10 0229738 3

KU-548-204

UNIVERSITY OF NOTTINGHAM
WITHDRAWN
FROM THE LIBRARY

UNIVERSITY OF NOTTINGHAM
WITHDRAWN
FROM THE LIBRARY

NOT TO
BE TAKEN
OUT OF
THE
LIBRARY

Interpreting Infrared, Raman, and Nuclear Magnetic Resonance Spectra

VOLUME 1   Variables in Data Interpretation of Infrared and Raman
Spectra

Interpreting Infrared, Raman, and Nuclear Magnetic Resonance Spectra

VOLUME 1   Variables in Data Interpretation of Infrared and Raman Spectra

RICHARD ALLEN NYQUIST

Nyquist Associates
Midland, MI

ACADEMIC PRESS

A Harcourt Science and Technology Company

San Diego   San Francisco   New York   Boston
London   Sydney   Tokyo

This book is printed on acid-free paper. ∞

Copyright © 2001 by Academic Press

All rights reserved.

No part of this publication may be reproduced or transmitted in any form or by
any means, electronic or mechanical, including photocopy, recording, or any
information storage and retrieval system, without permission in writing from
the publisher. Requests for permission to make copies of any part of the work
should be mailed to the following address: Permissions Department, Harcourt,
Inc., 6277 Sea Harbor Drive, Orlando, Florida, 32887-6777.

ACADEMIC PRESS
A Harcourt Science and Technology Company
525 B Street, Suite 1900, San Diego, CA 92101–4495, USA
http://www.academicpress.com

Academic Press
Harcourt Place, 32 Jamestown Road, London, NW1 7BY, UK
http://www.academicpress.com

Library of Congress Catalog Number: 00-108478

International Standard Book Number: 0-12-523475-9
Volume 1 International Standard Book Number: 0-12-523355-8
Volume 2 International Standard Book Number: 0-12-523470-8

Printed in the United States of America
01 02 03 04 05 QW 9 8 7 6 5 4 3 2 1

UNIVERSITY OF NOTTINGHAM
WITHDRAWN
FROM THE LIBRARY

CONTENTS

DEDICATION

This book is dedicated to all of the scientists I have been associated with for over 50 years, and to those scientists who find this book useful to them in increasing their skills in the interpretation of infrared (IR) and Raman spectra. This dedication especially includes my coauthors and other authors whose manuscripts or books were referenced in the present compilation.

ACKNOWLEDGMENTS

I thank the management of The Dow Chemical Company for providing me with a rewarding career in chemistry for over 41 years. I also thank the management of Sadtler Research Laboratories, a Division of Bio-Rad, for the opportunity to serve as an editorial consultant for several of their spectral collections of IR and Raman spectra.

I thank Marcia Blackson for typing the book manuscript. Her cooperation and editorial comments are appreciated.

NYQUIST'S BIOGRAPHY

In 1985, Richard A. Nyquist received the Williams-Wright Award from the Coblentz Society for his contributions to industrial IR spectroscopy. He was subsequently named an honorary member of the Coblentz Society for his contributions to vibrational spectroscopy, and in 1989, he was a national tour speaker for the Society of Applied Spectroscopy. The Association of Analytical Chemists honored Dr. Nyquist with the ANACHEM Award in 1993 for his contributions to analytical chemistry. He is listed in Who's Who in Science and Engineering, Who's Who in America, and Who's Who in the World. The Dow Chemical Company, from which Dr. Nyquist retired in 1994, honored him with the V.A. Stenger Award in 1981, and the Walter Graf European Award in 1994 for excellence in analytical chemistry. He has also been a member of ASTM, and received the ASTM Award of Appreciation for his contributions to the Practice of Qualitative Infrared Analysis. In 2000, Dr. Nyquist was awarded honorary membership in the Society of Applied Spectroscopy for his exceptional contributions to spectroscopy and to the society. Dr. Nyquist received his B.A. in chemistry from Augustana College, Rock Island, Illinois, his M.S. from Oklahoma State University, and his Ph.D. from Utrecht University, The Netherlands. He joined The Dow Chemical Company in 1953. He is currently president of Nyquist Associates, and is the author or coauthor of more than 160 scientific articles including books, book chapters, and patents. Nyquist has served as a consultant for Sadtler Research Laboratories for over 15 years. In 1997, Michigan Molecular Institute, Midland, Michigan, selected him as their consultant in vibrational spectroscopy.

PREFACE

My intention in compiling this book is to integrate IR and Raman data in order to aid analysts in the interpretation of spectral data into chemical information useful in the solution of problems arising in the real world.

There is an enormous amount of IR and Raman data available in the literature, but in my opinion there has not been enough emphasis on the effects of the physical environment of chemicals upon their molecular vibrations. Manipulation of the physical phase of chemicals by various experiments aids in the interpretation of molecular structure. Physical phase comprises solid, liquid, vapor, and solution phases.

In the solid crystalline phase, observed molecular vibrations of chemicals are affected by the number of molecules in the unit cell, and the space group of the unit cell. In the liquid phase, molecular vibrations of chemicals are affected by temperature, the presence of rotational conformers, and physical interaction between molecules such as hydrogen bonding and/or dipolar interaction. In the vapor phase, especially at elevated temperature, molecules are usually not intermolecularly hydrogen bonded and are free from dipolar interaction between like molecules. However, the rotational levels of the molecules are affected by both temperature and pressure. Induced high pressure (using an inert gas such as nitrogen or argon) will hinder the molecular rotation of molecules in the vapor phase. Thus, the rotational-vibrational band collapses into a band comparable to that observed in a condensed phase. Higher temperature will cause higher rotational levels to be observed in the vibrational-rotational bands observed in the vapor phase.

In solution, the frequencies of molecular vibrations of a chemical are affected by dipolar interaction and/or hydrogen bonding between solute and solvent. In addition, solute-solvent interaction also affects the concentration of rotational conformers of a solute in a solvent.

The number of intermolecular hydrogen-bonded molecules existing in a chain in solution depends upon the solute concentration. In addition, the number of molecules of a solute in solution existing in a cluster in the absence of intermolecular hydrogen bonding also depends upon solute concentration.

INTRODUCTION

Infrared (IR) and Raman (R) spectroscopy are essential tools for the study and elucidation of the molecular structures of organic and inorganic materials. There are many useful books covering both IR and R spectroscopy (1–14). However, none of these books emphasize the significance of changes in the molecular vibrations caused by changes in the physical state or environment of the chemical substance. One goal of this book is to show how changes in the physical environment of a compound aid in both the elucidation of molecular structure and in the identification of unknown chemical compositions. Studies of a variety of chemicals in various physical states have led to the development of the Nyquist Rule. The Nyquist Rule denotes how the in-phase- and out-of-phase- or symmetric and antisymmetric molecular vibrations (often called characteristic group frequencies) differ with changes to their physical environment. These group frequency shift differences aid the analyst in interpreting the data into useful chemical information. Another goal of this book is to gather information on the nature of solute-solvent interaction, solute concentration, and the effect of temperature. This knowledge also aids the analyst in interpretation of the vibrational data. Another goal of this work was to compile many of the authors' and coauthors' vibrational studies into one compendium.

REFERENCES

1. Herzberg, G. (1945). *Molecular Spectra and Molecular Structure II. Infrared and Raman Spectra of Polyatomic Molecules*, New Jersey: D. Van Nostrand Company, Inc.

2. Wilson, E. B. Jr., Decius, J. C., and Cross, P. C. (1955). *Molecular Vibrations*, New York: McGraw-Hill Book Company, Inc.

3. Colthup, N. B., Daly, L. H., and Wiberley, S. E. (1990). *Introduction to Infrared and Raman Spectroscopy*, New York: Academic Press.

4. Potts, W. J. Jr. (1963). *Chemical Infrared Spectroscopy*, New York: John Wiley & Sons, Inc.

5. Nyquist, R. A. (1984). *The Interpretation of Vapor-Phase Infrared Spectra: Group Frequency Data*, Philadelphia: Sadtler Research Laboratories, A Division of Bio-Rad.

6. Nyquist, R. A. (1989). *The Infrared Spectra Building Blocks of Polymers*, Philadelphia: Sadtler Research Laboratories, A Division of Bio-Rad.

7. Nyquist, R. A. (1986). *IR and NMR Spectral Data-Structure Correlations for the Carbonyl Group*, Philadelphia: Sadtler Research Laboratories, A Division of Bio-Rad.

8. Griffiths, P. R. and de Haseth, J. A. (1986). Fourier Transform Infrared Spectrometry, *Chemical Analysis*, vol. 83, New York: John Wiley & Sons, Inc.

9. Socrates, G. (1994). *Infrared Characteristic Group Frequencies Tables and Charts*, 2nd ed., New York: John Wiley & Sons, Inc.

10. Lin-Vien, D., Colthup, N. B., Fateley, W. G., and Grasselli, J. G. (1991). *The Handbook of Infrared and Raman Characteristic Frequencies of Organic Molecules*, San Diego, CA: Academic Press.

11. Nyquist, R. A., Putzig, C. L., and Leugers, M. A. (1997). *Infrared and Raman Spectral Atlas of Inorganic and Organic Salts*, vols. 1–3, San Diego, CA: Academic Press.

12. Nyquist, R. A. and Kagel, R. O. (1997). *Infrared Spectra of Inorganic Compounds*, vol. 4, San Diego, CA: Academic Press.

13. Nakamoto, K. (1997). *Infrared and Raman Spectra of Inorganic and Coordination Compounds, Part A: Theory and Applications in Inorganic Chemistry*, New York: John Wiley & Sons, Inc.

14. Nakamoto, K. (1997). *Infrared and Raman Spectra of Inorganic and Coordination Compounds, Part B, Applications in Coordination, Organometallic, and Bioinorganic Chemistry*, New York: John Wiley & Sons, Inc.

Theory of Vibrational Spectroscopy

Figures

*Numbers in parentheses indicate in-text page reference.

I. THEORY

For detailed discussion on the theoretical treatment of vibrational data (IR and Raman) the reader is referred to the following References (1–4). Extensive interpretation of IR vapor-phase spectra have been presented in Reference (5). The infrared and Raman methods are based on the fact that within any molecule the atoms vibrate within a few definite, sharply defined frequencies characteristic of the molecule. These vibrational frequencies occur in the region of the electromagnetic spectrum $13333\ cm^{-1}$ to $50\ cm^{-1}$ and beyond. Only those molecular vibrations producing a dipole-moment change are IR active, allowed in the IR, and only those molecular vibrations producing polarization of the electron cloud are Raman active, allowed in the Raman. In the vapor-phase, molecules are free to rotate in three-dimensional (3D) space. The molecular rotational moments of inertia are governed by molecular geometry, and the atomic mass of each atom in the molecule together with their relative spatial positions within the molecule. Therefore, in the vapor-phase, molecules undergo transitions between quantized rotation states as well as quantized vibrational transitions. The result is that a transition between the ground state and the first excited state of a normal mode is accompanied by a manifold of rotational transitions. This leads to a complex rotation-vibration band for every IR active molecular vibration. Overtones are also IR active for molecules without a center of symmetry, and they result from transitions between the ground state and the second excited state of a normal vibration. Combination tones result from simultaneous transitions from the ground state to the first excited state of two or more normal vibrations. Both combination tones and overtones are also accompanied by manifold rotational transitions. In the liquid or solution phase, rotation of the molecule in space is restricted. Therefore, the rotation-vibration bands are "pressure broadened," and do not exhibit the sharp manifold rotational translational lines.

The number of molecular vibrations allowed in the IR or R for a given molecule is governed by the number of atoms in the molecule together with its molecular geometry. For nonlinear molecules, the total number of normal vibrations is determined by the equation 3N-6, where N is the number of atoms in the molecule. Because the molecules are free to vibrate, rotate, and translate in 3D space, N is multiplied by 3. The number 6 is subtracted because the number of possible molecular vibrations is not determined by rotation and translation of the molecule. For linear molecules, the total number of normal vibrations is determined by equation 3N-5.

In order to determine the active IR and Raman normal vibrations for any molecule, one has to apply a method known as Group Theory (1–4). Application of Group Theory also allows the determination of which overtones and combination tones are active in either the IR or Raman.

Pure molecular rotation transitions are also IR active, and they occur in the IR spectrum in the region below $600\ cm^{-1}$ for small molecules having a permanent dipole, such as H_2O, NH_3, PH_3, etc.

In the vapor-phase, interpretation of the rotation-vibration bond contour is helpful in the elucidation of molecular structure. Band contours result from a combination of molecular symmetry and the moments of inertia I_A, I_B, and I_C about three mutually perpendicular axes.

II. EXAMPLES OF MOLECULAR STRUCTURE

A. LINEAR MOLECULES: $I_A = O$, $I_B = I_C$

Examples of these are:

  Hydrogen halides, HX;
  Carbon monoxide, CO;

Nitrogen oxide, NO;
Carbon dioxide, O=C=O;
Carbon disulfide, S=C=S; and
Acetylene, H−C≡C−H.

In the case of linear molecules, I_A is the moment of inertia along the molecular axis, and I_B and I_C are mutually perpendicular axes. The infrared active stretching vibrations produce a dipole-moment change along the molecular symmetry axis, and the resulting rotation-vibration band contour is called a parallel band. In this case, the P and R branches are predominate with no center Q branch.

Figure 1.1 shows the IR vapor-phase spectrum of hydrogen bromide. The P branch of HBr occurs in the region 2300–2550 cm^{-1} and the R branch occurs in the region 2500–2725 cm^{-1}. Figure 1.2 shows the IR vapor-phase spectrum of hydrogen chloride. The P branch of HCl occurs in the region 2600–2880 cm^{-1} and the R branch occurs in the region 2900–3080 cm^{-1}. The rotational subband spacings are closer together for HBr than those for HCl, and this is because the individual rotation subbands in the P and R branches are dependent upon the moments of inertia, and become more closely spaced as I_B and I_C become larger. Neither HBr nor HCl exhibits a central Q branch for the hydrogen-halogen stretching vibration.

Figure 1.3 shows an IR vapor-phase spectrum for carbon monoxide in the region 1950–2300 cm^{-1}. In this case the P branch occurs in the region 2000–2150 cm^{-1}, and the R branch occurs in the region 2150–2250 cm^{-1}. Both the P and R branches exhibit closely spaced rotation subbands of the CO stretching vibration, producing essentially a solid continuum of absorption lines.

Figure 1.4 shows an IR vapor-phase spectrum of nitrogen oxide. The NO stretching vibration is assigned to the parallel band occurring in the region 1760–1970 cm^{-1}. However, it should be noted that this is an exceptional case for linear molecules, because a central Q branch is observed near 1872 cm^{-1}. This exception results from the presence of an unpaired electron in nitrogen oxide causing a resultant electronic angular momentum about the molecular axis, which gives rise to a Q branch in the parallel band (1). The P branch subbands occur below 1872 cm^{-1}, and the R branch subbands occur above 1872 cm^{-1}. Otherwise, a band with this contour for linear molecules would be called a perpendicular band.

Figure 1.5 is a vapor-phase IR spectrum of carbon dioxide. The perpendicular band for CO_2 exhibits its Q branch near 670 cm^{-1} with a P branch near 656 cm^{-1} and an R branch near 680 cm^{-1}. This CO_2 bending vibration is doubly degenerate. The parallel band for the antisymmetric CO_2 stretching vibration occurs in the region 2300–2400 cm^{-1}. The P branch occurs near 2350 cm^{-1} and the R branch occurs near 2360 cm^{-1}.

Figure 1.6 is an IR vapor-phase spectrum of carbon disulfide. The P branch of the parallel band occurs near 1560 cm^{-1} and the R branch of the parallel band occurs near 1540 cm^{-1} for this antisymmetric CS_2 stretching vibration.

Figure 1.7 is an IR vapor-phase spectrum for acetylene. The perpendicular band occurring in the region 650–820 cm^{-1} is assigned to the in-phase $(\equiv C-H)_2$ bending vibration, which is doubly degenerate. The Q branch occurs near 730 cm^{-1}, the P branch in the region 650–720 cm^{-1}, and the R branch in the region 735–820 cm^{-1}. Rotational subbands are noted in both the P and R branches of this linear molecule.

Acetylene, carbon dioxide, and carbon disulfide have a center of symmetry; therefore, some vibrations are only IR active and some vibrations are only Raman active. Raman active vibrations for acetylene are the in-phase $(\equiv C-H)_2$ stretching vibration, the $C\equiv C$ stretching vibration, and the out-of-phase $(\equiv C-H)_2$ bending vibration. A Raman active fundamental for CO_2 and CS_2 is the symmetric CO_2 and CS_2 stretching vibration. Comparison of the frequency separation between the subbands of the P and R branches shows that it is less for CS_2 than for CO_2, and this is a result of a larger $I_B = I_C$ for CS_2 compared to CO_2.

B. SPHERICAL TOP MOLECULES: $I_A = I_B = I_C$

Molecules in this class have Td symmetry, and examples are methane and symmetrically substituted carbon tetrahalides. In this case IR active fundamentals exhibit P, Q and R branches in the vapor-phase comparable to that exhibited by perpendicular vibrations in linear molecules.

Figure 1.8 is an IR vapor-phase spectrum for methane. The antisymmetric CH_4 stretching vibration is triply degenerate, the perpendicular band exhibits the Q branch near $3020 \, cm^{-1}$, the P branch in the region $2850-3000 \, cm^{-1}$, and the R branch in the region $3040-3180 \, cm^{-1}$. The triply degenerate antisymmetric CH_4 bending vibration also appears as a perpendicular band. The Q branch occurs near $1303 \, cm^{-1}$, the P branch in the region $1200-1290 \, cm^{-1}$, and the R branch in the region $1310-1380 \, cm^{-1}$. Numerous subbands are noted in the P and R branches of both vibrations. The symmetric CH_4 stretching vibration is only active in the Raman ($2914 \, cm^{-1}$), and the symmetric CH_4 bending vibration, triply degenerate, is only Raman active ($1306 \, cm^{-1}$). The antisymmetric CH_4 stretching vibration, triply degenerate, and the symmetric CH_4 bending vibration, doubly degenerate, are both IR and Raman active.

Figure 1.9 shows an IR vapor-phase spectrum of carbon tetrafluoride. The antisymmetric CF_4 stretching vibration is triply degenerate, and its Q branch is noted near $1269 \, cm^{-1}$. The antisymmetric CF_4 bending vibration is triply degenerate. Its Q branch is noted near $630 \, cm^{-1}$, and the P and R branches are noted near $619 \, cm^{-1}$ and $650 \, cm^{-1}$, respectively. Because the moments of inertia are large, the subbands of the P and R branches are so narrowly spaced that the P and R branches appear as a continuum.

C. SYMMETRIC TOP MOLECULES

1. Prolate: $I_A < I_B = I_C$ (essentially rod shaped)

Molecules in this class have C_{3V} symmetry. Examples include the methyl halides, propyne, and 1-halopropynes.

Prolate symmetric top molecules exhibit both parallel and perpendicular bands in the vapor phase. Molecular vibrations mutually perpendicular to the highest symmetry axis exhibit perpendicular bands and molecular vibrations symmetric with respect to the highest symmetry axis exhibit parallel bands.

Figures 1.10, 1.11, and 1.12 are IR vapor-phase spectra of methyl chloride, methyl bromide, and methyl iodide, respectively.

The P and R branches of the parallel bands for these methyl halides are given here:

Branch	CH_3Cl cm^{-1}	CH_3Br cm^{-1}	CH_3I cm^{-1}	Assignment
P	2952	2980	2980	symmetric CH_3
R	2981	2985	2958	stretching
P	1346	1293	1261	symmetric CH_3
R	1366	1319	1235	bending
P	713	598	518	C–X stretching
R	748	622	540	

The parallel bands are nondegenerate.

The P, Q, and R branches of the perpendicular bands for these methyl halides are given here:

Branch	CH_3Cl cm^{-1}	CH_3Br cm^{-1}	CH_3I cm^{-1}	Assignment
P	~3000–3150	~3020–3140	~3020–3140	antisymmetric CH_3 stretching
Q	~1325–1600	~1300–1600	~1300–1600	antisymmetric CH_3 bending
R	~940–1120	~850–1050	~770–1000	CH_3 rocking

The perpendicular modes are doubly degenerate.

Figures 1.13 and 1.14 are IR vapor-phase spectra of 1-bromopropyne and 1- iodopropyne, respectively. Detailed assignments for these two compounds are given in Reference (6).

2. Oblate Top: $I_A = I_B < I_C$ (essentially disc shaped)

Molecules in this class have D_{6h} symmetry, and molecules with this symmetry include benzene, benzene-d_6, and the hexahalobenzenes, which contain only F_6, Cl_6, Br_6, or I_6. Oblate symmetric top molecules exhibit both parallel and perpendicular bands. Planar molecular vibrations exhibit parallel bands, and out-of-plane vibrations exhibit perpendicular bands. These complex molecules exhibit relatively simple IR vapor-phase IR spectra, because these molecules have a center of symmetry, and only a few normal modes are IR active.

Figure 1.15 is an IR vapor-phase spectrum of benzene. The type C perpendicular band with a Q branch at $\sim 670\,cm^{-1}$, with P and R branches near ~ 651 and $\sim 686\,cm^{-1}$, respectively, is assigned to the in-phase-out-of plane 6 hydrogen deformation. Benzene exhibits a type A band with P, Q, and R branches near 1019, 1039, and $1051\,cm^{-1}$, respectively. Both the type A and type C bands show P, Q, and R branches, but the spacings between P and R branches for type A bands for benzene are less than the spacings between the P and R branches of type C bands.

D. ASYMMETRIC TOP MOLECULES: $I_A \neq I_B \neq I_C$ (WHERE $I_A < I_B < I_C$, AND THE ROTATIONAL CONSTANTS ARE ORDERED A > B > C)

Molecules in this class will exhibit type A, B, and C bands providing the dipole moment change during the normal vibration is parallel to the a, b, or c axis, respectively. Mixed band contours described as type AB are exhibited by molecules where the dipole moment change during the normal vibration is not exactly parallel to the a or b symmetry axis.

Figure 1.16 is a vapor-phase IR spectrum of ethylene oxide. Type B bands exhibit no central peak, and classic type B bands are noted near 1269 and 875 cm^{-1}. For the 1269 cm^{-1} band the P and R branches are assigned near 1247 and 1292 cm^{-1}, respectively, and the Q^I and Q^{II} branches are assigned near 1263 and 1274 cm^{-1}, respectively. For the 875 cm^{-1} band, the P and R branches are assigned near 848 and 894 cm^{-1}, respectively, and the Q^I and Q^{II} branches are assigned near 869 and 881 cm^{-1}, respectively. The 1269 cm^{-1} band is assigned to ring breathing, the 875 cm^{-1} band to a ring deformation. The weak type C band with a Q branch near 820 cm^{-1} is assigned to CH$_2$ rocking (7).

III. PRESSURE EFFECT

With increasing pressure there are more frequent collisions of like molecules, or between a molecule and a diluent gas such as nitrogen or helium, and this has the effect of broadening the vapor-phase IR band contours resulting from molecular rotation-vibration. This effect is termed pressure broadening. The individual subbands (or lines) become increasingly broad due to restricted rotation in the vapor phase due to frequent molecular collisions. Under high pressure, the subbands are completely broadened so that the molecular vibrations with highly restricted molecular rotation produce IR band shapes with no apparent subbands. Thus, these pressure broadened vapor-phase IR bands for various molecular structures have shapes comparable to their IR band shapes observed in their neat liquid or solution phases. This is so because, in the neat liquid or solution phases, there are frequent collisions between like molecules or molecular collisions between solute molecules and between solute and solvent molecules for molecules in solution, which restrict molecular rotation of these molecules.

In order to measure the intensities of the IR vibrational bands for ethane and ethane-d$_6$ in the vapor-phase, the samples were pressurized up to 50 atm to broaden the bands (8).

IV. TEMPERATURE EFFECT

In the vapor-phase temperature can cause change in the band contours. The individual lines or subbands in the P and R branches are due to the relative population of the rotational states. An increase in temperature will change the population of rotation states, hence the change in IR band contour. Secondly, increased temperature in a closed volume cell increases the pressure, which induces the pressure broadening effect.

Hot bands are also temperature dependent, and change in temperature causes change in IR band intensity (3).

When molecules exist as rotational conformers (rotamers), they are also affected by changes in temperature, because the concentration of the different rotamers is dependent upon temperature. Therefore, it is essential to record IR spectra at different temperatures in cases where molecules exist as rotamers in order to determine which bands result from the same rotamer.

V. VAPOR-PHASE VS CONDENSED IR SPECTRA

The IR group frequencies of molecules are dependent upon physical phase. These frequency differences result from solute-solvent interaction via dipolar interaction or from weak intermolecular hydrogen bonding. Large frequency differences, as large as $400 \, cm^{-1}$, result from strong intermolecular hydrogen bonding between molecules or between solute and solvent. These effects are absent in the vapor phase, especially at temperatures above $180 \, ^\circ C$ at ordinary pressure. The vibrational frequency changes will be extensively discussed in later chapters.

VI. FERMI RESONANCE (F.R.) AND OTHER FACTORS

Some molecules exhibit two or more bands in a region where only one fundamental vibration is expected, excluding the presence of rotational conformers. In this case Fermi resonance between a fundamental and an overtone or combination tone of the same symmetry species interacts. The combination or overtone gains intensity at the expense of the intensity of the fundamental. The result is that one band occurs at a higher frequency and one band occurs at a lower frequency than expected due to this resonance interaction between the two modes. Langseth and Lord (9) have developed a method to correct for F.R. (10).[1] This equation is presented here:

$$W_o = \frac{W_a + W_b}{2} \pm \frac{W_a - W_b}{2} \cdot \frac{I_a - I_b}{I_a + I_b}$$

where, W_a and W_b are the observed vibrational frequencies, I_a and I_b their band intensities, and the two values of W_o calculated by the equation will be approximately the unperturbed frequencies. The amount of F.R. is dependent upon the unperturbed frequencies of the fundamental and the combination or overtone. If two bands of equal intensity are observed, each band results from an equal contribution from the fundamental vibration and an equal contribution from the combination tone or overtone. The combination or overtone may occur above or below the fundamental frequency. In the case in which the bands are of unequal intensity, both bands are still a mixture of both vibrations but the stronger band has more contribution from the fundamental than the weaker band. Correction for F.R. is necessary in

---

[1] Reference (10) also includes the same equation developed for the correction of two bands in Fermi resonance (8). In addition, the newer reference shows the development for the correction of Fermi resonance for cases where three vibrations are in Fermi resonance.

cases where one needs to perform a normal coordinate analysis, or when comparing IR data of certain classes of compounds where not all of the compounds show evidence for F.R.

Parameters such as bond force constants, bond lengths, bond angles, field effects, inductive effects, and resonance effects are independent of physical phase and these parameters are useful in the elucidation of molecular structure via IR spectra-structure interpretation.

REFERENCES

1. Herzberg, G. (1945). *Molecular Spectra and Molecular Structure II. Infrared and Raman Spectra of Polyatomic Molecules*, New Jersey: D. Van Nostrand Company, Inc.

2. Wilson, E. B. Jr., Decius, J. C., and Cross, P. C. (1955). *Molecular Vibrations*, New York: McGraw-Hill Book Company, Inc.

3. Potts, W. J. Jr. (1963). *Chemical Infrared Spectroscopy*, New York: John Wiley & Sons, Inc.

4. Colthup, N. B., Daly, L. H., and Wiberley, S. E. (1990). *Introduction to Infrared and Raman Spectroscopy*, 3rd ed., Boston: Academic Press.

5. Nyquist, R. A. (1984). *The Interpretation of Vapor-Phase Infrared Spectra: Group Frequency Data*, vol. 1, Philadelphia: Sadlter Research Laboratories, Division of Bio-Rad Laboratories.

6. Nyquist, R. A. (1965). *Spectrochim Acta*, **21**, 1245.

7. Lord, R. C. and Nolin, B. (1956). *J. Chem. Phys.*, **24**, 656.

8. Nyquist, I. M., Mills, I. M., Person, W. B., and Crawford, B. Jr. (1957). *J. Chem. Phys.*, **26**, 552.

9. Langseth, A. and Lord, R. C. (1948). *Kgl. Danske Videnskab Selskab Mat-fys. Medd*, **16**, 6.

10. Nyquist, R. A., Fouchea, H. A., Hoffman, G. A., and Hasha, D. L. (1991). *Appl. Spectrosc.*, **45**, 860.

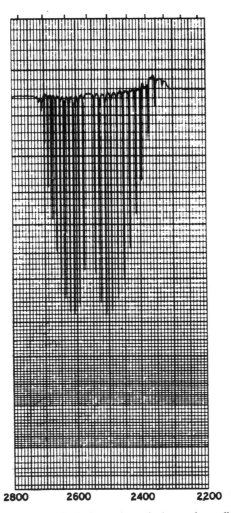

FIGURE 1.1 Infrared vapor-phase spectrum for hydrogen bromide (5-cm glass cell with KBr windows: 600 mm Hg HBr).

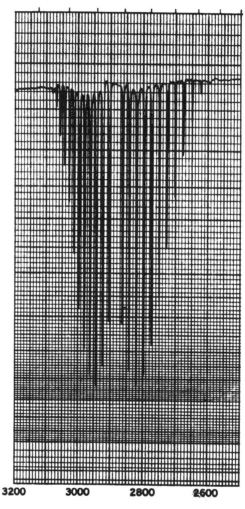

3200   3000   2800   2600

FIGURE 1.2   Infrared vapor-phase spectrum for hydrogen chloride (5-cm glass cell with KBr windows: 200 mm Hg HCl, total pressure 600 mm Hg with N_2).

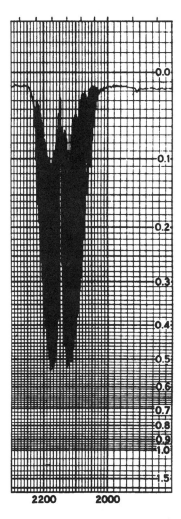

FIGURE 1.3 Infrared vapor-phase spectrum for carbon monoxide (5-cm glass cell with KBr windows: 400 mm Hg CO, total pressure 600 mm Hg with N_2).

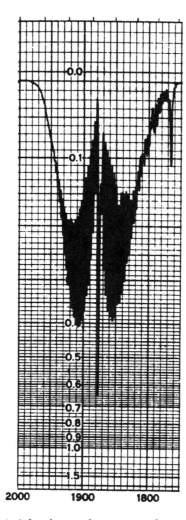

FIGURE 1.4   Infrared vapor-phase spectrum for nitrogen oxide, NO.

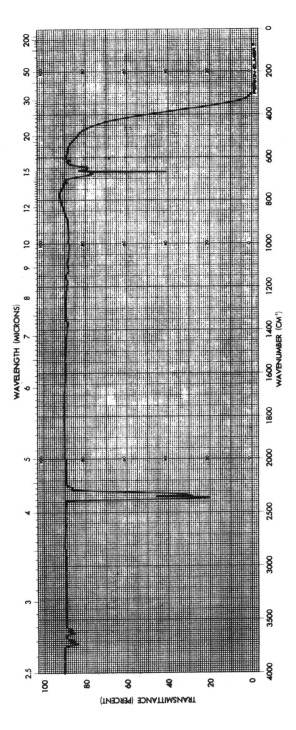

FIGURE 1.5 Infrared vapor-phase spectrum for carbon dioxide (5-cm glass cell with KBr windows: 50 and 200 mm Hg CO_2, total pressure 600 mm with N_2).

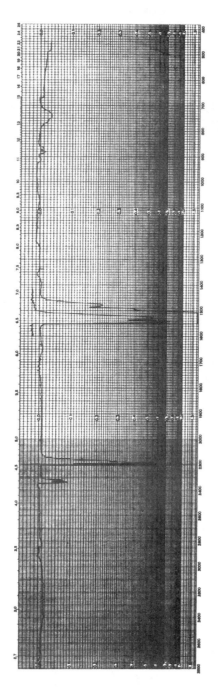

FIGURE 1.6   Infrared vapor-phase spectrum for carbon disulfide (5-cm glass cell with KBr windows: 2 and 100 mm Hg CS_2, total pressure 600 mm Hg with N_2).

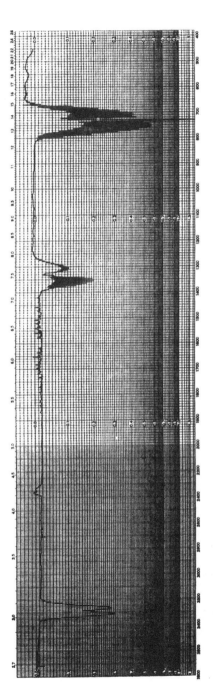

FIGURE 1.7   Infrared vapor-phase spectrum for acetylene (5-cm glass cell with KBr windows: 50 mm Hg C_2H_2; total pressure 600 mm Hg with N_2).

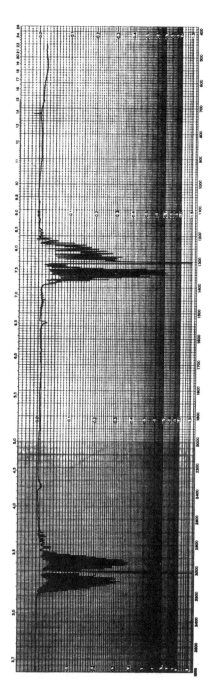

FIGURE 1.8   Infrared vapor-phase spectrum for methane (5-cm glass cell with KBr windows: 150 mm Hg CH$_4$, total pressure 600 mm Hg with N$_2$).

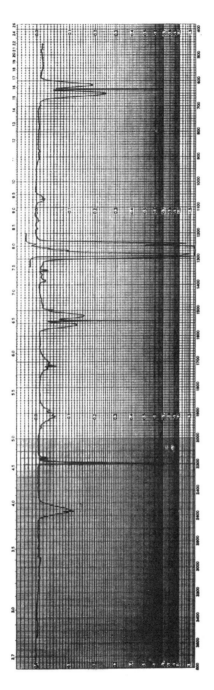

FIGURE 1.9 Infrared vapor-phase spectrum for carbon tetrafluoride (5-cm glass cell with KBr windows: 100 mm Hg CF_4, total pressure 600 mm Hg with N_2).

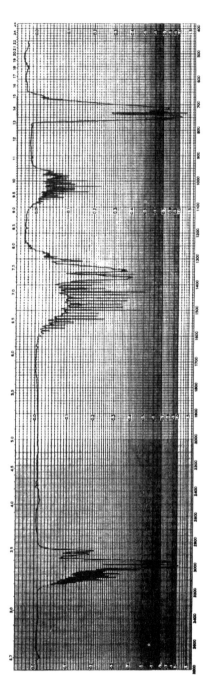

FIGURE 1.10   Infrared vapor-phase spectrum for methyl chloride (5-cm glass cell with windows: 200 mm Hg CH_3, Cl, total pressure 600 mm Hg with N_2).

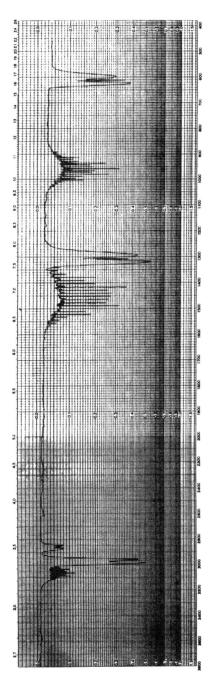

FIGURE 1.11 Infrared vapor-phase spectrum for methyl bromide (5-cm glass cell with KBr windows: 100 mm Hg CH₃Br, total pressure 600 mm Hg with N₂).

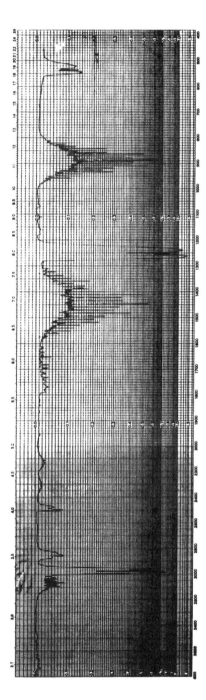

FIGURE 1.12 Infrared vapor-phase spectrum for methyl iodide (5-cm glass cell with KBr windows: 200 mm Hg CH_3I, total pressure 600 mm Hg with N_2).

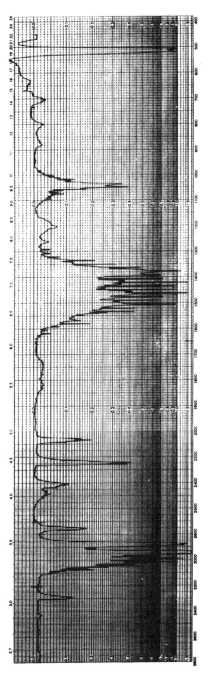

FIGURE 1.13 Infrared vapor-phase spectrum for 1-bromopropyne, C_3H_3Br.

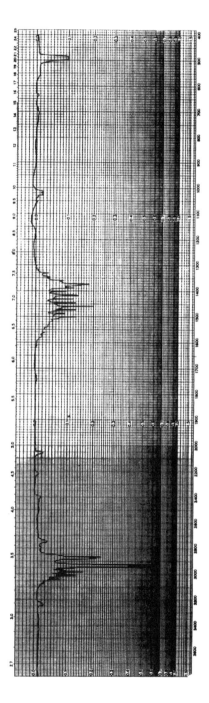

FIGURE 1.14   Infrared vapor-phase spectrum for 1-iodopropyne, C_3H_3I.

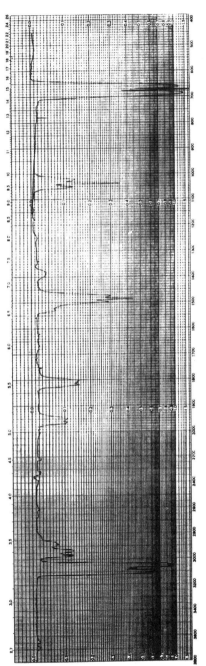

FIGURE 1.15 Infrared vapor-phase spectrum for benzene (5-cm glass cell with KBr windows: 40 mm Hg with C_6H_6, total pressure 600 mm Hg with N_2).

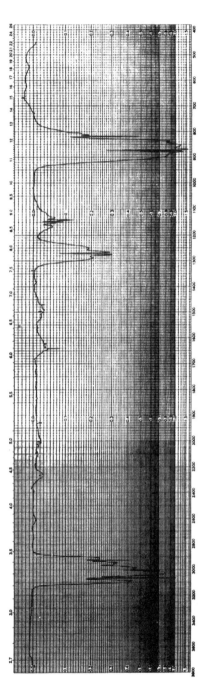

FIGURE 1.16 Infrared vapor-phase spectrum for ethylene oxide (5-cm glass cell with KBr windows: 50 mm Hg C_2H_2O, total pressure 600 mm Hg with N_2).

Experimental

Sample preparation is a very important part of the IR technique. Because chemicals can exist in the solid, liquid, vapor, or solution phases, different methods of preparation are required in order to be able to record their IR spectra. This chapter is not intended to include all methods of obtaining IR spectra. It includes only the methods used to acquire the data included in this book.

I. SOLIDS (EXCLUDING SINGLE CRYSTALS)

Solid samples are usually prepared as mulls or as KBr pellets. Both techniques require that the particle size be smaller than 2.5 μ. Larger particles scatter IR radiation via Rayleigh scattering in the IR region of interest. For example, if particles are present in decreasing concentration from 2.5 through 25 μ, the baseline of the spectrum will slope upward in the region 2.5 to 25 μ (4000–400 cm^{-1}, the region most commonly examined by chemists).

Thus, solid samples must be ground using a mortar and pestle (or a wiggle bug) to meet the preceding requirements. The closer one gets to a virtual horizontal baseline, the better the quality of the IR spectrum. The bands of the chemical solid start absorbing IR radiation at the point of the baseline where their vibrational frequencies occur in this region of the spectrum.

There is another factor that causes distortion of IR absorption bands. If the refractive index of solid particles and the surrounding medium differ appreciably, the Christiansen effect is encountered (1). The Christiansen effect develops because the refractive index of a chemical is a function of frequency that has a discontinuity in each frequency region of a strong absorption band. The refractive index falls rapidly on the high frequency side of the absorption maximum and on the low frequency side the refractive index falls rapidly from a high value to its value of no absorption. This effect causes peculiar absorption band distortion when there are many large particles present due to inadequate sample preparation. The Christiansen effect is minimized by a reduction of as many particles as possible smaller than $2\,\mu$ in size. However, the effect is never completely eliminated when recording spectra of powdered crystalline materials.

A. Mull Technique

After the sample has been properly ground, Nujol oil or Fluorolube is added and mixed in order to suspend the particles in the mulling agent. However, it is preferable to grind the solid in the presence of the mulling agent to help in the grinding process. The mulling agents suspend the solid particles, which helps to produce a closer match of the refractive index between these particles and the surrounding medium. Nujol is useful for recording IR spectra in the region $1333-400\,cm^{-1}$, and Fluorolube is useful for recording IR spectra in the region $4000-1333\,cm^{-1}$. In order to place these mulled suspensions in the IR sample compartment, sodium chloride and/or potassium bromide plates are required. A Fluorolube mull paste is then placed between two sodium chloride plates or two potassium bromide plates using pressure to obtain the proper thickness of this paste in order to obtain a quality IR spectrum in the region $4000-1333\,cm^{-1}$ that is essentially void of significant Fluorolube absorption bands. In the region $1333-400\,cm^{-1}$ the process is repeated using the Nujol mull suspension between potassium bromide plates, which minimizes absorption from Nujol oil.

B. Potassium Bromide (KBr) Disk Technique

After grinding the solid to proper particle size, KBr is mixed with the solid particles. A ratio somewhere between 50 to 100 KBr : 1 of solid sample usually produces a quality IR spectrum. The KBr and ground particles are then thoroughly mixed to produce a uniform mixture. Further grinding of the mixture should be avoided because this additional grinding will usually induce water into the mixture. The proper amount of this KBr preparation is placed in a special die, and a disk is pressed using approximately 9600 Kg pressure. The disk is placed in a holder and placed then in the sample compartment of the IR spectrometer.

Both the mull and KBr pressed disk technique can cause changes to the sample. Grinding or increased pressure upon the sample can cause changes in crystalline form of the sample. In addition, the pressed KBr disk can cause chemical reactions to occur between KBr and the sample (e.g., $R-NH_3^+Cl^- + KBr \rightarrow R-NH_3^+Br^- + KCl$). In the mull technique ion exchange can also occur between the window and the suspended sample (e.g., $R-NH_3{}^+F^- + KBr \rightarrow R-NH_3{}^+Br^-$). This latter reaction occurs when pressure between the plates together with plate rotation causes solid state interaction between the plates and the sample. Thus, one should be aware that the chemist can inadvertently alter the original sample during preparation, thereby causing problems in the solution of the chemical problem at hand.

The IR spectra of solids obtained after evaporating water solutions to dryness can also be recorded using the mull or KBr disk technique.

C. SOLUTIONS (SOLIDS, LIQUIDS, AND GASES)

Solids, liquids, and gases are often soluble in solvents such as carbon tetrachloride and carbon disulfide. Other solvents such as chloroform, methylene chloride, or dimethyl sulfoxide (DMSO) can also be used depending upon the particular problem. Since like dissolves like, the more polar solvents are used to dissolve the more polar compounds. Carbon tetrachloride and carbon disulfide are used to dissolve the less polar compounds.

Carbon tetrachloride is a useful solvent in the region 4000–1333 cm^{-1} and carbon disulfide is a useful solvent in the region 1333–400 cm^{-1}. This is because these solvents have the least absorption in these regions. Quality IR spectra can be recorded using samples prepared at 10% by weight in each of these solvents, and placing the solutions in 0.1-mm sodium chloride cell (4000–1333 cm^{-1} for CCl$_4$ solutions) and 0.1-mm potassium bromide cell (1333–400 cm^{-1} for CS$_2$ solutions). Comparable IR spectra will be recorded using 1% by weight using 1.0-mm cells; however, absorption from the solvent will be increased by a factor of 10. Variation of concentration and cell path length can be used to record the spectrum most useful in obtaining useful chemical information. For example, changing the concentration of a chemical, in say CCl$_4$, can help to distinguish between inter- and intramolecular hydrogen bonding. The OH or NH stretching frequencies remain essentially unchanged upon dilution, in the case of intramolecular hydrogen bonding, but increase markedly in frequency upon dilution in the case of intermolecular hydrogen bonding.

Solution spectra are very useful in performing quantitative analysis when both the sample concentration and cell path length are known, providing the absorbance of a band due to the presence of the critical analyte can be directly measured. In the worst case, interference from the presence of another analyte may have to be subtracted before the analysis can be performed on the critical analyte in question.

A solvent such as CS$_2$ is also useful for extracting some chemicals from water. After thorough shaking with water, the sample can be concentrated by partial solvent evaporation in a well-ventilated hood containing no source of ignition. The sample is then salted using dry NaCl powder in order to remove water. The CS$_2$ is then placed in a suitable KBr cell before the solution spectrum is recorded. This same solvent can be used to extract certain additives from polymer compositions.

II. LIQUID FILMS (SEE THE FOREMENTIONED I–C FOR LIQUIDS IN SOLUTION) AND CAST POLYMER FILMS

a. Liquid films between KBr plates are easily prepared by placing a drop or more on one plate and then placing the second plate on top with enough pressure to form the desired film thickness. Very volatile liquids are better prepared as solutions.

b. IR spectra of polymers are often recorded of freestanding film or of films cast on a suitable IR plate. Freestanding films are prepared by heating polymeric material above its melting point between heated plates in a suitable press. The film is allowed to cool to ambient temperature before removing it from the metal plates. The freestanding film is then placed in the IR spectrometer.

c. IR spectra are often recorded of polymeric substances cast from boiling solution onto a preheated KBr plate, often under a nitrogen atmosphere in order to avoid oxidation of the polymer. After the solvent has evaporated, the plate and film are allowed to cool to ambient temperature before placing in the IR sample compartment. Failure to preheat the KBr plate will cause it to shatter when first in contact with the hot solution. Moreover, removal of the heated cast film after solvent evaporation to a ambient environment without prior cooling will also cause the KBr plate to shatter. Organic solvents such as 1,2-dichlorobenzene, toluene, and dimethyl formamide are often used to cast polymer films. The solvent used is dependent upon the particular polymer or copolymer, and its solubility in that solvent.

Silver chloride plates are often used as the substrate for casting films of water-soluble polymers. The water is removed by heat after first placing the water solution onto the AgCl plate placed under an IR heat lamp. It is essential that AgCl plates be stored in the dark to prevent darkening of the plates due to the formation of silver oxide upon exposure to light.

It is essential that all plates be cleaned after being used in these experiments. The polymer film can be removed from the plates using the same solvents used to cast the films. It is again necessary to avoid sudden temperature change to the plates during the cleaning operation in order to avoid plate shattering in cases such as KBr, NaCl, CaF_2, etc.

III. VAPORS AND GASES

Quality IR spectra can be recorded using partial pressures and appropriate cell lengths equipped, say, with potassium bromide windows. It should be noted that certain inorganic compounds react with KBr and form other inorganic salts on the surface of the KBr plate. Their reactions are readily detected because their reaction products can not be removed when the sample is evacuated under vacuum from the cell.

The IR spectra of chemicals with low vapor pressure can be recorded using a variable long path length vapor cell either at ambient temperature or at elevated temperature. It is best if these variable path length cell walls are coated with a substance such as poly (tetrafluorothylene) to help prevent adsorption of the chemical on the surface of the metallic cell body. It is sometimes necessary to heat the cell body under vacuum to remove adsorbed chemical molecules. Another method to clean out adsorbed molecules present in the cell is to flush the cell with nitrogen or dry air.

It is also possible to perform quantitative analysis of compounds whose IR spectra have been recorded in the vapor phase. An often-used method is to record the partial pressure of the chemical in the vapor, and then bring the total pressure up to 600-mm Hg using dry nitrogen. The constant total pressure of 600-mm Hg helps eliminate the effects of pressure broadening on the absorption bands.[1] This requires of course that the appropriate vacuum line and dry nitrogen be available to the chemist. A longer path length setting is required as vapor pressure of a chemical falls. Of course, some chemicals react with the mirrors, and this will limit this application for these particular compounds.

It is possible to detect low parts per million of chemicals in air using the variable long path cells. The interpretable regions of the IR spectrum are significantly improved by spectral

---

[1] The windows (KBr, NaCl, etc) of most glass-bodied cells are adhered to the glass body by a material such as paraffin wax. Pressures higher than 760-mm Hg will blow the windows from the glass body. Thus, 600-mm Hg sample pressure is a reasonable total pressure to safely achieve in these laboratory experiments.

subtraction of absorption due to the presence of H_2O and CO_2. This can be done electronically or by dual cells placed in separate beams of a double beam spectrometer. In the case of FT-IR spectrometers, it is an easy task to remove these absorption bands due to the presence of air. In a case where there is no IR radiation being transmitted in a particular region of the IR spectrum, detection of a compound present in air (or in any phase) is not possible in these spectral regions, because the spectrometer is "dead" in these regions. Spectral subtraction will not change the "dead" regions of the spectrum.

A simple method of obtaining IR vapor spectra of chemicals with high vapor pressure is to connect the sample container using a rubber stoppered hose to a 0.1-mm (or 0.2-mm, etc.) liquid cell. The stopper is opened and the chemical vapor is allowed to flush out the 0.1-mm cell. The exit port of the cell is then stoppered, the sample connection is closed and removed, and the entrance part of the cell is also stoppered. The cell is now filled with the sample in the vapor phase at ambient temperature and pressure. Of course, this operation should be performed in a well-ventilated hood.

Gas chromatography has been coupled to infrared spectroscopy (GC/FT-IR) to form a powerful analytical technique capable of solving many real world problems. This technique requires that the chromatographed vapors pass sequentially through a gold-coated light pipe heated to a temperature of over 200 °C. The light pipe path-length must be short enough so that only one chromatographed component is in the light pipe at one time. This technique has a major pitfall in that not all chemical compounds are stable at the high temperatures encountered utilizing this technique. For example, phthalic acid present as one component in a mixture would not be detected as one of the chromatographed fractions. This is because at these elevated temperatures, water splits out of phthalic acid to form phthalic anhydride, which is the compound detected using this technique. Other types of chemical reactions can occur if the chemicals contact hot metal surfaces (excluding gold) during their path through the GC/FT-IR system.

In order to identify unequivocally a vapor-phase IR spectrum of a chemical, an IR vapor-phase standard spectrum of this compound recorded under comparable conditions must be available for comparison. The reasons for this are presented here. A compound such as acetic acid exists as a hydrogen-bonded cyclic dimer in the condensed phase and in the vapor-phase at temperatures 150 °C and below. At elevated temperature, acetic acid exists as isolated CH_3CO_2H molecules. In this monomeric state, the OH stretching frequency exhibits a weak-medium sharp band near 3580 cm^{-1} in the vapor phase, and the C=O stretching frequency exhibits a strong bond near 1791 cm^{-1}. These features are uniquely different from the condensed phase IR spectra of acetic acid. This monomeric situation is even more complicated in situations where intramolecular hydrogen bonding can occur between the proton of the carboxylic acid group and a basic site in the molecule. For example, pyruvic acid (2-oxo-propionic acid) exhibits two bands in the vapor phase at 95 °C (2). A weak band near 3580 cm^{-1} is assigned to an unassociated OH group of CO_2H. The weak-medium bond near 3465 cm^{-1} results from the intramolecular hydrogen bond OH group to the free pair of electrons on the ketone carbonyl group to form a 5-membered cyclic ring as illustrated here:

Other situations occur where molecules that are intermolecularly hydrogen bonded in the condensed phase form intramolecular hydrogen bonds in the vapor phase. In addition, the regions for group frequencies in the condensed or solution phases have shifted from those in the vapor phase. Therefore, one must have at hand a collection of vapor-phase group frequency data, available to enable one to interpret these GC/FT-IR spectra by spectra-structure correlations (3). A compilation of the vapor-phase group frequency data has been developed from editorial work performed by Nyquist on the 10,000 vapor-phase spectra published by Sadtler Research Laboratories, G. Division of Bio-Rad Laboratories, Inc. The collection of these Sadtler spectra are a valuable asset for those employing the GC/FT-IR technique to solve real world problems.

Raman spectra of solids and liquids are routinely recorded utilizing dispersive or Fourier transform systems.

For example, Raman spectra of liquid ethynyl benzene and ethynyl benzene-d were recorded utilizing a Hylger spectrometer and 4358 Å radiation (7 Å/mm) filtered through rhodamine/ nitrite filters. Depolarization measurements were made (4). These depolarization measurements aid in distinguishing between in-plane vibrational modes and out-of-plane vibrational modes in the case of ethynyl benzene. The in-plane modes are polarized and the out-of-plane modes are depolarized.

More recently, Raman spectra of inorganics in water solution have been recorded utilizing the Dilor XY Raman triple spectrograph operating in the double subtractive mode and fitted with 1200 g/mm gratings. The detector is a 3-stage Peltier-cooled EG&G silicon CCD model 15305 equipped with a Thomson 1024×256 chip, operated at $-60\,^{\circ}C$ (5).

Sample fluorescence limits the application of the Raman technique because fluorescence is a first-order phenomenon, and the Raman effect is a second-order phenomenon. This fluorescence problem has been overcome recently by the development of FT/Raman. In this case, near-IR is used as the source of excitation of the molecules. Coleyshaw *et al.* reported on the quality of FT-Raman spectra as related to the color of the minerals. They report that white-, gray-, yellow-, pink-, orange-, and red-colored minerals yield good FT-Raman spectra, but they had little success with blue-, green-, or dark-colored minerals (6). This is because these colors absorb red light.

A Nicolet model 800 FT-IR spectrometer/Nicolet FT-Raman accessory equipped with a CaF_2 beamsplitter, Ge detector, and a CVI model C-95 Nd/YAG laser can be used successfully in recording the Raman spectra of many organic and inorganic compounds. The beauty of this combination device is that it can be used to record either IR or Raman spectra. Other manufacturers also produce FT-Raman systems.

REFERENCES

1. Potts, W. J. Jr. (1963). *Chemical Infrared Spectroscopy*, New York: John Wiley & Sons, Inc.

2. Welti, D. (1970). *Infrared Vapor Spectra*, New York: Hyden & Son Ltd.

3. Nyquist, R. A. (1984). *The Interpretation of Vapor-Phase Infrared Spectra: Group Frequency Data*, vols. 1 and 2, Philadelphia: Sadlter Research Laboratories, Division of Bio-Rad Laboratories.

4. Evans, J. C. and Nyquist, R. A. (1960). *Spectrochim. Acta*, **16**, 918.

5. Nyquist, R. A., Putzig, C. L., and Leugers, M. A. (1997). *Infrared and Raman Spectral Atlas of Inorganic Compounds and Organic Salts*, vol. 1, Boston: Academic Press.

6. Coleyshaw, E. E., Griffith, W. P., and Bowell, R. J. (1994). *Spectrochim. Acta*, **50A**, 1909.

Alkyl Carbon–Hydrogen Vibrations

*Numbers in parentheses indicate in-text page reference.

This chapter discusses alkyl carbon–hydrogen molecular vibrations and, in some cases, looks at how these molecular vibrations are affected by their surrounding chemical environment. However, in some cases the alkyl carbon–hydrogen vibrations will be included in the section that discusses their most distinguishing molecular vibrations.

The series of *n*-alkanes, C_nH_{2n+2} were prepared as 0.5 wt. % solutions in CCl_4, $CDCl_3$, and 54.6 mol % $CHCl_3/CCl_4$. Table 3.1 lists the IR frequency data for the νasym. CH_3 and νsym. CH_3 stretching frequencies for C_5H_{12} to $C_{18}H_{38}$ (1). Figure 3.1 shows plots of νasym. CH_3 vs the molecular weight (M.W.) of each *n*-alkane and Fig. 3.2 shows plots of νsym. CH_3 vs M.W. of each *n*-alkane. A study of the IR data and figures show that νasym. CH_3 generally decreases as the number of carbon atoms increases in the order C_5H_{12} to $C_{18}H_{38}$ by 0.11 to 0.22 cm^{-1} in going from solution in CCl_4 to solution in $CDCl_3$.

The νsym. CH_3 mode for this series of n-alkanes shows that it generally decreases by approximately $0.8\,cm^{-1}$ in CCl_4 solution and approximately $1.2\,cm^{-1}$ in $CDCl_3$ solution progressing in the series C_5H_{12} to $C_{18}H_{38}$ in going from solution in CCl_4 to solution in $CDCl_3$.

Table 3.2 lists the IR absorbance data for C_5H_{12} to $C_{18}H_{38}$ for the νasym. CH_3 and νsym. CH_3 modes, Fig. 3.3 shows a plot of (νasym. CH_3)/(νasym. CH_2) vs A(νsym. CH_3)/A(νsym. CH_2) in CCl_4 solution for C_5H_{12} to $C_{18}H_{38}$, and Fig. 3.4 shows a plot of (νsym. CH_3)/(νsym. CH_2) vs A(νasym. CH_3)/A(νasym. CH_2) in CCl_4 solution for C_5H_{12} to $C_{18}H_{38}$. Both plots show an essentially linear relationship. The slight deviation from linearity is most likely due to overlapping interferences in the measurement of these peak height absorbances.

Table 3.2a lists absorbance ratios; all of these absorbance ratios for the νCH_3 and νCH_2 modes generally decrease progressing in the series C_5H_{12} to $C_{18}H_{38}$.

Table 3.3 lists IR data for νasym. CH_2 and νsym. CH_2 for the n-alkane at 0.5 wt. % in CCl_4, $CDCl_3$, and 54.6 mol % $CDCl_3/CCl_4$ solutions. Figure 3.5 shows plots of νasym. CH_2 vs νsym. CH_2 in each of the solvent systems, and Fig. 3.6 shows plots of νasym. CH_3 vs νsym. CH_3 in all three solvent systems. These plots show that these relationships are not linear over the entire n-alkane series. The plots do point out in general that as νasym. CH_2 decreases in frequency, νsym. CH_2 also decreases in frequency progressing in the series C_5H_{12} to $C_{18}H_{38}$, and that the νasym. CH_3 and νsym. CH_3 frequencies show the same trend.

A study of Tables 3.1 and 3.3 show some interesting trends in CCl_4, $CDCl_3$, or 54.6 mol % solutions progressing in the series C_5H_{12} to $C_{18}H_{38}$. The νasym. CH_3 frequency decrease in going from solution in CCl_4 to solution in $CDCl_3$ is small ($\sim 0.1\,cm^{-1}$) while for νsym. CH_3 the frequency decrease is more in $CDCl_3$ solution ($1.2\,cm^{-1}$) than in CCl_4 solution ($0.7\,cm^{-1}$). In addition, the frequency difference for νasym. CH_3 in CCl_4 and in $CDCl_3$ increases from 0.11 to $0.22\,cm^{-1}$ and for νsym. CH_2 in CCl_4 and in $CDCl_3$ decreases from 0.45 to 0.85 progressing in the series C_5H_{12} to $C_{18}H_{38}$. The νasym. CH_2 frequency decrease in going from solution in CCl_4 to solution is small ($0.1\,cm^{-1}$) while for νsym. CH_2 the frequency decrease is larger in going from solution in CCl_4 to a solution in $CDCl_3$ ($0.6\,cm^{-1}$) progressing in the series C_5H_{12} to $C_{18}H_{38}$. Moreover, these data show that the νsym. CH_2 mode changes in frequency by a factor of approximately 5 times more than νasym. CH_2, νasym. CH_3, and νsym. CH_3. In addition, the νasym. CH_2 frequency increases in frequency while the νsym. CH_2 frequency decreases in frequency in going from solution in CCl_4 to $CDCl_3$. In general these last two trends generally decrease progressing in the series C_5H_{12} to $C_{18}H_{38}$.

Table 3.4 lists the frequency difference between νasym. CH_3 and νsym. CH_3 and between νasym. CH_2 and νsym. CH_2 in the three solvent systems. These data show that the frequency separation is much larger for the two νCH_3 vibrations ($\sim 85^{-1}$) than for the two νCH_2 vibrations ($\sim 69\,cm^{-1}$). Figure 3.7 shows plots of νasym. $CH_3 - \nu$sym. CH_3 vs νasym. $CH_2 - \nu$sym. CH_2, which clearly shows the behavior of the frequency separation of the νCH_3 and νCH_2 vibrations in the three solvent systems.

SUMMARY

For n-alkanes, $C_{5-18}H_{12-38}$, the νasym. CH_3 occurs in the region 2957.26–$2959.55\,cm^{-1}$ in CCl_4 and in the region 2957.48–$2959.66\,cm^{-1}$ in $CDCl_3$, and νsym. CH_3 occurs in the region 2872.35–$2873.12\,cm^{-1}$ in CCl_4 and in the region 2871.50–$2872.67\,cm^{-1}$ in $CDCl_3$. Moreover,

$v$asym. CH_2 occurs in the region 2926.61–2927.73 cm^{-1} in CCl_4 and in the region 2926.92–2928.14 cm^{-1} in $CDCl_3$, and $v$sym. CH_2 occurs in the region 2854.59–2861.82 cm^{-1} in CCl_4 and in the region 2854.55–2861.21 cm^{-1} in $CDCl_3$. In addition, these four vibrations decrease in frequency progressing in the series C_5H_{12} through $C_{18}H_{38}$.

The n-alkanes are nonpolar molecules, and one would expect that there would be minimal solute-solvent interaction between n-alkane molecules and solvent molecules such as CCl_4 and $CDCl_3$. The n-alkanes in going from solution CCl_4 to solution in $CDCl_3$ solution show a frequency increase of 0.11 to 0.22 cm^{-1} for $v$asym. CH_3 and for $v$sym. CH_3 it decreases by -0.45 to -0.85 cm^{-1}, with decreases for $v$sym. CH_3 progressing in the order C_5H_{12} to $C_{18}H_{38}$. In addition, the $v$asym. CH_2 frequency difference is 0.41 to 0.23 cm^{-1} and for $v$sym. CH_2 is -0.61 to -0.04 cm^{-1}. Again the $v$asym. CH_2 mode increases in frequency and the $v$sym. CH_2 mode decreases in frequency in going from solution in CCl_4 to solution in $CDCl_3$. Thus, both $v$asym. CH_3 and $v$sym. CH_2 increase in frequency and both $v$sym. CH_3 and $v$sym. CH_2 decrease in frequency in going from solution in CCl_4 to solution in $CDCl_3$ at 0.5 wt. % solutions. These data confirm that the effects of these solvents are minor in these four molecular stretching vibrations. It is noteworthy that the $v$sym. CH_2 vibration shifts progressively to lower frequency in the order C_5H_{12} to $C_{18}H_{38}$, and that it decreases in frequency by a factor of at least 7 times more than the $v$sym. CH_3 vibration.

Apparently, the n-alkane protons form weak intermolecular hydrogen bonds with the free pair of electrons on the Cl atoms of the CCl_4 and/or $CDCl_3$ solvent system, and an explanation is needed to determine the frequency behavior of these four molecular vibrations in going from solution in CCl_4 to solution in $CDCl_3$. As these $v$asym. CH_3, $v$asym. CH_2, $v$sym. CH_3, and $v$sym. CH_2 modes vibrate, the protons obtain a weak positive charge and the carbon atom obtains a weak negative charge. This is the so-called dipole moment change during these molecular vibrations. Therefore, the n-alkane protons would form weak intermolecular hydrogen bonds with the free pair of electrons on the Cl atoms of the CCl_4 and/or $CDCl_3$ solvent system. The Cl atoms of CCl_4 would be expected to be more basic than those for $CDCl_3$ due to the fact that the D atom attracts electrons from the Cl atoms. In addition, there is intermolecular bonding between D and Cl such as $CCl_3D:ClCCl_2D:ClCCl_3$. Therefore, one would expect a stronger C–H:Cl bond to be formed between the protons in n-alkanes and the Cl atoms in CCl_4 than for Cl atoms in $CDCl_3$. Hydrogen bonding also weakens the O–H or C–H bond, and the vibration vOH:X or vC–H:X is expected to decrease in frequency—this is what we noted in the $v$sym. CH_3 and $v$sym. CH_2 modes. However, the opposite was observed for $v$asym. CH_3 and $v$asym. CH_2, where both modes increased in frequency in going from solution in CCl_4 to solution in $CDCl_3$. This frequency increase for the $v$asym. modes needs an explanation. Because the $v$asym. CH_3 and $v$asym. CH_2 modes increase in frequency, it requires more energy for these two modes to vibrate in going from solution in CCl_4 to solution in $CDCl_3$. The two CH_3 groups are isolated by $(CH_2)_n$ groups, and as the Cl atoms in $CDCl_3$ are weaker bases than the Cl atoms in CCl_4 a weaker C–H$\ldots$ClCDCl$_2$ bond is expected. Consequently, the $v$asym. CH_3 and $v$asym. CH_2 increases in frequency when CCl_4 is replaced by $CDCl_3$.

On the other hand there are $(CH_2)_n$ units present in the *n*-alkane series which are capable of forming *n* units of

in either CCl_4 or $CDCl_3$ solution. It is noted that *v*sym. CH_3 increasingly decreases in frequency progressing in the series C_5H_{12} to $C_{18}H_{38}$ in going from solution in CCl_4 to solution in $CDCl_3$. This indicates that the C–H : Cl$CDCl_3$ bond strength is increased as *n* is increased for $CH_3(CH_2)_nCH_3$. The inductive effect of additional CH_2 groups apparently weakens the CH_3 bonds, causing *v*sym. CH_2 to decrease in frequency as the number of CH_2 groups are increased in the *n*-alkane. As the number of CH_2 groups are increased, the decrease in *v*sym. CH_2 for *n*-alkanes decreases progressing in the series C_5H_{12} to $C_{10}H_{22}$ and is relatively constant from C_9H_{20} to $C_{18}H_{38}$. This suggests that the effect of the number of CH_2 groups forming intermolecule hydrogen bonds with a $CDCl_3$ chain is minimized after eight CH_2 groups are present in that the effect of the number of $(CH_2)ClCDCl_3$ intermolecular hydrogen bonds formed between the *n*-alkane and $CDCl_3$ is minimized in the series $C_{11}H_{24}$ to $C_{18}H_{38}$. It is also possible that the Cl atoms in $CDCl_3$ are closer in space to the C–H bonds compared to that for CCl_4, and this fact would also contribute to lower *v*sym. CH_2 frequencies.

In the series C_5H_{12} to $C_{18}H_{38}$ there is a comparatively large change in the *v*sym. $(CH_2)_n$ mode. There is a decrease of $7.23\,cm^{-1}$ CCl_4 and $6.66\,cm^{-1}$ in $CDCl_3$. This is attributed to the increasing number of CH_2 groups stretching in-phase progressing in the *n*-alkane series.

The smooth correlation of the absorbance values of the CH_2 and CH_3 groups as the ratio of the CH_3 groups to CH_2 groups decreases is just what is predicted. There is apparently no significant difference in the dipole moments of either CH_2 or CH_3 stretches progressing in the *n*-alkane series.

OTHER *n*-ALKANE VIBRATIONS

The CH_2 bend asym. CH_3 bend, and the sym. CH_3 bend occurs near 1467, 1458, and $1378.5\,cm^{-1}$ in CCl_4 solution (Table 3.5).

1,2-EPOXYALKANES

Table 3.6 lists IR vapor-phase data for the alkyl (R) vibrations of 1,2-epoxyalkanes (2). The *v*asym. CH_3 mode occurs in the region $2953–2972\,cm^{-1}$, the *v*asym. CH_2 mode in the region $2920–2935\,cm^{-1}$, the *v*sym. CH_2 mode in the region $2870–2932\,cm^{-1}$, and the *v*sym. CH_3 bending mode in the region $1363–1388\,cm^{-1}$.

SODIUM DIMETHYLPHOSPHONATE $(CH_3)_2P(O)_2Na$

The IR and Raman dating for sodium dimethylphosphonate are listed in Table 3.7 (3). The *v*asym. CH_3 and *v*sym. CH_3 modes are assigned at 2985 and $2919\,cm^{-1}$, respectively. The asym.

$(CH_3)_2$ bending modes are assigned at 1428 and 1413 cm^{-1} and the sym. $(CH_3)_2$ bending modes are assigned at 1293 and 1284 cm^{-1}.

METHYLTHIOMETHYL MERCURY, DIMETHYLMERCURY (4), AND METHYLTHIOCHLOROFORMATE (5)

Table 3.8 lists assignments for the CH_3-Hg and CH_3-S groups for the preceding 3 compounds. These assignments should aid the reader in assigning vibrations for these CH_3 groups in other compounds.

CYCLOALKANES

Table 3.9 lists IR vapor-phase data for cycloalkanes (6). Raman data for the ring breathing and ring deformation modes are also presented for cyclobutane and cyclopentane. The νasym. CH_2 mode occurs in the region 2930–3100 cm^{-1}. The νsym. CH_2 mode occurs in the region 2880–3020 cm^{-1}. Both νCH_2 vibrations decrease in frequency as the ring becomes larger. This is the result of lesser ring strain with increasing ring size. The CH_2 bend, CH_2 wag, CH_2 twist, CH_2 rock, ring breathing, and ring deformation vibration assignment are also presented.

MISCELLANEOUS ALKYL AND CYCLOALKYL COMPOUND

Vibrational assignments for cycloalkyl groups are presented in Table 3.10, for alkyl groups of monomers and polymers in Table 3.11, for cyclopropane derivatives in Table 3.12, and for octadecane, octadecane-D_{38}, tetracosane, and tetracosane-D_{50} in Table 3.13.

REFERENCES

1. Nyquist, R. A. and Fiedler, S. L. (1993). *Appl. Spectrosc.*, **47**, 1670.
2. Nyquist, R. A. (1986). *Appl. Spectrosc.*, **40**, 275.
3. Nyquist, R. A. (1968). *J. Mol. Struct.*, **2**, 111.
4. Nyquist, R. A. and Mann, J. R. (1972). *Spectrochim. Acta*, **28A**, 511.
5. Nyquist, R. A. (1967–68). *J. Mol. Struct.*, **1**, 1.
6. Nyquist, R. A. (1984). *The Interpretation of Vapor-Phase Infrared Spectra: Group Frequency Data*, Philadelphia: Sadtler Research Laboratories, A Division of Bio-Rad.

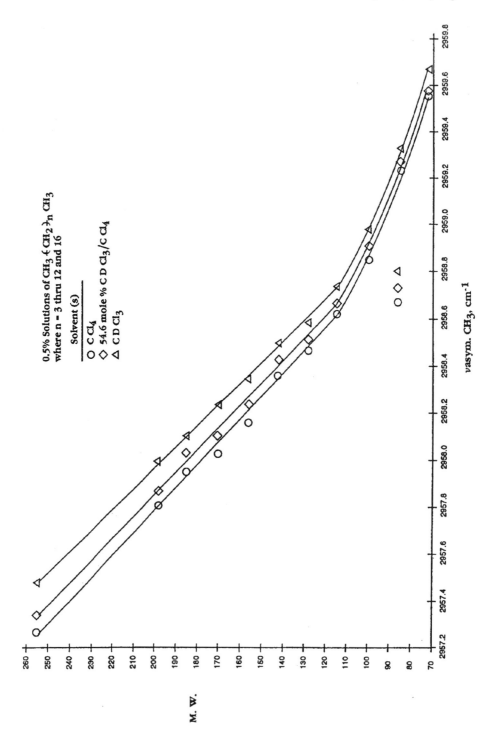

FIGURE 3.1   Plots of νasym. CH_3 vs the molecular weight of each n-alkane.

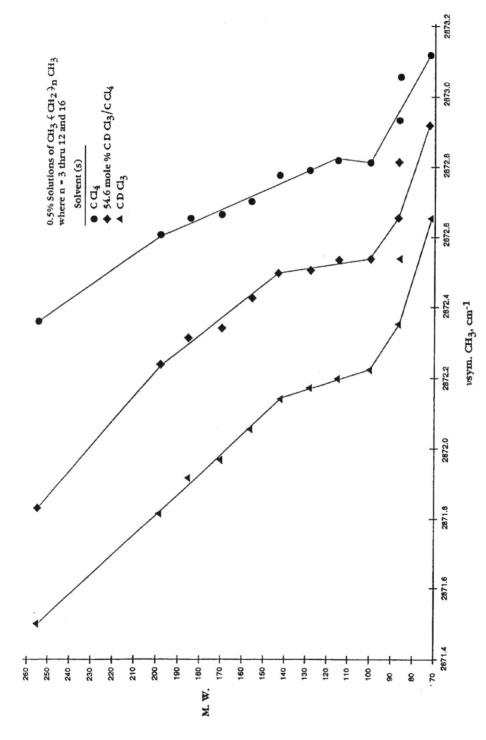

FIGURE 3.2 Plots of νsym. CH_3 vs the molecular weight of each n-alkane.

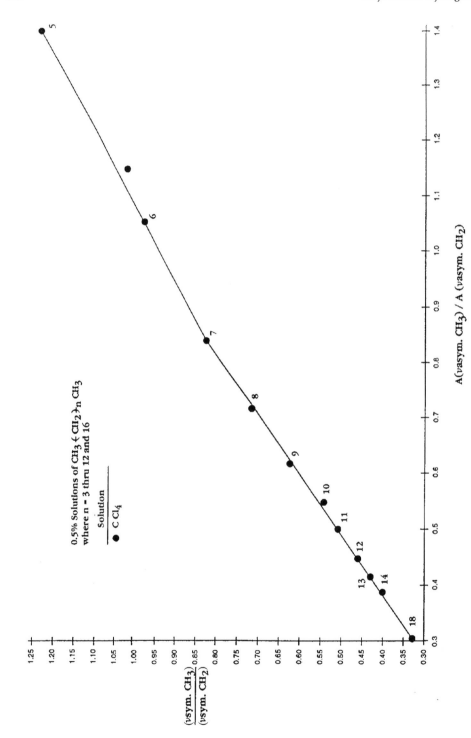

FIGURE 3.3 A plot of the absorbance ratio $A(\nu asym.\ CH_3)/A(\nu asym.\ CH_2)$ vs the absorbance ratio $A(\nu sym.\ CH_3)/A(\nu sym.\ CH_2)$ in CCl_4 solution for C_5H_{12} to $C_{18}H_{38}$.

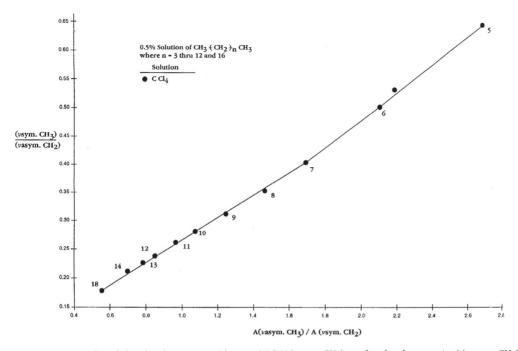

FIGURE 3.4 A plot of the absorbance ratio A($v$sym. CH_3)/A($v$sym. CH_2) vs the absorbance ratio A($v$asym. CH_3)/A($v$asym. CH_2) in CCl_4 solution for C_5H_{12} to $C_{18}H_{38}$.

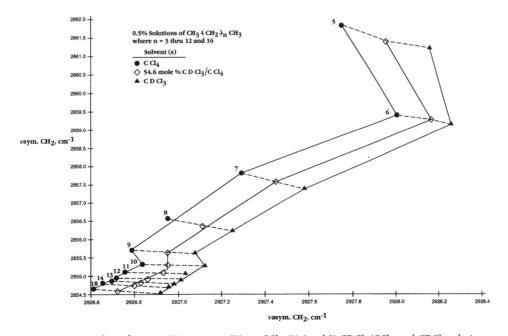

FIGURE 3.5 Plots of $v$asym. CH_2 vs $v$sym. CH_2 in CCl_4, 54.6 mol % $CDCl_3$/CCl_4, and $CDCl_3$ solutions.

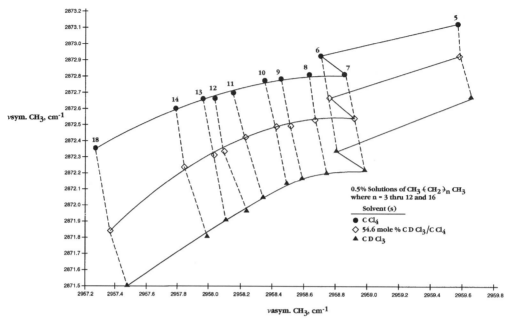

FIGURE 3.6   Plots of νasym. CH_3 vs νsym. CH_3 in CCl_4, 54.6 mol % $CDCl_3/CCl_4$, and $CDCl_3$ solutions.

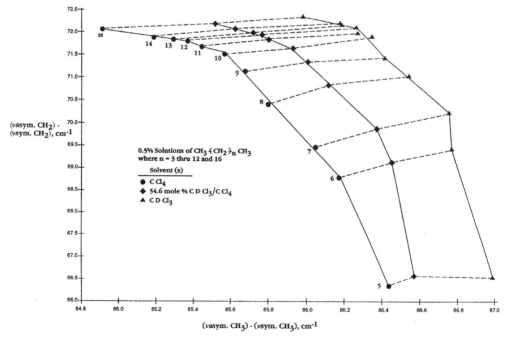

FIGURE 3.7   Plots of the frequency separation (νasym. CH_3-νsym. CH_3) vs the frequency separation (νasym. CH_2-νsym. CH_2) for each n-alkane in CCl_4, 54.6 mol % $CDCl_3/CCl_4$, and $CDCl_3$ solutions.

TABLE 3.1 IR data for a. and s.CH$_3$ stretching for n-alkanes in 0.5 mol % solutions in CCl$_4$, 54.6 mol % CHCl$_3$/CCl$_4$

Compound	a.CH$_3$ str. CCl$_4$ cm^{-1}	a.CH$_3$ str. 54.6 mol % CDCl$_3$/CCl$_4$ cm^{-1}	a.CH$_3$ str. CDCl$_3$ cm^{-1}	s.CH$_3$ str. CCl$_4$ cm^{-1}	s.CH$_3$ str. 54.6 mol % CDCl$_3$/CCl$_4$ cm^{-1}	s.CH$_3$ str. CDCl$_3$ cm^{-1}
Pentane	2959.6	2959.6	2959.7	2873.1	2872.9	2872.7
Hexane	2958.7	2958.7	2958.8	2872.9	2872.7	2872.3
Hexane	2959.2	2959.3	2959.3	2873.1	2872.8	2872.5
Heptane	2958.9	2958.9	2959	2872.8	2872.5	2872.2
Octane	2958.6	2958.7	2958.7	2872.8	2872.5	2872.2
Nonane	2958.5	2958.5	2958.6	2872.8	2872.5	2872.2
Decane	2958.4	2958.4	2958.5	2872.8	2872.5	2872.1
Undecane	2958.1	2958.2	2958.3	2872.7	2872.4	2872.1
Dodecane	2958	2958.1	2958.2	2872.7	2872.3	2872
Tridecane	2957.9	2958	2958.1	2872.7	2872.3	2871.9
Tetradecane	2957.8	2957.9	2858	2872.6	2872.2	2871.8
Octadecane	2957.3	2957.3	2957.5	2872.4	2871.8	2871.5
delta cm^{-1}	−2.3	−2.3	−2.2	−0.7	−1.1	−1.2

Compound	a.CH$_3$ str. [CCl$_4$]- [54.6 mol % CDCl$_3$/CCl$_4$] cm^{-1}	a.CH$_3$ str. [CCl$_4$]- [CDCl$_3$] cm^{-1}	s.CH$_3$str. [CCl$_4$]- [54.6 mol % CDCl$_3$/CCl$_4$] cm^{-1}	s.CH$_3$ str. [CCl$_4$]- [CDCl$_3$] cm^{-1}
Pentane	0.02	0.11	−0.2	−0.45
Hexane	0.03	0.08	−0.24	−0.51
Heptane	0.05	0.12	−0.27	−0.58
Octane	0.04	0.11	−0.29	−0.62
Nonane	0.04	0.12	−0.29	−0.62
Decane	0.06	0.14	−0.29	−0.64
Undecane	0.08	0.19	−0.28	−0.65
Dodecane	0.16	0.2	−0.33	−0.7
Tridecane	0.08	0.16	−0.34	−0.74
Tetradecane	0.06	0.2	−0.36	−0.79
Octane	0.08	0.22	−0.52	−0.85

TABLE 3.2 IR absorbance data for *n*-alkanes: a. and s.CH_2 and CH_3 stretching and CH_2 bending

Compound	A[a.CH_3 str.]	A[a.CH_2 str.]	A[s.CH_3 str.]	A[s.CH_2 str.]
Pentane	1.09	0.78	0.502	0.407
Hexane	0.879	0.759	0.408	0.402
Hexane	0.903	0.86	0.419	0.43
Heptane	0.769	0.925	0.372	0.45
Octane	0.67	0.938	0.329	0.458
Nonane	0.637	1.027	0.319	0.512
Decane	0.563	1.028	0.287	0.527
Undecane	0.538	1.083	0.283	0.558
Dodecane	0.478	1.073	0.256	0.56
Tridecane	0.496	1.197	0.271	0.632
Tetradecane	0.473	1.23	0.263	0.66
Octadecane	0.374	1.234	0.221	0.672

	A[CH_2 bend] CCl_4	A[CH_2 bend] 54.6 mol % $CDCl_3/CCl_4$	A[CH_2 bend] $CDCl_3$
Pentane	0.137	0.141	0.105
Hexane	0.127	0.115	0.082
Hexane	0.152	0.123	0.09
Heptane	0.147	0.111	0.084
Octane	0.128	0.096	0.073
Nonane	0.143	0.103	0.063
Decane	0.136	0.092	0.06
Undecane	0.134	0.093	0.057
Dodecane	0.132	0.084	0.05
Tridecane	0.141	0.081	0.051
Tetradecane	0.141	0.081	0.047
Octadecane	0.132	0.08	0.037

TABLE 3.2A   Absorbance ratios for CH_3 and CH_2 groups for n-alkanes

Compound	A[a.CH_3 str.] /A[s.CH_3 str.]	A[a.CH_2 str.] /A[s.CH_2 str.]	A[a.CH_3 str.] /A[a.CH_2 str.]	A[s.CH_3 str.] /A[s.CH_2 str.]
Pentane	2.171	1.916	1.397	1.233
Hexane	2.154	1.912	1.143	1.015
Hexane	2.155	2	1.05	0.974
Heptane	2.067	2.064	0.831	0.827
Octane	2.036	2.048	0.714	0.718
Nonane	1.997	2.006	0.62	0.623
Decane	1.962	1.951	0.548	0.545
Undecane	1.901	1.941	0.497	0.507
Dodecane	1.867	1.916	0.445	0.457
Tridecane	1.83	1.894	0.414	0.428
Tetradecane	1.798	1.864	0.385	0.398
Octadecane	1.692	1.836	0.303	0.329

	A[a.CH_3 str.] /A[s.CH_2 str.]	A[s.CH_3 str.] /A[a.CH_2 str.]	A[s.CH_3 str.] /A[CH_2 bend]
Pentane	2.678	0.644	4.781
Hexane	2.187	0.531	4.976
Hexane	2.1	0.5	4.656
Heptane	1.689	0.402	4.429
Octane	1.463	0.351	4.507
Nonane	1.244	0.311	5.063
Decane	1.068	0.279	4.783
Undecane	0.964	0.261	4.965
Dodecane	0.854	0.238	5.12
Tridecane	0.785	0.226	5.313
Tetradecane	0.717	0.214	5.596
Octadecane	0.557	0.179	5.972

TABLE 3.3 IR data for a. and s.CH$_2$ stretching for n-alkanes in 0.5% solutions in Cl$_4$, 54.6 mol % CDCl$_3$/CCl$_4$, and CDCl$_3$

Compound	a.CH$_2$ str. CCl$_4$ cm^{-1}	a.CH$_2$ str. 54.6 mol % CDCl$_3$/CCl$_4$ cm^{-1}	a.CH$_2$ str. CDCl$_3$ cm^{-1}	s.CH$_2$ str. CCl$_4$ cm^{-1}	s.CH$_2$ str. 54.6 mol % CDCl$_3$/CCl$_4$ cm^{-1}	s.CH$_2$ str. CDCl$_3$ cm^{-1}
Pentane	2927.7	2928	2928.1	2861.8	2861.4	2861.2
Hexane	2928	2928.2	2928.3	2959.4	2859.3	2859.1
Hexane	2927.9	2928	2928.1	2859.1	2859	2858.7
Heptane	2927.3	2927.5	2927.6	2857.8	2857.6	2857.4
Octane	2927	2927.1	2927.3	2856.6	2856.3	2856.2
Nonane	2926.8	2927	2927.1	2855.7	2855.6	2855.6
Decane	2926.8	2927	2927.1	2855.3	2855.3	2855.3
Undecane	2926.8	2926.9	2927	2855.1	2855.1	2855.1
Dodecane	2926.7	2926.9	2927	2854.9	2854.9	2854.9
Tridecane	2926.7	2926.8	2927	2854.8	2854.8	2854.8
Tetradecane	2926.7	2926.8	2927	2854.8	2854.7	2854.7
Octadecane	2926.6	2926.7	2926.9	2854.6	2854.6	2854.6
delta cm^{-1}	−1.1	−1.3	−1.2	−7.2	−6.8	−6.6

	a.CH$_2$ str. [CCl$_4$]- [mol % CDCl$_3$/CCl$_4$] cm^{-1}	a.CH$_2$ str. [CCl$_4$]- [CDCl$_3$] cm^{-1}	s.CH$_2$ str. [CCl$_4$]- [mol % CDCl$_3$/CCl$_4$] cm^{-1}	s.CH$_2$ str. [CCl$_4$]- [CDCl$_3$] cm^{-1}
Pentane	0.22	0.41	−0.41	−0.61
Hexane	0.16	0.25	−0.08	−0.26
Hexane	0.15	0.23	−0.17	−0.38
Heptane	0.16	0.29	−0.25	−0.44
Octane	0.16	0.31	−0.21	−0.33
Nonane	0.16	0.28	−0.09	−0.1
Decane	0.12	0.3	−0.04	−0.06
Undecane	0.17	0.27	−0.02	−0.04
Dodecane	0.14	0.29	−0.02	−0.04
Tridecane	0.13	0.28	−0.01	−0.04
Tetradecane	0.14	0.3	−0.03	−0.06
Octadecane	0.12	0.31	−0.01	−0.06

TABLE 3.4   The frequency separation between the a. and s.CH$_3$ and the a. and s.CH$_2$ stretching vibrations for n-alkanes in CCl$_4$, CDCl$_3$/CCl$_4$, and CDCl$_3$ solutions

Compound	[a.CH$_3$ str.]- [s.CH$_3$ str.] CCl$_4$ cm^{-1}	[a.CH$_3$ str.]- [s.CH$_3$ str.] 54.6 mol % CDCl$_3$/CCl$_4$ cm^{-1}	[a.CH$_3$ str.]- [s.CH$_3$ str.] CDCl$_3$ cm^{-1}	[a.CH$_2$ str.]- [s.CH$_2$ str.] CCl$_4$ cm^{-1}	[a.CH$_2$ str.]- [s.CH$_2$ str.] mol % CDCl$_3$/CCl$_4$ cm^{-1}	[a.CH$_2$ str.]- [s.CH$_2$ str.] CDCl$_3$ cm^{-1}
Pentane	86.43	86.55	86.99	66.32	66.54	66.52
Hexane	86.18	86.45	86.77	66.6	68.84	69.11
Heptane	86.05	86.37	86.75	69.48	69.89	70.21
Octane	85.8	86.13	86.53	70.4	70.77	71.04
Nonane	85.67	86	86.41	71.11	71.36	71.42
Decane	85.57	85.92	86.35	71.52	71.68	71.88
Undecane	85.44	85.8	86.28	71.67	71.86	71.98
Dodecane	85.37	85.76	86.27	71.8	71.96	72.13
Tridecane	85.29	85.71	86.19	71.87	72.01	72.19
Tetradecane	85.19	85.61	86.18	71.89	72.06	72.25
Octadecane	84.91	85.51	85.98	72.02	72.15	72.34
delta cm^{-1}	−1.52	−1.04	−1.01	5.7	5.61	5.82

TABLE 3.5   The CH$_3$ and CH$_2$ bending frequencies and frequency separations for n-alkanes

Compound	CH$_2$ bend cm^{-1}	a.CH$_3$ bend cm^{-1}	s.CH$_3$ bend cm^{-1}	[CH$_2$ bend]- [s.CH$_3$ bend] cm^{-1}	[a.CH$_3$ bend]- [s.CH$_3$ bend] cm^{-1}
Petane	1467.55	1458	1379.23	88.32	78.77
Hexane	1467.22	1458.37	1378.48	88.74	79.89
Hexane	1467.35	1458.29	1378.59	88.76	79.7
Heptane	1467.39	1458.08	1378.66	88.73	79.42
Octane	1467.36	1458.85	1378.62	88.74	80.23
Nonane	1467.33	1458.27	1378.57	88.76	79.7
Decane	1467.39	1458.56	1378.39	89	80.17
Undecane	1467.63	1458.27	1378.6	88.76	79.67
Dodecane	1467.39	1458.27	1378.62	88.77	79.65
Tridecane	1467.36	1458.27	1378.64	88.72	79.63
Tetradecane	1467.35	1458.56	1378.61	88.74	79.95
Octadecane	1467.33	1457.97	1378.64	88.69	79.33

TABLE 3.6   IR vapor-phase data for the alkyl group of 1,2-epoxyalkanes

1,2-epoxyalkane R	a.CH$_3$ str. cm^{-1}	A	s.CH$_2$ str. cm^{-1}	A	a.CH$_2$ str. cm^{-1}	A	a.CH$_3$ bending cm^{-1}	A	s.CH$_3$ bending cm^{-1}	A
CH$_3$	2969	0.835	2930	0.594			1457	0.368	1370	0.25
							1445	0.353		
C$_2$H$_5$	2972	1.568	2932	0.804	2920	0.769	1465	0.595	1379	0.158
C$_3$H$_7$	2965	~1.8	2920	0.6	2934	1.37	1467	0.473	1381	0.226
iso-C$_4$H$_9$	2970	~1.8	2920	0.729	2935	1.158	1469	0.672	1388[*1]	0.428
									1371[*2]	0.415
iso-C$_3$H$_7$	2970	~1.7	2881	0.533			1470	0.574	1382[*1]	0.205
									1366[*2]	0.32
t-C$_4$H$_9$	2965	~1.8	2870	0.561			1467	0.435	1381[*1]	0.168
									1366[*2]	0.783
C$_8$H$_{17}$	2960	0.649	2870	0.49	2930	0.88	1467	0.28	1378	0.087
C$_9$H$_{19}$	2959	0.52	2870	0.37	2929	0.87	1465	0.238	1378	0.07
C$_{12}$H$_{25}$	2958	0.546	2878	0.435	2930	0.847	1468	0.302	1377	0.07
C$_{14}$H$_{29}$	2958	0.482	2870	0.386	2920	0.842	1465	0.295	1374	0.07
C$_{16}$H$_{33}$	2953	0.47			2920	0.82	1462	0.27	1372	0.052
Range	2953–2972		2870–2932		2920–2935		1445–1470		1363–1388	

[*1] [in-phase].
[*2] [out-of-phase].

TABLE 3.7   IR and Raman data for $(CH_3)_2PO_2Na$ in water and in the solid phase

$(CH_3)_2P(O)_2Na$ Assignments	IR solid phase cm^{-1}	IR H_2O soln. cm^{-1}	Raman H_2O soln. cm^{-1}	5 a1	4 a2	4 b1	5 b2
a.[$(CH_3)_2$ str.]	2985		2988	1	1	1	1
s.[$(CH_3)_2$ str.]	2919		2921	1			1
a.[$(CH_3)_2$ bend]	1428	1429	1422	1	1	1	1
	1413	~1412	1413				
s.[$(CH_3)_2$ bend]	1293	1309		1			1
	1284	1301					
	912		915	1			
	860sh						
$(CH_3)_2$ rock	851	878			1	1	1
	839sh						
$(CH_3)_2$ torsion	?	?			1	1	

TABLE 3.8   IR data and assignments for CH_3HgSCH_3, $CH_3SC(C=O)Cl$, and $(CH_3)_2Hg$

$CH_3-Hg-S-CH_3$ cm^{-1}	$CH_3-S-C(=O)Cl$ cm^{-1}	$CH_3-S-P(=O)Cl_2$ cm^{-1}	$(CH_3)_2Hg$ cm^{-1}	Assignment
[CH_3-Hg group]			[CH_3-Hg group]	
2984			2970	a.CH_3 str.
2919			2910	a.CH_3 str.
1408			1397 or	a.CH_3 bend
			1443	
1177			1182	s.CH_3 bend
763			700, 788	CH_3 rock
?			?	CH_3 torsion
[CH_3-S group]	[CH_3-S group]	[CH_3-S group]		
2984	3025	3011		a.CH_3 str.
2919	2940	2938		s.CH_3 str.
1432	1430	1431		a.CH_3 bend
1309	1320	1321		s.CH_3 bend
956	976	972		CH_3 rock
?	?	?		CH_3 torsion
533				Hg−C str.
333				Hg−S str.
190				Hg−S−C bend
120				Hg−S−CH_3 torsion ?

? not observed.

TABLE 3.9 IR vapor-phase data for cycloalkanes

Compound	a.CH₂ str.	s.CH₂ str.	CH₂ bend	CH₂ wag	CH₂ twist	CH₂ rock	Ring breathing	Ring def.	(A)s.CH₂ str. /(A)a.CH₂ str.
Cyclopropane	3130 (0.200)	3035 (0.480)	1439 (0.022)	1050 (0.175)				899 (0.206)	0.46
	3100 (1.250)	3020 (0.570)	1430 (0.018)	1024 (0.325)				862 (0.512)	
		2998 (0.325)	1414 (0.020)	1000 (0.160)				832 (0.162)	
Cyclobutane	[2974, vs]	[2945, s]	[1443, m]	[1260, s]	[1224, m]		[1188]		
	[2965, vs]								
	2988 (0.850)								
Cyclopentane	2960 (1.250)	2885 (0.400)	1479 (0.050)				[1005, vs, R]	[926, s, R]	0.32
	2945 (0.500)		1460 (0.061)				[889, stg, R]	[901, w, R]	
			1445 (0.050)					914 (0.020)	
								895 (0.040)	
								870 (0.020)	
Cyclohexane	2945 (0.950)	2875 (0.352)	1475 (0.074)	1271 (0.015)	1041 (0.020)		879 (0.010)	910 (0.020)	0.49
	2935 (1.250)	2860 (0.619)	1459 (0.180)	1261 (0.019)			861 (0.020)	900 (0.020)	
		2850 (0.340)	1444 (0.076)	1245 (0.012)			842 (0.012)		
Cycloheptane	2935 (1.250)	2860 (0.230)	1464 (0.060)	1354 (0.005)	950 (0.005)	730 (0.001)	812 (0.011)		0.18
Cyclooctane	2930 (1.250)	2862 (0.619)	1460 (0.143)	1355 (0.044)	1042 (0.020)	767 (0.020)	852 (0.020)	952 (0.010)	0.49
Cyclodecane	2930 (1.250)	2880 (0.370)	1486 (0.110)	1354 (0.040)	1286 (0.040)	720 (0.040)			0.29
		2865 (0.340)	1454 (0.120)						0.29
Cyclododecane	2940 (1.250)	2870 (0.520)	1471 (0.120)	1350 (0.041)	1310 (0.020)	718 (0.060)	1030 (0.020)	1080 (0.011)	0.42
			1453 (0.115)						
	a.CD₂ str.	s.CD₂ str.	CD₂ bend	CD₂ wag	CD₂ twist	CD₂ rock			(A)s.CD₂ str. /(A)a.CD₂ str.
Cyclododecane-D₂₄	2202 (1.250)	2102 (0.610)	1160 (0.040)	1160 (0.040)	1118 (0.050)	550 (0.042)	983 (0.085)	1095 (0.180)	0.49
Cyclododecane /cyclododecane-D₂₄	1.335	1.365	1.268 1.253	1.164	1.172	1.305	1.048	0.986	

TABLE 3.10   Raman and IR data for compounds containing cycloalkyl groups

Compound	Raman neat a.CH$_2$ str.	Raman neat s.CH$_2$ str.	Raman neat s.CH$_2$ str. in F.R.	Raman neat CH$_2$ bend	IR v.p. a.CH$_2$ str.	IR v.p. s.CH$_2$ str.
Cyclopropane (vap.)		3039 (83, p)	3020 (37, p)	1452 (2, p)	3100	3020
Cyclopropane carboxylic acid		3020 (37, p)		1458 (5, p)	3102	3032
Chlorocyclobutane	2985 (26, p)	2955 (44, p)		1433 (4, p)		
Cyclobutane carboxylic acid					2995	2890
Cyclopentane	2943 (43, p)	2868 (45, p)		1446 (6, p)	2960	2880
Chlorocyclopentane	2971 (46, p)	2919 (33, p)	2876 (21, p)	1446 (6, p)	2980	2890
Bromocyclopentane	2967 (43, p)	2915 (29, p)		1448 (5, p)	2970	2890
Cyclopentanecarbonitrile[1]	2970 (40, p)	2875 (29, p)		1446 (6, p)		
Cyclopentane carboxylic acid	2967 (41, p)	2875 (29, p)		1449 (8, p)	2970	2890
Cyclopentyl alcohol	2962 (84, p)	2875 (45, p)		1448 (9, p)		
Cyclohexane	2924 (40)	2852 (52)		1446 (10)	2935	2860
Chlorocyclohexane					2948	2870
Bromocyclohexane					2942	2865
Cyclohexyl alcohol	2940 (42, p)	2855 (45, p)		1440 (7, p)	2940	2864
Cyclohexyl amine	2937 (71, p) 2920 (59, p)	2855 (76, p)		1440 (10, p)	2938	2862
1,2,4-Trivinylcyclohexane	2932 (4)	2852 (4)		1443 (1)		
Cycloheptane	2927 (42, p)	2853 (37, p)		1441 (8, p)		
Cyclodecane	2914 (84, p)			1442 (10, p)	2930	2880

[1] [CN str., 2234 (40, p)].

TABLE 3.11 Raman data for the CH stretching and bending modes of the alkyl group, and C–C–C=O skeletal bending for monomers and polymers

Compound	a.CH_3O str	a.CH_3 str	s.CH_3 str	a.CH_2 str	s.CH_2 str	s.CH_3 str	CH_2 bend	a.CH_3 bend	s.CH_3 bend	CC=O bend
Methyl acrylate	3001 (4)		2957 (4)	2924 (2)		2858 (4)	1443 (1)	1451 (1)	1391 (0)	601 (4)
Poly(methyl methacrylate)			2952 (9)			2844 (1)		1455 (6)	1392 (0)	602 (4)
Ethyl methacrylate								1453 (3)		
Propyl acrylate			2941 (7)		2882 (5)			1452 (5)		
Propyl methacrylate		2963 (4)	2931 (9)		2882 (5)			1453 (4)		604 (4)
Poly(propyl acrylate)			2938 (9)		2881 (5)			1452 (5)		
Poly(propyl methacrylate)			2938 (9)		2881 (5)			1452 (3)		602 (3)
Dibutyl phosphonate		2965 (5)	2938 (8)	2913 (9)	2876 (9)			1452 (3)		[2434 (1) P–H str.]
Dibutyl fumarate		2964 (5)	2939 (7)	2914 (8)	2877 (7)			1451 (3)		
Dibutyl phthalate			2938 (7)	2914 (7)	2876 (6)			1450 (3)		
Poly(butyl acrylate)			2937 (9)	2919 (9)	2875 (7)			1451 (4)		
Poly(butyl methacrylate)		2963 (9)		2915 (8)	2876 (6)					603 (3)
Hexyl methacrylate		2961 (4)	2930 (9)	2901 (7)	2876 (6)			1441 (3)		604 (2)
Bis(isoctyl) fumarate		2963 (8)		2909 (8)	2878 (9)			1450 (4)		
Vinyl(isooctyl ether)			2933 (7)	2913 (7)	2874 (9)			1450 (3)		
Poly(isodecyl methacrylate)		2959 (7)	2932 (9)		2875 (9)					605 (1)
Undecyl acrylate				2900 (8)	2854 (8)			1439 (4)		
Undecyl methacrylate			2928 (9)		2854 (8)			1439 (5)		604 (2)
Vinyl undecyl ether				2900 (8)	2854 (8)			1439 (4)		
Dodecyl acrylate				2897 (8)	2853 (9)			1440 (4)		
Dodecyl methacrylate			2929 (9)	2896 (9)	2853 (9)			1439 (4)		605 (1)
Poly(dodecyl methacrylate)			2929 (8)	2804 (9)	2853 (9)			1440 (4)		
Vinyl dodecyl ether				2895 (8)	2853 (9)	2874 (8)		1439 (3)		
Strontium stearate			2923 (3)	2884 (9)	2852 (5)			1447 (4)		
Zinc stearate				2882 (9)	2849 (5)			1440 (3)		
Poly(vinyl stearate)				2882 (9)	2848 (8)			1438 (3)		
Vinyl octadecyl ether				2890 (7)	2852 (9)			1439 (3)		
Poly(octadecyl acrylate)				2862 (9)	2847 (7)			1438 (3)		
Octadecyl methacrylate		2964 (8)	2928 (7)	2894 (8)	2852 (9)			1439 (4)		605 (1)
Poly(octadecyl methacrylate)		2965 (7)	2928 (7)	2892 (8)	2852 (9)	2875 (9)		1440 (4)		605 (0)
Isobutyl methacrylate		2964 (9)	2931 (9)	2922 (8)	2877 (9)			1453 (4)		603 (3)
Poly(isobutyl methacrylate)			2937 (9)	2901 (7)	2875 (9)			1453 (4)		602 (3)
Vinyl isobutyl ether		2962 (5)	2943 (4)	2911 (6)	2876 (9)			1464 (2)		
Poly(vinyl isobutyl ether)		2957 (6)	2939 (9)	2913 (8)	2872 (9)			1462 (3)		

2-Ethylhexyl acrylate	2963 (5)	2939 (9)		2877 (9)		1451 (3)	
2-Ethylhexyl methacrylate	2983 (5)	2931 (9)	2900 (7)	2877 (7)		1450 (4)	
3,3,5-Trimethylhexyl methacrylate	2960 (7)	2930 (9)					
Ethylene dimethacrylate	2962 (5)	2931 (8)			1380 (2)		603 (3) [801 (9) ring breathing]
Cyclohexyl acrylate	2954 (9)			2861 (8)		1447 (4)	
3,5,5-Trimethylcyclohexyl acrylate	2956 (9)	2931 (9)		2873 (9)		1463 (4) 1450 (4)	
1,1,1-Trimethylolethane tri-acrylate	2966 (2)		2902 (1)			1464 (1)	
Trimethylolpropane triethoxyacrylate	2951 (4)			2885 (4)			
Poly(2-Hydroxyethyl methacrylate)	2944 (9)					1455 (9)	604 (9)
2-Hydroxypropyl acrylate	2940 (4)			2885 (2)		1458 (2)	
Poly(2-Hydroxypropyl methacrylate)	2939 (9)					1455 (8)	604 (5)
Propargyl acrylate			2953 (1)				[2132 (9) CC str.]
Propargyl methacrylate		2933 (1)				1439 (1)	[2132 (9) CC str.]
1,H,1H,3H-terafluoro-propyl methacrylate	2971 (5)	2937 (7)				1439 (1)	
2,2,2-Trifluoromethyl methacrylate	2978 (5)	2937 (7)				1453 (2)	603 (3)
2,2,2-Trichloroethyl acrylate	2980 (2)					1455 (0)	
Trichloromthyl methacrylate		2929 (0)					
Tribromoneopentyl acrylate			2921 (2)	2883 (1)		1465 (2) 1434 (2)	
Tribromoneopentyl methactylate	2969 (5)	2928 (2)				1467 (4)	606 (3)
N,N-dimethylaminoethyl methacrylate	2954 (9)	2931 (9)				1447 (4)	612 (1)
Trimethylammonium methosulfate methacrylate	2963 (2)					1460 (3)	[1063 (9) s.SO3 str.]
3-Sulfopropyl potassium salt methacrylate	2983 (3)	2933 (5)				1446 (1)	601 (1) 619 (1)
Poly(phenyl methacrylate)						1448 (3)	615 (2) i.p. ring

TABLE 3.12 IR vapor-phase data for cyclopropane derivatives

Compound	a.CH₂ str.	s.CH₂ str.	CH₂ bend	CH₂ wag	Ring Breathing	CH₂ twist	CH₂ rock	Ring Deformation	Other
Cyclopropane	3130 (0.200) 3035 (0.480); 3100 (1.250)	1439 (0.022); 3020 (0.570); 2998 (0.325)	1050 (0.175); 1430 (0.018); 1414 (0.020)	1024 (0.325); 1000 (0.160)	[1188]		899 (0.206)	862 (0.512); 832 (0.162)	
Bromocyclopropane	3110 (0.170); 3101 (0.210); 3090 (0.220)	3030 (0.280); 3020 (0.300); 3010 (0.250)	1446 (0.172)	1045 (0.329); 1035 (0.430); 1025 (0.240)	1279 (1.240); 1271 (1.040); 1260 (1.140)	1191 (0.035)	868 (0.100)	811 (0.160); 800 (0.180)	C—Br str. 560 (0.280); 550 (0.240); 544 (0.220)
Cyclopropane-carbonitrile	3125 (0.210); 3118 (0.279); 3105 (0.210)	3050 (0.355); 3040 (0.430); 3030 (0.410)	1445 (0.210); 1440 (0.155); 1434 (0.171)	1074 (0.490); 1060 (0.630); 1050 (0.620)	950 (1.103); 940 (1.115); 930 (1.215)	1205 (0.080); 1199 (0.060); 1186 (0.089)	735 (0.255); 722 (0.290); 711 (0.180)	824 (0.375); 814 (0.435); 805 (0.415)	CN. str. 2260 (0.540); 2255 (0.625); 2245 (0.530)
1-Phenylcyclopropane-carbonitrile	3100 (0.250)	3038 (0.400)	1447 (0.220)	1030 (0.282)	942 (0.310)	1191 (0.010)	masked	860 (0.068)	2235 (0.250); Phenyl o.p. ip. H5 def. 751 (0.940); o.p. Phenyl def. 694 (1.250)
Cyclopropane-carboxylic acid	3108 (0.040)	3035 (0.090)	1425 (0.330); 1413 (0.360); 1405 (0.320)	1060 (0.080); 1035 (0.160); 1018 (0.100)	938 (0.110); 929 (0.150); 919 (0.140)	1181 (0.160)	755 (0.100)	825 (0.165)	O—H str. 3590 (0.291), 3580 (0.340), 3575 (0.260); C=O str. 1779 (0.950), 1772 (1.250), 1768 (1.050); C—O str. 1135 (1.150), 1130 (1.250), 1125 (1.125); γ C=O 585 (0.355), 570 (0.290), 562 (0.290); 2(C=O)
Methyl cyclopropane-carboxylate	3102 (0.052) [a.CH₃ str.] [2960 (0.200)]	3024 (0.169) [s.CH₃ str.] [2910 (0.050)]		1035 (0.100)	905 (0.172)			827 (0.100)	C=O str. 1751 (0.945); o.p. Phenyl def. 1179 (1.240), 1091 (0.295); 2(C=O) 3490 (0.015)
Cyclopropylbenzene	masked Phenyl CH str. 3084 (1.050)	3035 (1.250) & Phenyl CH str.	1465 (0.130)	1028 (0.340)	898 (0.270)	1178 (0.015)	masked	812 (0.130)	Phenyl o.p. ip. H5 def. 750 (0.730); o.p. Phenyl def. 695 (1.030)

TABLE 3.13 IR data for octadecane, octadecane-D$_{38}$, tetracosane, and tetracosane-D$_{50}$

Compound	a.CH$_3$ str.	a.CH$_2$ str.	s.CH$_3$ str.	s.CH$_2$ str.	a.CH$_3$ bend	CH$_2$ bend	s.CH$_3$ bend	CH$_2$ wag	CH$_2$ twist	CH$_3$ rock?	CH$_2$ rock
Octadecane	2970 (0.471)	2930 (1.250)		2863 (0.621)	1465 (0.120)		1385 (0.030)	1354 (0.042)	1310 (0.1030)	1105 (0.010)	720 (0.021)
Octadecane-D$_{38}$	a.CD$_3$ str. 2215 (0.650)	a.CD$_2$ str. 2199 (1.250)	s.CD$_3$ str.	s.CD$_2$ str. 2100 (0.590)	a.CD$_3$ bend 1085 (0.111)	CD$_2$ bend	s.CD$_3$ bend 1060 (0.050)	CD$_2$ wag 975 (0.020)	CD$_2$ twist 960 (0.020)	CD$_3$ rock? 865 (0.015)	CD$_2$ rock 530 (0.020)
Octadecane /octadecane-D$_{38}$	1.341	1.332		1.363	1.349		1.316	1.389	1.365	1.277	1.358
Tetracosane	a.CH$_3$ str. 2970 (0.241)	a.CH$_2$ str. 2930 (1.250)	s.CH$_3$ str.	s.CH$_2$ str. 2860 (0.435)	a.CH$_3$ bend 1460 (0.077)	CH$_2$ bend	s.CH$_3$ bend 1385 (0.015)	CH$_2$ wag 1351 (0.020)	CH$_2$ twist 1300 (0.015)	CH$_3$ rock?	CH$_2$ rock 715 (0.012)
Tetracosane-D$_{50}$	a.CD$_3$ str.	a.CD$_2$ str. 2200 (1.250)	s.CD$_3$ str.	s.CD$_2$ str. 2100 (0.470)	a.CD$_3$ bend 1085 (0.101)	CD$_2$ bend	s.CD$_3$ bend 1053 (0.040)	CD$_2$ wag 985 (0.020)	CD$_2$ twist 960 (0.020)	CD$_3$ rock? 875 (0.010)	CD$_2$ rock 530 (0.020)
Tetracosane /Tetracosane-D$_{50}$		1.332		1.362	1.346		1.315	1.371	1.354		1.349

Alkenes and Other Compounds Containing C=C Double Bonds

Tables

*Numbers in parentheses indicate in-text page reference.

The IR spectra of a variety of chemicals containing carbon-carbon double bonds (C=C) together with spectra-structure correlations are readily available in book form to aid chemists in the identification of these important polymer building blocks (1). Chapter 4 contains IR and Raman data for these compounds in different environments. Discussions of both chemical and physical effects upon group frequencies associated with carbon-carbon double bonds are included.

Table 4.1 lists vibrational assignments for C=C stretching ($vC=C$), vinyl twist, vinyl CH_2 wag, vinyl $CH=CH_2$ wag, and the first overtone of vinyl CH_2 wag in the vapor-phase (2). The $vC=C$ mode for 1-alkenes ($R-CH=CH_2$) occur in the region 1641–1650 cm^{-1}. The vinyl twist mode occurs in the region 991–1008 cm^{-1}. Branching in the 3-carbon atom increases the vinyl twist frequency. The vinyl $C=CH_2$ wag occurs in the region 910–918 cm^{-1} and its first overtone occurs in the region 1829–1835 cm^{-1}. The vinyl wag vibration occurs in the region 572–685 cm^{-1}. This vibration increases in frequency with increased substitution on the 3-carbon atom (e.g., $R-CH_2-CH=CH_2$, 572–627 cm^{-1}; $R_2CH-CH=CH_2$, 653–678 cm^{-1}, and $R_3C-CH=CH_2$, 681–685 cm^{-1}).

Absorbance ratios and frequency separations are also presented for some vibrational bands.

IN-PLANE VIBRATIONS

Table 4.2 lists IR vapor-phase frequencies and assignments for a variety of compounds containing C=C double bonds. Acrylonitrile and divinylsulfone exhibit $vC=C$ at 1613 and 1620 cm^{-1}, respectively. The allyl derivatives exhibit $vC=C$ in the region 1641–1653 cm^{-1}. It is apparent that CN and SO_2 groups joined to the vinyl group have the effect of lowering the C=C stretching frequency (3).

The $v$asym. $CH_2=$, $v$sym. $CH_2=$, and $CH_2=$ bending modes occur in the regions 3082–3122, 2995–3045, and 1389–1430 cm^{-1}, respectively. The CN and SO_2 groups raise the $v$asym. $CH_2=$ frequencies. Compare acrylonitrile (3122 cm^{-1}) and divinylsulfone (3110 cm^{-1}) vs those for 3-butenoic acid and the allyl derivatives (3082–3100 cm^{-1}).

OUT-OF-PLANE VIBRATIONS

The $CH=CH_2$ twist frequencies for these compounds are assigned in the region 962–997 cm^{-1}. The lowest frequency, 962 cm^{-1}, is exhibited by divinylsulfone, and the two vinyl groups are joined to the sulfur atom of the SO_2 group. All of the other compounds exhibit $CH=CH_2$ twist in the region 970–997 cm^{-1}, and in these cases the vinyl group is joined to a carbon atom.

The $C=CH_2$ wag vibration and its first overtone occur in the regions 911–971 cm^{-1} and 1835–1942 cm^{-1}, respectively. The CN and SO_2 groups cause the $C=CH_2$ wag mode to occur at higher frequency than the other vinyl compounds, which exhibit this molecular vibration in the region 911–934 cm^{-1}.

As noted here, the $CH=CH_2$ wag mode for vinyl groups joined to $R-CH_2$, $(R-)_2CH$, and $R-_3C$ groups for 1-alkenes occur in the regions 572–627, 653–678, and 681–685 cm^{-1}, respectively. With the exceptions of allyl formate (638 cm^{-1}), allyl benzene (648 cm^{-1}) and allyl naphthalene (655 cm^{-1}), $CH=CH_2$ wag occurs in the region 551–558 cm^{-1}. Acrylonitrile and divinylsulfone exhibit $CH=CH_2$ wag at 680 and 715 cm^{-1}, respectively, and occur at higher frequency than for compounds of form $(R-)_3CCH=CH_2$. On the other hand, allylbenzene and allylnaphthalene exhibit $CH=CH_2$ wag at 648 and 655 cm^{-1}, respectively, and occur at higher frequency than for 1-alkenes of form $R-CH_2-CH=CH_2$ (572–627 cm^{-1}). The reason for the $CH=CH_2$ wag frequency behavior with chemical structure is as follows: The $CH=CH_2$ wag includes bending of the $C-C=C$ bonds, and as the $C(-)_2$ becomes increasingly branched it becomes more difficult for the $C-C=C$ group to bend out-of-plane together with the three hydrogen atoms joined to vinyl group bending out-of-plane in the same direction as the 3-carbon atom. Therefore, $CH=CH_2$ wag increases in frequency as the 3-carbon atom is increasingly branched. Apparently, in the case of acrylonitrile and divinylsulfone, it is more difficult for the $C=C-CN$ and $C=C-S$ bonds to bend than for $(R-)_3C-C=C$ bond bending in the complex $CH=CH_2$ wag vibration.

BAND ABSORBANCE RATIOS AND FREQUENCY SEPARATIONS FOR VINYL TWIST, VINYL C=CH$_2$, AND VINYL CH=CH$_2$ WAG

Table 4.2a and Table 4.1 list the numbers and frequencies for the normal vibrations listed in this section (3). For 1-alkenes the ratio of the absorbances (A) for (A) $CH=CH_2$ twist/(A)$C=CH_2$ wag is in the range 0.341–0.617. This shows that the $C=CH_2$ wag mode has more intensity than the $CH=CH_2$ twist mode. Table 4.2a shows the same trend except for acrylonitrile and allyl alcohol where these two modes have essentially identical intensity.

In Table 4.1 it is noted that the frequency separation between $CH=CH_2$ twist and $C=CH_2$ wag varies between 80 and 94 cm^{-1}, and in Table 4.2a it varies between 41 and 80 cm^{-1} with the exception of 19 and 9 cm^{-1} for acetonitrile and divinylsulfone, respectively.

The frequency separation between $CH=CH_2$ twist and $CH=CH_2$ wag varies between 319 and 419 cm^{-1} and with most compounds it varies between 319 and 344 cm^{-1}. These latter 1-alkenes are substituted in the 3-position with a methyl group (see Table 4.1). For the 1-alkenes not substituted in the 3-position, this frequency separation, varies between 370 and 419 cm^{-1}. In contrast, the frequency separation between $CH=CH_2$ twist and $CH=CH_2$ wag is 290 and 247 cm^{-1} for acrylonitrile and divinylsulfone, respectively. Other compounds studied extend this frequency separation to include 339 to 437 cm^{-1} for allyl naphthalene and allyl carbamate, respectively.

In the case of the frequency separation between C=CH$_2$ wag and CH=CH$_2$ wag it varies between 226 and 339 cm^{-1} for 1-alkenes (see Table 4.1). The frequency separation for the 1-alkenes not substituted in the 3-position varies between 286 and 239 cm^{-1}, while for those substituted in the 3-position it varies between 226 and 257 cm^{-1} (see Table 4.2a).

RAMAN DATA FOR X−CH=CH$_2$ COMPOUNDS

Table 4.2b contains Raman data and assignments for X−CH=CH$_2$ compounds and for cis- and trans-crotononitrile. The numbers in parentheses are for the relative Raman band intensities. They vary between 0 and 9, nine being the most intense band in the spectrum, and 0 being the least intense band in the spectrum. As these are whole numbers, it does not differentiate between the variances in intensity for bands whose intensities lie between any of the whole numbers (4).

In the case of cis- and trans-crotononitrile the νC=C modes occur at 1639 and 1629 cm^{-1}, respectively, and the Raman band is stronger in the case of the trans isomer than for the cis isomer (the trans to cis ratio is 5:4). Thus, the polarization of the electron cloud during the νC=C vibration is larger for the trans isomer than it is for the cis isomer.

Comparison of the Raman data for 1-octene and 1-decene shows that νC=C occurs at 1642 cm^{-1}; however, it is noted that the band intensity is 8 for 1-octene and 5 for 1-decene. The empirical structure for 1-octene is CH$_3$−(CH$_2$)$_5$−CH=CH$_2$ and for 1-decene is CH$_3$−(CH$_2$)$_7$−CH=CH$_2$. The CH$_2$ groups are in a 5:7 in this case while the band intensity is in an 8:5 for νC=C. This is what is expected, because the more CH$_2$ groups present in 1-alkenes the stronger the relative intensity for νC=C. Perhaps the ratios would be exactly 5:7 and 7:5 if the exact band intensities were measured.

CONJUGATED VINYL GROUPS

Vinyl-containing compound where the vinyl group is joined directly to a carbon atom of an aromatic ring exhibits νC=C in the region 1629–1633^{-1} (Table 4.2b) while 1-alkenes exhibit νC=C at higher frequency (1641–1653 cm^{-1}). This decrease in frequency is attributed to resonance of the C=C group with the aromatic ring. Resonance weakens the strength of the C=C bond, which causes νC=C to vibrate at lower frequency.

Table 4.2b show that the lowest νC=C frequencies are for those compounds containing the Si−CH=CH$_2$ group (νC=C, 1590–1603 cm^{-1}). The Raman band intensities for νC=C are relatively in the weak-medium class. The νC=C frequency for vinyl phenyl sulfone is also low (1607 cm^{-1}) where the vinyl group is joined to sulfur (3).

ROTATIONAL CONFORMERS OR FERMI RESONANCE (F.R.)

Table 4.2b shows that the vinyl ethers of form R−O−CH=CH$_2$ exhibit a Raman band in the region 1626–1639 cm^{-1} and a Raman band in the region 1610–1620 cm^{-1}. In the case of vinyl phenyl ether, Raman bands are observed at 1644 and 1593 cm^{-1}. The 1593 cm^{-1} most likely results from an in-plane bend stretching mode of the phenyl group. In all cases of R−O−CH=CH$_2$, the Raman band in the region 1610–1620 cm^{-1} has more intensity than the

Raman band in the region 1626–1639 cm^{-1}. In the IR, two (or sometimes three) bands are observed in the νC=C stretching region of the spectrum (5). An explanation is required to explain the existence of two bands in this region when only one O−CH=CH$_2$ group is present in these molecules. There are two possibilities for this observation (presuming the spectra represent pure materials), and these are for the presence of rotational conformers or from band splitting due to Fermi resonance (F.R.) of νC=C with the first overtone of a lower lying fundamental of the CH=CH$_2$ group. Criteria for F.R. to occur between a fundamental and a combination or overtone of a lower lying fundamental or two lower lying fundamentals are presented in what follows (4).

The combination or overtone must be of the same symmetry species as the νC=C vibration and the combination or overtone must involve molecular motion within the CH=CH$_2$ group. In addition, νC=C and the combination or overtone must occur at similar frequencies in order for a significant amount of F.R. to occur between νC=C and its combination or overtone. It does not matter if the combination or overtone occurs above or below the unperturbed νC=C vibration. If the combination or overtone is identical in frequency to the unperturbed νC=C frequency, both Raman or IR bands will have equal intensity. In other words, in this case both bands result from an equal contribution of νC=C and an equal amount of the combination or overtone. Because combination or overtones usually have intensities at least an order of magnitude less than most fundamental vibrations, an explanation is required for the often two strong bands or a strong and medium band in the event of F.R. This occurs because the fundamental contributes intensity to the combination or overtone, which causes the perturbed fundamental to be weaker than its unperturbed intensity and the perturbed combination or perturbed overtone to be more intense than its unperturbed intensity. It is usual practice to assign the most intense band of the Fermi doublet and the weaker band to the combination or overtone. However, the truth of the matter is that both bands are in F.R., and both modes contribute to both the band frequencies and the intensities. The stronger of the bands has more contribution from the unperturbed fundamental and the weaker of the bands has the least contribution from the fundamental and the most contribution from the unperturbed combination or overtone.

It so happens that C=CH$_2$ wag for vinyl phenyl ether is assigned at 850 cm^{-1}, and its first overtone would be expected to occur above 1700 cm^{-1}. This is due to the fact that C=CH$_2$ wag exhibits negative anharmonicity (occurs at higher frequency than twice the fundamental frequency) (5). A very weak band is noted at 1710 cm^{-1} in the case of vinyl phenyl ether, and it is reasonably assigned as 2(C=CH$_2$ wag). The Raman band at 1644 cm^{-1} and the strong 1643 IR band are assigned to a νC=C mode. A weak-medium IR band is noted at 1615 cm^{-1}. Obviously, from both the intensities and frequencies of these two IR bands neither fits the criteria for F.R. On this basis we assign the 1643 cm^{-1} band to νC=C for the gauche conformer and the 1615 cm^{-1} bond to the cis conformer for vinyl phenyl ether (5,6). Therefore, the vinyl alkyl ethers exhibit gauche νC=C in the region 1636–1639 cm^{-1} and cis νC=C in the region 1610–1620 cm^{-1}.

Table 4.2c lists IR vapor-phase data for vinyl alkyl ether (3). The gauche νC=C conformer is assigned in the region 1630–1648 cm^{-1} and the cis νC=C conformer in the region 1611–1628 cm^{-1}. Comparison of the vapor-phase IR data vs the Raman liquid phase data for vinyl isobutyl ether [gauche νC=C, VP(1645) vs liquid (1638 cm^{-1}) and cis νC=C, νP(1618 cm^{-1} vs liquid (1612 cm^{-1})], and for vinyl octadecyl ether [gauche νC=C, νP(1641 cm^{-1})] cis νC=C, νP(1613 cm^{-1}) vs liquid (1610 cm^{-1}] indicates that both gauche νC=C and cis νC=C occur at higher frequency in the vapor-phase (3).

In the vapor-phase, νasym. $CH_2=$ for vinyl alkyl ethers occur in the region (3122–3135 cm^{-1}), $\nu CH=$ occurs in the region (3060–3070 cm^{-1}), νsym. $CH_2=$ in the region (2984–3015 cm^{-1}), $CH_2=$ bend in the region (1400–1418 cm^{-1}), $CH=$ rock in the region (1311–1321 cm^{-1}), $CH=CH_2$ twist in the region (961–965 cm^{-1}), $C=CH_2$ wag in the region (812–825 cm^{-1}), cis$-CH=CH_2$ wag in the region (687–701 cm^{-1}), νasym. $C=C-O-C$ in the region (1203–1220 cm^{-1}), and νsym. $C=C-O-C$ in the region (837–888 cm^{-1}).

1-ALKENE (CH=CH$_2$) CH AND CH$_2$ STRETCHING VIBRATIONS

The carbon hydrogen stretching frequencies for 1-alkenes all take place within the plane of the $C=C$ group (Table 4.2b).

The Raman band in the region 3120–3122 cm^{-1} with the relative intensity between 0 and 2 is assigned to νasym. $CH_2=$, the Raman band in the region 3043–3046 cm^{-1} with the relative bond intensity between 1 and 4 is assigned to $\nu CH=$, the Raman band in the region 3022–3023 cm^{-1} with a relative intensity between 1 and 2 is assigned to νsym. $CH_2=$, and the Raman band in the region 1320–1329 cm^{-1} with a relative intensity between 2 and 9 is assigned $CH=$ in-plane rocking. The $CH=$ in-plane rocking mode for vinyl phenyl ether is assigned at 1311 cm^{-1}.

C=CH$_2$ WAG FREQUENCIES VS (SIGMA ρ-SIGMA$'$)

The term (sigma ρ-sigma$'$) defines the inductive effect of G for $G-CH=CH_2$. Table 4.3 lists the $C=CH_2$ wag frequencies and the inductive value of group G. The positive values withdraw sigma electrons from the vinyl group, the negative values contribute sigma electrons to the vinyl group. A plot of the $C=CH_2$ wag frequencies vs $\sigma\rho-\sigma'$ shows a smooth relationship (5).

CH=CH$_2$ TWIST FREQUENCIES VS σ' AND pK$_a$ OF G FOR G$-$CH$_2$CO$_2$H

Table 4.3a lists IR CS_2 solution and IR vapor-phase data for $CH=CH_2$ twist frequencies, pK_a of $G-CH_2-CO_2H$ and σ' of G. Table 4.3b shows that in general the $CH=CH_2$ twist mode decreases in frequency as the pK_a value of $G-CH_2CO_2H$ decreases in value. The σ' values of G do not correlate as well with $CH=CH_2$ twist as do the pK_a values. These parameters are useful in assigning $CH=CH_2$ twist vibrations in unknown materials containing this group (5). It should be noted than the vapor-phase $CH=CH_2$ twist frequencies occur at higher frequency than they do in CS_2 solution.

ALLYL HALIDES

Table 4.4 lists the IR data and assignments for the cis and gauche conformers of the allyl halides. In all cases, cis νC=C (1645–1652 cm^{-1}) occurs at higher frequency than gauche νC=C (1630–1643 cm^{-1}) (7). However, the frequency separation for gauche CH=CH$_2$ twist and gauche C=CH$_2$ wag is nearly constant (48.2–48.9 cm^{-1}). An interesting correlation exists for the frequency separation between gauche CH=CH$_2$ twist and gauche CH=CH$_2$ wag in that it increases in the order F(347 cm^{-1}), Cl(395.2 cm^{-1}), Br(446.2 cm^{-1}), and I(489.7 cm^{-1}). The frequency separation between gauche C=CH$_2$ wag and gauche CH=CH$_2$ wag increases in the same order: F(292.7 cm^{-1}), Cl(347 cm^{-1}), Br(397.3 cm^{-1}), and I(440.8 cm^{-1}). The gauche CH=CH$_2$ wag frequency is the most affected progressing in the order F through I for these allyl halides, because the frequency separation between F and Cl is 52.3 cm^{-1}, Cl and Br is 52.3 cm^{-1}, and Br and I is 46.5 cm^{-1}. In this case the inductive effect of the halogen atoms decreases in the order F to I while the mass increases in the same order. The C—C= bond strength decreases in the order F to I due to the decreasing inductive effect of the halogen atoms, which causes CH=CH$_2$ wag to occur at a lower frequency because the C—C= bond is more easily bent during this complex CH=CH$_2$ wag fundamental.

CYCLOALKENES AND CYCLOALKADIENES

Table 4.5 lists IR data and C=C or (C=C)$_2$ stretching assignments for cyclopentene, cyclohexene, 1,4-cyclohexadiene, and 1,3-cyclohexadiene in mole % CHCl$_3$/CCl$_4$ solutions (8). Only three data points are listed in the table, but 20 data points were taken in the original experiment. Figure 4.1 shows the plot of the νC=C or ν(C=C)$_2$ modes vs mole % CDCl$_3$/or CHCl$_3$/solvent system.

The νC=C mode for cyclopentene (1613.9 to 1611.2 cm^{-1}) and cyclohexene (1652.8 to 1650.9) decreases in frequency as the mole % CHCl$_3$ increases. These data suggest that the strength of the intermolecular proton bond formed between the C=C π system and the H or D atom of CHCl$_3$ or CDCl$_3$ increases as the mole % CHCl$_3$ or CDCl$_3$ increases. In addition the slopes of the plots for cyclopentene and cyclohexene are essentially identical. The relatively large frequency difference between that for cyclopentene and cyclohexene does not represent the basicity of the C=C bond. The major factor in determining the νC=C frequency is the bond angles of the carbon atoms joined to cis C=C (9). The νC=C frequency changes randomly in the order cyclopropene (1656 cm^{-1}), cyclobutene (1566 cm^{-1}), cyclopentene (1613.9 cm^{-1}), and cyclohexene (1652.8 cm^{-1}) (9,10).

Both 1,4- and 1,3-cyclohexadienes contain two C=C double bonds, and in each case the C=C bonds couple and split into out-of-phase (C=C)$_2$ stretching and in-phase stretching (8).

In the case of 1,3-cyclohexadiene ν_{ip}(C=C)$_2$ occurs at 1577.9 cm^{-1} and ν_{op}(C=O)$_2$ at 1603.3 cm^{-1} in the neat phase, while in solution with CDCl$_3$ ν_{ip}(C=C)$_2$ occurs at 1577.1 cm^{-1} and ν_{op}(C=C)$_2$ occurs at 1608 cm^{-1}. In this case, ν_{ip}(C=C)$_2$ decreases in frequency by 0.8 cm^{-1} while ν_{op}(C=C)$_2$ increases in frequency by 4.7 cm^{-1} in going from the neat phase to solution in CDCl$_3$. In the case of 1,4-cyclohexadiene, the ν_{ip}(C=C)$_2$ frequency occurs at 1672.3 cm^{-1} and ν_{op}(C=C)$_2$ occurs at 1676.8 cm^{-1} in the neat phase, while in CHCl$_3$ solution

v_{ip} occurs at 1676.8 cm^{-1} and $v_{op}(C=C)_2$ occurs at 1637.9 cm^{-1}. In this case, it is the $v_{ip}(C=C)_2$ frequency that increases 4.5 cm^{-1} while the $v_{op}(C=C)$ frequency decreases 1.3 cm^{-1}. However, Figure 4.1 shows that the higher frequency band in each set increases in frequency while the lower frequency band in each set decreases in frequency going from the neat phase to decreasing concentration in CHCl$_3$ or CDCl$_3$ solutions (8).

In the case of 1,3-cyclohexadiene, the two C=C groups are conjugated CH=CH—CH=CH$_2$ while in the case of 1,4-cyclohexadiene the two C=C groups are not conjugated. For the sake of comparison, the $v_{ip}(C=C)_2$ and $v_{op}(C=C)_2$ modes for s-trans-butadiene in the liquid phase occur at 1638 and 1592 cm^{-1}, respectively (10). The compound, cis 2-tert-butylbutadiene, exhibits $v_{ip}(C=C)_2$ and $v_{op}(C=C)_2$ at 1610 and 1645 cm^{-1}, respectively. Therefore, the $v_{ip}(C=C)_2$ and $v_{op}(C=C)_2$ frequency order is the same for 1,3-cyclohexadiene and cis tert-butylbutadiene. In this case, the two C=C groups have to be in a cis-cis configuration due to ring restraint, and this is in good agreement with the cis configuration assignment for cis-tert-butylbutadiene.

Table 4.5a lists IR vapor-phase data for cis-cycloalkene derivatives. Cyclopentene exhibits vC=C at 1621 cm^{-1} in the vapor phase and 1614 cm^{-1} in CCl$_4$ solution, and cyclohexene exhibits vC=C at 1651 cm^{-1} in the vapor phase and 1653 cm^{-1} in CCl$_4$ solution. In this case the vC=C mode for the solution-phase data read directly from the computer are considered more accurate than the manually read vapor-phase data. If these data are valid, the correlation of vC=C occurring at higher frequency in the vapor than in solution is an exception in the case of cyclohexene.

A study of Table 4.5a shows that vC=C for the 5-membered rings occur at the lower frequency than those for the 6-membered rings as already discussed here. Comparison of the vC=C vibrations for 2-cyclopentene-1-one and 2-cyclohexene-1-one (1600 cm^{-1} vs 1624 cm^{-1}) shows that vC=C for the 5-membered ring still occurs at lower frequency than it does for the 6-membered ring. However, both vC=C modes are lower in frequency than for cyclopentene (lower by 21 cm^{-1}) and for cyclohexene (lower by 27 cm^{-1}). The reason for this is that the C=C and C=O bonds are conjugated, which causes both bonds to become weaker and the connecting C—C bond to become stronger. Therefore, vC=O also occurs at lower frequency in the case of 2-cyclopentene-1-one (1745 cm^{-1}) vs (1765 cm^{-1}) for cyclopentanone in the vapor phase and 2-cyclohexene-1-one (1710 cm^{-1}) vs (1732 cm^{-1}) for cyclohexanone in the vapor phase. In the vC=O cases, they are lower by 20 and 22 cm^{-1} for the 5- and 6-membered rings, respectively. Therefore, conjugation of C=C with C=O decreases both modes in the same order of magnitude.

An IR band in the region 699–750 cm^{-1} is assigned to cis CH=CH wag in the six cis compounds studied. Another band is noted in the region 635–658 cm^{-1}. It is not certain whether these bands result from a different vibrational mode or from cis CH=CH wag of another conformer.

ALKYL ACRYLATES AND ALKYL METHACRYLATES

It is relatively easy to distinguish between alkyl acrylates and alkyl methacrylates by studying the vC=C and vC=O frequencies. The acrylates exhibit two bands in the vC=C region of the spectrum, and it has been suggested that they result from cis and trans conformers (10,11). Table 4.6 lists IR data and assignments for alkyl acrylates in CHCl$_3$ and CCl$_4$ solutions (11). In CCl$_4$

solution, cis $vC=C$ is assigned in the region 1619.2–1620.4 cm^{-1}, and in CHCl$_3$ solution, in the region 1618.5–1619.9 cm^{-1}. In CCl$_4$ solution, trans $vC=C$ occurs in the region 1635.3–1637.2 cm^{-1}, and in CHCl$_3$ solution in the region 1635.1–1636.6 cm^{-1}. In all cases cis $vC=C$ occurs at lower frequency than trans $vC=C$, and both cis and trans $vC=C$ occur at lower frequency in CHCl$_3$ than in CCl$_4$ solution.

Table 4.6a lists IR data and assignments for alkyl methacrylates in CCl$_4$ and CHCl$_3$ solution (11). In CCl$_4$ solution $vC=C$ for the alkyl methacrylates occurs in the region 1635.7–1637.3 cm^{-1}, in CHCl$_3$ solution, $vC=C$ occurs in the region 1635.7–1637.3 cm^{-1}, and in CHCl$_3$ solution, $vC=C$ occurs in the region 1635.8–1637.3 cm^{-1}. In all cases $vC=C$ occurs at lower frequency in CHCl$_3$ than in CCl$_4$ solution by 0.8 to 2.3 cm^{-1}.

The $vC=O$ frequencies for the alkyl methacrylates occur in the region 1719.5–1726 cm^{-1} (in CCl$_4$) and 1709.5–1718 cm^{-1} (in CHCl$_3$). Thus, $vC=O$ occurs at lower frequency by 8 to 10.6 cm^{-1} in going from CCl$_4$ to CHCl$_3$ solution. The $vC=O$ frequencies for alkyl acrylates occur in the region 1722.9–1734.1 cm^{-1}1 (in CCl$_4$) and in the region 1713.8–1724.5 cm^{-1} (in CHCl$_3$). Thus, the $vC=O$ occurs at lower frequency by 3.4–10.8 cm^{-1} in going from CCl$_4$ to CHCl$_3$ solution (11). In the vapor phase $vC=O$ for methyl methacrylate and methyl acrylate occur at 1741 and 1751 cm^{-1}, respectively (2). These data show that the $vC=O$ frequencies occur at higher frequency in the vapor phase than in solution.

It should be noted from the preceding $vC=O$ for the methacrylates occur at lower frequency than for the acrylates. This shift to lower frequency is attributed to the inductive contribution of the CH$_3$ group to the carbonyl group, which weakens the C=O bond.

It should be also noted that the alkyl group also causes a shift in the $vC=O$ frequency. For example, in the case of alkyl acrylates in CCl$_4$ solution, $vC=O$ occurs at 1734.1, 1727.3, 1727.1, and 1722.9 cm^{-1} for the methyl, butyl, 2-ethylhexyl, and tert-butyl analogs, respectively. In the case of the same series of alkyl methacrylates in CCl$_4$ solution, $vC=O$ occurs at 1718, 1710.5, and 1709.5, respectively. These $vC=O$ frequency decreases also are attributed to the increased inductive contribution of the alkyl group progressing in the series methyl to tert-butyl.

In the case of alkyl acrylates, the CH$_2$= bend occurs in the region 1401.6–1407.2 cm^{-1} (in CCl$_4$) and in the region 1404–1410.3 cm^{-1} (in CHCl$_3$). Thus CH$_2$= bend increases in frequency by 2.4 to 3.4 cm^{-1} on going from CCl$_4$ to CHCl$_3$ solution.

Alkyl acrylates exhibit CH$_2$=CH twist in the region 982.6–984.9 cm^{-1} (in CCl$_4$) and 983.4–985 cm^{-1} (in CHCl$_3$). The C=CH$_2$ wag frequencies occur in the region 966.3–968.4 cm^{-1} (in CCl$_4$) and 966.5–970.3 cm^{-1} (in CHCl$_3$).

TRANS-ALKENES

Table 4.7 lists IR vapor-phase frequency data for trans-alkenes and compounds containing a trans carbon-carbon double bond (3). Compounds containing a trans carbon-carbon double bond have the following configuration:

When the R groups or X groups are identical the νC=C is not allowed in the IR, because these compounds have a center of symmetry. Therefore, there is no dipole moment change during a cycle of νC=C. However, in this case the νC=C mode is Raman active. However, even in the case where the trans-alkenes do not have a center of symmetry, the dipole moment change during a cycle of trans νC=C is small, and the IR band is very weak. Raman spectroscopy is required to readily detect the C=C in trans-alkenes.

Alkyl crotonates exist in a trans configuration, and in the vapor phase trans νC=C occurs near 1664 cm^{-1}. A weak band at 1649 cm^{-1} and a weak-medium band is noted at 1621 cm^{-1} in the vapor-phase spectrum of cinnamonitrile, suggesting that both the cis and trans isomers are present.

The trans CH=CH twist is assigned in the region 961–972 cm^{-1} for most of these compounds included in Table 4.7. Halogen atoms joined to C=C lowers the trans CH=CH twist mode. For example, it occurs at 897 cm^{-1} for trans-1,2-dichloroethylene and at 930 cm^{-1} for 1,3-dichloropropene; Table 4.8 lists data for a variety of trans disubstituted ethylenes in CS$_2$ solution in most cases (5). The trans CH=CH twist mode occurs in the region 896–975 cm^{-1}.

ALKYLCINNAMATES

Table 4.9 lists IR vapor-phase data for alkyl cinnamates (3). The νC=C vibration occurs in the region 1640–1642 cm^{-1} in the vapor-phase, and it has medium band intensity. The CH=CH twist occurs in the region 972–982 cm^{-1}, and its intensity is always less than that exhibited by νC=C as demonstrated by the absorbance ratio [(A)HC−CH twist]/[(A)νC=C] in the range 0.42–0.67. The νC=O mode for these alkyl cinnamates will be discussed in Chapter 15.

CINNAMYL ESTERS

Table 4.10 lists IR vapor-phase data for cinnamyl esters (3). In this case, νC=C occurs in the region 1652–1660 cm^{-1}, and it occurs at higher frequency than in the case of alkyl cinnamates (compare with data in Table 4.10). In the case of alkyl cinnamate, ϕ−CH=CH−C(=O)−O−R, the C=O and C=C groups are conjugated causing both νC=C and νC=O to occur at lower frequency than those for cinnamyl esters, ϕ−CH=CH−CH$_2$−O−(=O)−R. In the case of the cinnamyl esters the absorbance ratio [(A)CH=CH/(A)νC=C] is in the range 6.9–20.5, and this is the reverse of that exhibited by the alkyl cinnamates. Chapter 15 discusses νC=O for these cinnamyl esters.

Table 4.10 also lists vibrational data and assignment for the R−C=O−O− portion of these esters.

1,1-DISUBSTITUTED ETHYLENES

Table 4.11 lists IR data for C=CH$_2$ wag and its first overtone for 1,1-disubstituted ethylenes (5). The fundamental is recorded in CS$_2$ solution and its overtone in CCl$_4$ solution. The C=CH$_2$ wag mode occurs in the region 711–1004 cm^{-1}, and its first overtone occurs in the region 1400–2020 cm^{-1}. The C=CH$_2$ wag frequency exhibits negative anharmonicity because the first overtone occurs at more than twice the C=CH$_2$ wag frequency.

Infrared and Raman spectra for 1-bromo-1-chloroethylene are shown in Figure 4.2, and these data were used to assign its 12 fundamental vibrations (13). Comparison of these data with those for 1,1-dichloroethylene (14,15) and 1,1-dibromoethylene (16) show that 9 fundamentals decrease in the order Cl$_2$, ClBr, and Br$_2$, and this trend is often observed in other halogenated analogs (see Table 4.11a).

Table 4.12 lists Raman data and assignments for the vinyl esters of carboxylic acids in the neat phase (4). With the exception of vinyl cinnamate (νC=C at 1636 cm^{-1}) all of the other vinyl esters exhibit νC=C in the region 1644–1648 cm^{-1}. In all cases where the aliphatic group of the ester is saturated the νC=C mode has much higher relative intensity (RI) than the (RI) for νC=O. The ratio [(RI)νC=O]/[(RI)νC=C] is in the range 0.11–0.29. In cases where the carboxylate portion of the ester is conjugated with the C=C group this RI ratio varies from 0.22 to 2.

The frequency separation between νC=C and νC=O varies between 98 and 110 cm^{-1} for compounds of form R$-$C(=O)$-$O$-$CH=CH$_2$, and between 84 and 92 cm^{-1} in cases where the C=O is conjugated with C=C or an unsaturated ring. This difference in the lower frequency separation is attributed to resonance between the C=C and C=O groups (C=C$-$C=O).

STYRENES AND α-METHYL STYRENES

Table 4.13 lists Raman data and assignments for styrene, α-methyl styrene monomers in the condemned phase (4). The νC=C vibration occurs in the region 1624–1635 cm^{-1}. The α-methylstyrene (νC=C, 1631 cm^{-1}) and 1,3-di(α-methyl) styrene (1631 cm^{-1}) exhibit νC=C frequencies comparable to styrene. However, νC=C of these compounds occur lower in frequency than for 1-octene (1642 cm^{-1}), and this is the result of conjugation between νC=C and the phenyl group. Other Raman bands listed in this table are assigned to in-plane phenyl ring vibrations.

Table 4.14 lists IR group frequency data and assignments for styrene and ring-substituted styrenes (12). Only data for styrenes that had corresponding ring-substituted phenols whose pK_a values were included were taken from Reference 12. Figure 4.3 gives the C=CH$_2$ wag frequencies plotted against the pK_a value of the corresponding ring-substituted phenol. Examination of Fig. 4.3 shows that styrenes substituted with atoms or groups in at least the 2,6-positions correlate in a manner different from styrenes not substituted in the 2,6-positions. The pK_a values are affected by contributions from both inductive and resonance effects of substituent atoms or groups joined to the phenyl ring. In the case of 2,6-disubstituted styrenes, the vinyl and 2,6-disubstituted phenyl group are not coplanar; therefore it is not possible for the resonance effects of the 2,6-atoms or groups to affect the C=CH$_2$ wag frequencies. Thus, it is the inductive effects of the atoms or groups substituted in the 2,6-positions that are affecting the $C=CH_2$ wag frequencies (assuming that the effects of intramolecular forces between atoms or groups in the 2,6-positions and the vinyl group are negligible). The $C=CH_2$ wag frequencies are affected by both inductive and resonance effects in cases where styrenes are not substituted in the 2,6-positions.

Figure 4.4 is a plot of the frequency separation between $CH=CH_2$ twist and $C=CH_2$ wag for styrene and ring-substituted styrenes vs the same pK_a values. Again, separate correlations exist for the planar and nonplanar styrenes with the exception of 2,4,6-trimethyl styrene, where the frequency separation between $CH=CH_2$ twist and $C=CH_2$ wag is less than for the corresponding 2,6-disubstituted phenols whose pK_a values are lower than ~ 7.2.

Table 4.15 lists IR data for α-halostyrenes, α-alkylstyrenes, and related compounds (12). Figure 4.5 shows a plot of the CH_2 wag frequencies for styrene and ring-substituted styrenes vs the $C=CH_2$ wag frequency for α-methylstyrene and correspondingly ring-substituted α-methyl-styrenes. This plot suggests that the factors affecting $C=CH_2$ wag for both styrenes and α-methyl styrenes are comparable. However, the CH_2 wag frequencies for α-methyl styrene occur at lower frequency than the correspondingly substituted styrenes by $10-18\,cm^{-1}$. The α-methylstyrenes are not planar for those substituted with Cl or CH_3 in the 2-position, while for styrenes it takes substitution in the 2,6-positions in order to sterically prevent the vinyl and phenyl groups from being coplanar.

BUTADIENES, PROPADIENES, CONJUGATED CYCLIC DIENES

Table 4.16 lists IR vapor-phase data and assignments for butadienes and propadienes (3,17). Compounds containing the 1,3-butadiene structure are of form $C=C-C=C$, and the two $C=C$ bonds couple during their $v(C=C)_2$ vibrations into in-phase $v(C=C)_2$ and out-of-phase $v(C=C)_2$. In the cases of 1,3-butadiene and 2-methyl-1,3-butadiene ip $v(C=C)_2$ occur at 1684 and $1649\,cm^{-1}$, respectively, while op $v(C=C)_2$ occur at 1594 and $1602\,cm^{-1}$, respectively. Other vibrational assignments are presented for these two molecules.

Propadienes have the basic skeletal structure $C=C=C$, and exhibit ip $(C=C)_2$ and op $v(C=C)_2$ vibrational mode. In the case of propadiene, the ip $v(C=C)_2$ is not IR active due to its molecular symmetry. With substitution of an atom or group in the 1-position of propadiene, these molecules have only a plane of symmetry, and ip $v(C=C)$ occurs in their region $1072-1101\,cm^{-1}$ progressing in frequency in the order CH_3, I, Br, and Cl. The op $v(C=C)$ mode is allowed in IR for propadiene, and its monosubstituted derivatives, and it occurs in the region $1953-1663\,cm^{-1}$ (18).

The compound 2,5-norbornadiene has the following basic structure:

In this case, the $C=C$ bonds are each in a fused 5-membered ring and they are conjugated. Therefore, 2,5-norbornadiene and related compounds exhibit ip and op $v(C=C)_2$ modes. The compounds 2,5- norbornadiene and 2,5-norbornadiene-yl acetate exhibit ip $v(C=C)_2$ at 1639 and $1651\,cm^{-1}$, respectively, and op $v(C=C)$ at 1546 and $1543\,cm^{-1}$, respectively. In the case of

1,3-cyclohexadiene the ring is larger than in the case of the 2,5-norbornadienes. The 6-membered ring is not fused in this case, but the two C=C groups are conjugated. The ip $v(C=C)_2$ and op modes are assigned at 1600 and 1701 cm^{-1}, respectively. The order of ip and op $v(C=C)_2$ is reversed when comparing it to those for the 2,5-norbornadienes.

The ip and op cis $(CH=CH)_2$ wag occur at 748 and 658 cm^{-1} for 1,3-cyclohexadiene, respectively. In the case of the 1,5-norbornadienes ip and op cis $(CH=CH)$ wag occur at 654 and 728–735 cm^{-1}, respectively. Again the ip and op cis (CH=CH wag) mode assignments are in the reverse order for these two sets of compounds.

ALKYL GROUPS OF 1-ALKENES AND VINYL ALKYL ETHERS

Table 4.17 contains the IR vapor-phase data and assignments for the carbon hydrogen vibrations for the 1-alkenes (also see Table 4.1). These assignments are given here rather than in Chapter 3 for the convenience of those interested in the vibrational spectra of 1-alkenes.

Table 4.18 contains the IR vapor-phase data and assignments for the alkyl group of vinyl alkyl ethers (also see Table 4.2b). The assignments are placed here rather than in Chapter 3 for the convenience of those interested in the vibrational spectra of vinyl alkyl ethers.

REFERENCES

1. Nyquist, R. A. (1989). *The Infrared Spectra, Building Blocks of Polymers*, Philadelphia: Sadtler Research Laboratories. A Division of Bio-Rad.

2. Nyquist, R. A. (1984). *The Interpretation of Vapor-Phase Infrared Spectra: Group Frequency Data*, Philadelphia: Sadlter Research Laboratories, A Division of Bio-Rad.

3. Nyquist, R. A. (ed.) (1984). *A Collection of 9200 Spectra*, Philadelphia: Sadtler Standard Infrared Vapor Phase Spectra, Sadtler Research Laboratories, A Division of Bio-Rad.

4. (1987). *Sadlter Standard Raman Spectra*, Philadelphia: Sadtler Research Laboratories, A Division of Bio-Rad.

5. Potts, W. J. and Nyquist, R. A. (1959). *Spectrochim. Acta*, **15**, 679.

6. Owen, N. L. and Sheppard, N. (1964). *Trans, Faraday Soc.*, **60**, 634.

7. McLachlan, R. D. and Nyquist, R. A. (1968). *Spectrochim. Acta*, **24A**, 103.

8. Nyquist, R. A. (1992). *Appl. Spectrosc.*, **47**, 560.

9. Colthup, N. B., Daly, L. H., and Wiberley, S. E. (1990). *Introduction to Infrared and Raman Spectroscopy*, 3rd ed., New York: Academic Press.

10. Lin-Vien, D., Colthup, N. B., Fateley, W. G., and Grasselli, J. G. (1991). *The Handbook of Infrared and Raman Frequencies of Organic Molecules*, Boston: Academic Press, Inc.

11. Nyquist, R. A. and Streck, R. (1994). *Vib. Spectrosc.*, **8**, 71.

12. Nyquist, R. A. (1986). *Appl. Spectrosc.*, **40**, 196.

13. Nyquist, R. A. and Thompson, J. W. (1977). *Spectrochim. Acta*, **33A**, 63.

14. Winter, R. (1970). *Z. Naturforsch.*, **25A**, 1912.

15. Joyner, P. and Slockler, G. (1952). *J. Chem. Phys.*, **20**, 302.

16. Scherer, J. R. and Overend, J. (1960). *J. Chem. Phys.*, **32**, 1720.

17. Nyquist, R. A., Lo, Y.-S., and Evans, J. C. (1964). *Spectrochim. Acta*, **20**, 619.

18. Nyquist, R. A., Lo, Y.-S., and Evans, J. C. (1964). *Spectrochim. Acta*, **20**, 619.

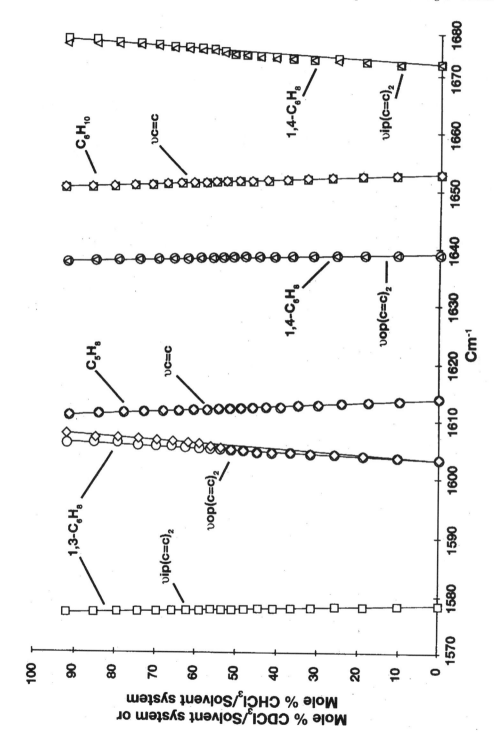

FIGURE 4.1   Shows plots of the $\nu C{=}C$ or $\nu(C{=}C)_2$ modes for 1,3-cyclohexadienes, 1,4-cyclohexadienes, cyclopentene, and cyclohexene vs mole % $CDCl_3$/or $CHCl_3$/solvent system.

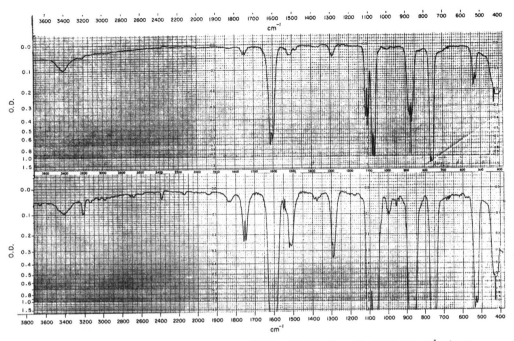

. Upper: Infrared vapor-phase spectrum of CH_2=CBrCl in the region 3800–400 cm^{-1} using a 10-cm KBr cell. The vapor pressure is 10 mm Hg. Lower: Same as upper spectrum except the vapor pressure is 100 mm Hg.

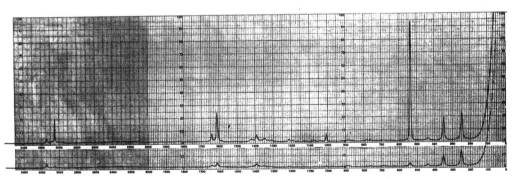

Raman liquid-phase spectrum of CH_2CBrCl. Upper: Parallel polarization. Lower: Perpendicular polarization.

FIGURE 4.2   Infrared and Raman spectra for 1-bromo-1-chloroethylene.

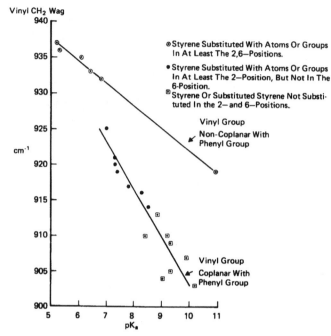

FIGURE 4.3 Plots of the C=CH$_2$ wag frequencies for styrenes and ring substituted styrenes vs the pK$_a$ values for correspondingly substituted phenols.

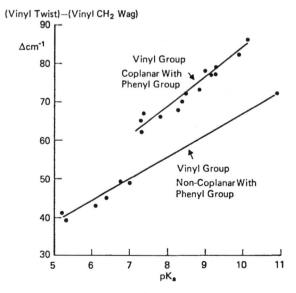

FIGURE 4.4 A plot of the frequency separation between CH=CH$_2$ twist and C=CH$_2$ wag for styrene and ring-substituted styrenes.

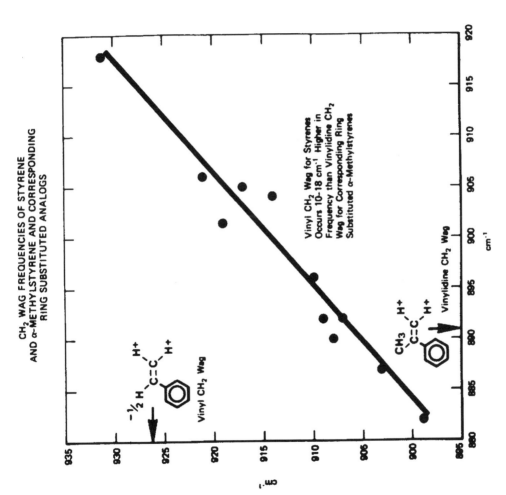

FIGURE 4.5 A plot of the C=CH₂ wag frequencies for styrene and ring-substituted styrenes *vs* the C=CH₂ wag frequency for α-methyl-α-methylstyrenes.

TABLE 4.1 IR vapor-phase data and assignments for 1-alkenes

Compound	Empirical structure	$2(CH_2$ wag)	C=C str.	$CH=CH_2$ twist	$C=CH_2$ wag	$CH=CH_2$ wag	[(A)$CH=CH_2$ twist]/[(A)$C=CH_2$ wag]	[$CH=CH_2$ twist]-[$C=CH_2$ wag]	[$CH=CH_2$ twist]-[$CH=CH_2$ wag]
Propene	$CH_3CH=CH_2$	1830 (0.122)	1650 (0.123)	991 (0.245)	911 (1.250)	572 (0.100)		80	419
1-Butene	$CH_3CH_2CH=CH_2$	1834 (0.029)	1648 (0.118)	995 (0.150)	911 (0.440)	625 (0.058)	0.341	84	370
4-Methyl-1-pentene	$(CH_3)_2CHCH_2CH=CH_2$	1835 (0.040)	1645 (0.130)	998 (0.161)	918 (0.350)	627 (0.050)	0.461	80	371
3-Methyl-1-butene	$(CH_3)_2CHCH=CH_2$	1830 (0.028)	1643 (0.070)	997 (0.140)	910 (0.285)	660 (0.059)	0.491*	87*	337*
3-Methyl-1-pentene	$CH_3CH_2(CH_3)CHCH=CH_2$*	1829 (0.040)	1641 (0.140)	1000 (0.230)	914 (0.389)	671 (0.071)	0.591*	86*	329*
3,4-Dimethyl-1-pentene	$(CH_3)_2CH(CH_3)CHCH=CH_2$*	1831 (0.021)	1643 (0.068)	1000 (0.110)	916 (0.245)	671 (0.035)	0.449*	84*	329*
3,7-Dimethyl-1-octene	$(CH_3)_2CH(CH_2)_3(CH_3)CHCH=CH_2$	1830 (0.020)	1643 (0.080)	998 (0.095)	914 (0.225)	678 (0.040)	0.422	84	320
3,3-Dimethyl-1-butene	$(CH_3)_3CCH=CH_2$	1833 (0.020)	1648 (0.091)	1001 (0.096)	915 (0.150)	682 (0.061)	0.641	86	319
3,3-Dimethyl-1-pentene	$CH_3CH_2(CH_3)_2CCH=CH_2$	1831 (0.020)	1648 (0.080)	1008 (0.117)	915 (0.270)	681 (0.050)	0.433	93	327
3,3-Dimethyl-1-hexene	$CH_3(CH_2)_2(CH_3)_2CCH=CH_2$	1830 (0.022)	1643 (0.061)	1005 (0.090)	911 (0.200)	685 (0.050)	0.451	94	320
Vinylcyclohexane	$C_5H_{10}CHCH=CH_2$*	1830 (0.020)	1641 (0.071)	997 (0.087)	911 (0.141)	653 (0.025)	0.617*	86*	344*

Compound	Empirical structure	[(A)$CH=CH_2$ twist]/[(A)$CH=CH_2$ wag]	[(A)$C=CH_2$ wag]/[(A)$CH=CH_2$ wag]	[$C=CH_2$ wag]-[$CH=CH_2$ wag]
Propene	$CH_3C=CH_2$	2.45	12.5	339
1-Butene	$CH_3CH_2CH=CH_2$	2.59	7.59	286
4-Methyl-1-pentene	$(CH_3)_2CHCH_2CH=CH_2$	3.22	7.01	291
3-Methyl-1-butane	$(CH_3)_2CHC=CH_2$*	2.37*	4.83*	250*
3-Methyl-1-pentene	$CH_3CH_2(CH_3)CHC=CH_2$*	3.24*	5.48*	243*
3,4-Dimethyl-1-pentene	$(CH_3)_2CH(CH_3)CHCH=CH_2$*	3.14*	4.83*	245*
3,7-Dimethyl-1-octene	$(CH_3)_2CH(CH_2)_3(CH_3)CHCH=CH_2$	2.38	5.63	236
3,3-Dimethyl-1-butene	$(CH_3)_3CCH=CH_2$	1.57	7.01	234
3,3-Dimethyl-1-pentene	$CH_3CH_2(CH_3)_2CH=CH_2$	2.34	5.41	234
3,3-Dimethyl-1-hexene	$CH_3(CH_2)_2(CH_3)_2CH=CH_2$	1.8	4.01	226
Vinylcyclohexane	$C_5H_{10}CHCH=CH_2$*	3.48*	5.64*	257*

* New data assigned from Sadtler Collection of IR Vapor Phase Spectra.

TABLE 4.2   The IR vapor-phase frequencies and assignments for the X–CH=CH_2 group

Compounds	a.CH_2= str.	s.CH_2= str.	CH_2= bend	2(C=CH_2 wag)	C=C str.	CH=CH_2 twist	C=CH_2 wag	CH=CH_2 wag
Acrylonitrile	3135 (0.060) 3122 (0.040) 3015 (0.059) [CN str.] 2240 (0.018)	3055 (0.060) 3045 (0.038) 3035 (0.040) [a.CH_2 str.] 2942 (0.039)	1425 (0.131) 1411 (0.090) 1401 (0.124)	1918 (0.112) 1911 (0.089) 1898 (0.101)	1626 (0.101) 1613 (0.070) 1601 (0.078)	970 (1.250)	951 (1.250)	700 (0.119) 680 (0.320) 659 (0.148)
Divinylsulfone	3110 (0.040) [a.SO_2 str.] 1354 (1.250)	3035 (0.060) [s.SO_2 str.] 1149 (0.970)	1389 (0.250) [a.CSC str.] 770 (1.040)	1942 (0.032) [s.CSC str.] 715 (0.270)	1620 (0.020) 1660 (0.062) [SO_2 bend] 589 (0.145)	962 (0.420) [SO_2 wag] 485 (0.185)	971 (0.565)	715 (0.270)
Allyl alcohol	3082 (0.265) H-O str. 3662 (0.285)	3002 (0.300) C-O str. 1021 (1.250)	1415 (0.340)	1855 (0.059)	1652 (0.070) 1645 (0.077)	985 (0.850) [C-O str.]	918 (0.875)	552 (0.095)
Allyl amine	3082 (0.211) [a.NH_2 str.] 3420 (0.005)	3002 (0.221) [s.NH_2 str.] 3355 (0.011)	1414 (0.135) [NH_2 bend] 1635 (0.230)	1841 (0.045) [NH_2 wag] 780 (1.220)	masked [a.CH_2 str.] 2920 (0.340)	997 (0.270)	919 (0.620)	551 (0.085)
Allyl cyanide	3099 (0.210) [CN str.] 2262 (0.021)	3040 (0.130) [a.CH_2 str.] 2941 (0.133)	1421 (0.375) [s.CH_2 str.] 2840 (0.020)	1870 (0.051) [CH_2 bend] 1439 (0.285)	1650 (0.210) [CH= rock] 1290 (0.069)	989 (0.582) [s.CH_2 str.] 2862 (0.300)	931 (1.250) [CH_2 bend] 1438 (0.122)	558 (0.250)
3-Butenoic acid	3094 (0.064) [H-O str.] 3580 (0.390)	2995 (0.154) [C=O str.] 1785 (1.040)	1420 (0.080) [COH bend] 1362 (0.230)	[gamma C=O] 640 (0.270)	1653 (0.106) [a.CH_2 str.] 2940 (0.044)	970 (0.110)	929 (0.150)	555 (0.201)
Allyl carbamate	3099 (0.051) [a.NH_2 str.] 3582 (0.111)	3030 (0.030) [s.NH_2 str.] 3462 (0.119)	[C=O str.] 1772 (1.139)	[NH_2 bend] 1589 (0.410)	1650 (0.052) [a.NCO str.] 1322 (1.245)	992 (0.100) [s.NCO str.] 1108 (0.300)	928 (0.151) [C-O str.] 1058 (0.481)	555 (0.079)
Allyl formate	3100 (0.071) [C-O str.] 1750 (1.240)	[C-O str.] 1168 (1.250)	1430 (0.080) [a.CH_2 str.] 2928 (0.290)	1865 (0.028)	1643 (0.045)	990 (0.151)	934 (0.268)	638 (0.081)
Allyl benzene	[a.CH_2 str.] 2920 (0.320)	[s.CH_2 str.] 2858 (0.090)	[CH_2 bend] 1449 (0.097)	1835 (0.061) [i.p. o.p. 5H Ring def.] 739 (0.770)	1641 (0.239) [o.p. Ring def.] 698 (1.240)	991 (0.290)	911 (0.835)	648 (0.211)
1-Allyl naphthalene	[a.CH_2 str.] 2925 (0.110)	[s.CH_2 str.] 2862 (0.040)	[CH_2 bend] 1447 (0.060)	1835 (0.035) [i.p. o.p. 3H Ring def.] 785 (0.590)	1641 (0.110) [i.p. o.p.4H Ring def.] 775 (1.245)	994 (0.100)	915 (0.225)	655 (0.050)

TABLE 4.2A   IR absorbance ratio and assignments for X–CH=CH$_2$ compounds

Compound	[(A)HC=CH twist] /[(A)C=CH$_2$ wag]	[(A)HC=CH twist] /[(A)CH=CH$_2$ wag]	[(A)C=CH$_2$ wag] /[(A)CH=CH$_2$ wag]	[HC=CH twist]- [C=CH$_2$ wag]	[HC=CH twist]- [CH=CH$_2$ wag]	[C=CH$_2$ wag] [CH=CH$_2$ wag]
Acrylonitrile	1	3.91	3.91	19	290	271
Divinylsulfone	0.74	1.56	2.09	9	247	256
Allyl alcohol	0.97	8.94*	9.21*	67	433	366
Allyl cyanide	0.47	2.33	5.01	58	431	373
3-Butenoic acid	0.73	0.55	0.75	41	415	374
Allyl carbamate	0.66	1.27	1.91	64	437	373
Allyl formate	0.56	1.86	3.31	56	352	296
Allyl benzene	0.35	1.37	3.96	80	343	263
Allyl naphthalene	0.44	2	4.51	79	339	260

*exceptionally high value.

TABLE 4.2B Raman data and assignments for X–CH=CH₂ compounds

Compound	C=C str. cm⁻¹ (A)	a.CH₂= str. cm⁻¹ (A)
1- Octene	1642 (8)	3081 (1)
1-Decene	1642 (5)	3081 (1)
1,2,4-Trivinyl cyclohexane	1641 (9)	3081 (2)
cis-Crotononitrile	1639 (4)	
trans-Crotononitrile	1629 (5)	
4-Vinyl benzoic acid	1629 (9)	
4-Vinyl benzene sulfonic acid, Na salt	1631 (9)	
2-Vinyl naphthalene	1632 (6)	
4-(Vinyl) phenylacetonitrile	1633 (9)	
2-Vinyl-4,6-diamino-S-triazine	1640 (9)	
Vinyl trimethoxysilane	1603 (3)	
Vinyl tris(betamethoxy ethoxy) silane	1600 (4)	
Divinyl diphenyl silane	1590 (2)	

Compound	C=C str.	C=C str. cis cm⁻¹ (A)	a.CH₂= str. cm⁻¹ (A)	CH= str. cm⁻¹ (A)	s.CH₂= str. cm⁻¹ (A)	CH= rock cm⁻¹ (A)	delta C=C str. cm⁻¹	(A)C=C str. /(A)C=C str. cis
Vinyl isobutyl ether	1638 (2)	1612 (3)	3122 (1)	3046 (3)	3023 (2)	1322 (7)	26	0.67
Vinyl isooctyl ether	1638 (1)	1611 (2)	3121 (0)	3046 (2)	3022 (1)	1320 (4)	27	0.5
Vinyl decyl ether	1637 (1)	1611 (2)	3121 (0)	3046 (2)	3022 (1)	1321 (5)	26	0.5
Vinyl dodecyl ether	1638 (0)	1610 (1)	3121 (0)	3046 (2)	3022 (1)	1320 (4)	28	0.5
Vinyl octadecyl ether	wk	1610 (0)		3046 (1)		1321 (2)		
Diethyleneglycol divinyl ether	1639 (3)	1620 (4)	3120 (2)	3045 (4)		1323 (9)	19	0.75
1,4-butandiol divinyl ether	1639 (3)	1616 (4)	3120 (1)	3044 (4)	3022 (2)	1322 (9)	23	0.75
Triethylene glycol divinyl ether	1636 (3)	1620 (4)	3120 (2)	3044 (4)		1329 (9)	19	0.75
Vinyl 2-(2-ethoxyethyl) ether	1639 (1)	1620 (2)	3120 (1)	3045 (2)		1323 (4)	19	0.5
Vinyl 4-hydroxybutyl ether	1639 (3)	1617 (4)	3120 (0)	3043 (2)		1322 (9)	22	0.75
Vinyl phenyl ether	1644 (3)	1593 (2)				1311 (2)	51	1.5
Vinyl phenyl sulfone	1607 (2)							
1,3-Diisopropenyl benzene	1631 (8)							

TABLE 4.2C   IR vapor-phase data for vinyl alkyl ethers

Vinyl ether	a.CH₂ str.	CH= str.	s.CH₂= str.	C=C str. gauche	C=C str. cis	CH₂= bend	CH= rock	CH=CH₂ twist	C=CH₂ wag	CH=CH₂ wag	CH=CH₂ wag cis	a.C=C-O-C str.	s.C=C-O-C str.	C-O-C bend
Methyl	3139 (0.075)	3070 (0.055)	3023 (0.140)	1655 (0.490)	1614 (0.640)	1400 (0.040)	1335 (0.180)	962 (0.191)	836 (0.210)	728 (0.035)	701 (0.025)	1231 (1.090)	900 (0.180)	600 (0.020)
	3129 (0.080)		3015 (0.171)	1630 (0.890)	1602 (0.510)		1321 (0.240)	940 (0.115)	811 (0.350)			1220 (1.245)	888 (0.170)	585 (0.015)
	3119 (0.080)		2998 (0.190)				1310 (0.190)	976 (0.193)				1205 (0.944)	874 (0.190)	575 (0.018)
	3139 (0.075)		2997 (0.434)				1330 (0.309)							
Ethyl	3135 (0.070)	3070 (0.060)	2984 (0.370)	1648 (0.632)	1625 (0.746)		1321 (0.329)	963 (0.245)	830 (0.279)		700 (0.031)	1204 (1.240)		
	3129 (0.081)				01611 (0.580)		1311 (0.359)	955 (0.225)	812 (0.353)		695 (0.028)			
											685 (0.031)			
Butyl	3125 (0.070)	3060 (0.059)		1640 (0.580)	1620 (0.612)		1321 (0.294)	965 (0.179)	818 (0.229)					
Octadecyl	3124 (0.015)			1641 (0.080)	1613 (0.090)		1323 (0.051)	962 (0.021)	812 (0.038)			1209 (0.140)		
Isobutyl	3125 (0.092)	3061 (0.050)		1645 (0.570)	1618 (0.650)		1320 (0.270)	964 (0.175)	814 (0.245)		700 (0.020)	1209 (1.245)		
2-Ethylhexyl	3130 (0.090)	3060 (0.070)	3015 (0.070)	1640 (0.572)	1612 (0.743)	1417 (0.060)	1321 (0.310)	965 (0.160)	815 (0.250)		700 (0.020)	1204 (1.250)	837 (0.250)	
2-Methoxyethyl	3122 (0.084)	3060 (0.062)		1637 (0.683)	1621 (0.699)		1319 (0.410)	962 (0.280)	824 (0.270)		695 (0.020)	1203 (1.250)		
2-Chloroethyl	3130 (0.041)	3070 (0.021)		1642 (0.520)	1628 (0.629)		1319 (0.231)	961 (0.165)	825 (0.186)		687 (0.082)	1204 (1.230)		
Bis[2-(vinyloxy) ethyl]ether	3130 (0.062)	3060 (0.030)		1640 (0.590)	1618 (0.605)	1418 (0.060)	1320 (0.320)	965 (0.180)	820 (0.210)		698 (0.020)	1205 (1.240)		

TABLE 4.3 $CH_2=$ wag frequencies vs (sigma ρ-sigma$'$) for group G

Group G	$C=CH_2$ wag CS_2 soln.	(sigma ρ-sigma$'$)
CO_2H	970	
CF_3	965	0.14
$(C=O)OCH_3$	964	
$(C=O)NH_2$	964	
$(C=O)OC_2H_5$	961	0.2
CN	960	0.07
$(C=O)CH_3$	960	0.25
$Si(CH_3)_3$	949	0.11
$CHCl_2$	937	
CH_2Cl	929	0.01
R	908	−0.13
I	905	−0.1
Br	898	−0.22
Cl	894	−0.24
$CH_3(C=O)O$	873	
F	863	−0.44
C_6H_5O	851	−0.41
CH_3O	813	−0.5

TABLE 4.3A IR carbon disulfide solution data for the $CH=CH$ twist frequency for compounds of form $G-CH_2CO_3H$ vs σ' of G and pK_a of $G-CH_2CO_2H$

Group G	twist CS_2 soln. cm^{-1}	$HC=CH_2$ twist vapor phase cm^{-1}	pK_a of $G-CH_2CO_2H$	σ' of G
$(CH_3)_3$	1009			−0.12
$(CH_3)_3C$	999			−0.07
$(CH_3)_2CH$	996		4.78	−0.056
$(CH_3)CH_2$	990		4.82	−0.052
CH_3	986		4.88	−0.045
C_6H_5	989		4.26	0.101
$4-Cl-C_6H_4$	983		4.19	
$ClCH_2$ [gauche conformer]	983	[995.5; 987.7]	4.08	0.17
$ClCH_2$ [cis conformer]		982.6		
ICH_2 [gauche conformer]	982	[990.6; 980.6]	4.06	
$BrCH_2$ [gauche conformer]	981	987.5	4.01	
$R(C=O)NH$	972		3.65	
CH_3O	963		3.53	0.23
CN	960	970	2.44	0.58
C_6H_5O	944		3.13	0.38
I	943		3.15	0.38
Br	936	941	2.87	0.45
Cl	938	941	2.86	0.47
F	~925	929	2.68	0.5

TABLE 4.4   IR data and assignments for the cis and gauche conformers of allyl halides

Allyl fluoride	Allyl chloride	Allyl bromide	Allyl iodide	Assignment
1630	1642.7	1638.4	1631.8	gauche C=C str.
1651.7	1649	1647	1645	cis C=C str.
989.3	985.2	983.9	980.9	gauche CH=CH$_2$ twist
935	937	935	932	gauche C=CH$_2$ wag
925				cis C=CH$_2$ wag
642.3	590	537.7	491.2	gauche CH=CH$_2$ wag
552.1	549.4		540	cis CH=CH$_2$ wag
48.3	48.2	48.9	48.9	[gauche CH=CH$_2$ twist]-
				[gauche C=CH$_2$ wag]
347	395.2	446.2	489.7	[gauche CH=CH$_2$ twist]-
				[gauche CH=CH$_2$ wag]
292.7	347	397.3	440.8	[gauche C=CH$_2$ wag]-
				[gauche CH=CH$_2$ wag]

TABLE 4.5   IR data and assignments for cyclopentene, cyclohexene, 1,4-cyclohexadiene, and 1,3-cyclohexadiene [C=C and (C=C)$_2$ stretching]

Mole % CHCl$_3$ /C$_5$H$_8$	C=C str.	Mole % CHCl$_3$ /C$_6$H$_{10}$	C=C str.	Mole % CHCl$_3$ /1,4-C$_6$H$_8$	i.p. C=C str.	o.p. C=C str.	Mole % CDCl$_3$ /1,3-C$_6$H$_8$	o.p. C=C str.	i.p. C=C str.
0	1613.9	0	1652.8	0	1672.3	1639.2	0	1603.3	1577.9
51.6	1612.2	52.5	1651.9	50.8	1674.3	1638.6	50.7	1605.4	1577.5
91.4	1611.2	92.5	1650.9	91.9	1676.8	1637.9	92	1608	1577.1
[delta cm^{-1}]	[−2.7]		[−1.9]		[4.5]	[−1.3]		[4.7]	[−0.8]

TABLE 4.5A   IR vapor-phase data and assignments for cis-cycloalkene derivatives

Compound	a.CH$_2$= str.	a.CH$_2$ str.	a.CH$_2$ str.	s.CH$_2$ str.	C=C str.	CH$_2$ bend	cis HC=CH wag	cis HC=CH wag
Cyclopentene	3070 (0.320)	2964 (0.950)	2920 (0.800)	2864 (0.620)	1621 (0.048)	1449 (0.050)	720 (0.179) 699 (0.595) 678 (0.199)	649 (0.095)
5-Norbornene-2-carbonitrile	3042 (0.110)		[CN str.] 2245 (0.050)		1635 (0.020)	1454 (0.050)	721 (0.445)	
Cyclohexene	3032 (0.410)	2935 (1.250)		2865 (0.550)	1651 (0.050)	1463 (0.191)	732 (0.065) 719 (0.200) 701 (0.060)	657 (0.160) 635 (0.244) 615 (0.095)
4-Methylcyclohexene	3035 (0.400)				1651 (0.045)		728 (0.050)	658 (0.240)
2-Cyclopentene-1-one	3075 (0.041)	2980 (0.060)	2940 (0.120)	2875 (0.030)	1600 (0.030)	1440 (0.041)	750 (0.230)	[C=O str.] 1745 (1.230)
2-Cyclohexene-1-one	3042 (0.110)	2942 (0.439)		2890 (0.150)	1624 (0.030)	1430 (0.100)	730 (0.160)	1710 (1.230)

TABLE 4.6 IR C=O stretching frequency data for alkyl acrylates [CHCl$_3$ and CCl$_4$ solutions]

Mole % CHCl$_3$/CCl$_4$	Methyl acrylate	2-Hydroxy-butyl acrylate	Allyl acrylate	2- Hydroxy-propyl acrylate	Butyl acrylate	2-Ethyl-hexyl acrylate	tert-Butyl acrylate
	C=O str.	C=O str.	C=O str.	C=O str.	C=O str.	C=O str.	C=O str.
[vapor]	1751				1741		
0	1734.1	1730.6	1730.5	1729.4	1727.3	1727.1	1722.9
100	1724.5	1721.6	1727.1	1721.3	1716.7	1716.3	1713.8
[delta cm^{-1}]	16.9; 26.5; 9.6	9	3.4	8.1	13.8; 24.3; 10.6	10.8	9.1
	s-trans C=C str.	s-trans C=C str.	s-trans C=C str.	s-trans C=C str.	s-trans C=C str.	s-trans C=C str.	s-trans C=C str.
0	1635.3	1636.5	1635.4	1637	1637.2	1635.7	1635.6
100	1635.1	1636.1	1635.4	1636.6	1635.8	1636.3	1635.8
[delta cm^{-1}]	−0.2	−0.4	0	−0.4	−1.4	0.6	0.2
	s-cis	s-cis	s-cis	s-cis	s-cis	s-cis	s-cis
0	1620.4	1619.3			1619.7		1619.2
100	1619.9				1619.2		1618.5
[delta cm^{-1}]	−0.5				−0.5		−0.7
	CH$_2$= bend	CH$_2$= bend	CH$_2$= bend	CH$_2$= bend	CH$_2$= bend	CH$_2$= bend	CH$_2$= bend
0		1406.4	1404.6	1406.6	1407.2	1406.5	1401.6
100		1409.4	1407	1409.5	1410.3	1409.9	1404
[delta cm^{-1}]		3	2.4	2.9	3.1	3.4	2.4
	HC=CH twist	CH=CH twist	CH=CH twist	CH=CH twist	CH=CH twist	CH=CH twist	CH=CH twist
0	984.9	982.6	984.7	983.3	983.5	983.6	984.5
100	985	983.4	984.4	983.9	984.3	984.5	985.6
[delta cm^{-1}]	0.1	0.8	−0.3	0.6	0.8	0.9	1.1
	C=CH$_2$ wag	C=CH$_2$ wag	C=CH$_2$ wag	C=CH$_2$ wag	C=CH$_2$ wag	C=CH$_2$ wag	C=CH$_2$ wag
0	968.1	982.6			968.4		966.3
100	970.3	983.4			969		966.5
[delta cm^{-1}]	2.2	0.8			0.6		0.2

TABLE 4.6A IR data and assignments for alkyl methacrylates in CHCl$_3$ and CCl$_4$ solutions [C=O and C=C stretching]

Mole % CHCl$_3$/CCl$_4$	2-Ethylhexyl C=O str. cm^{-1}	Ethyl C=O str. cm^{-1}	Butyl C=O str. cm^{-1}	Hexyl C=O str. cm^{-1}	Isobutyl C=O str. cm^{-1}	Allyl C=O str. cm^{-1}	2-Hydroxyethyl C=O. str. cm^{-1}	Glycidyl C=O str. cm^{-1}	Methyl C=O str. cm^{-1}
									1741
[vapor]	1719.5	1719.6	1720	1720.1	1720.6	1723.3	1723.9	1724.8	1726
0	1709.5	1710.5	1710.5	1710.4	1710	1714.7	1715	1717.1	1718
100	−10	−9.1	−9.5	−9.7	−10.6	−8.6	−8.9	−7.7	[−15;
[delta cm^{-1}]									−23; −8]
	C=C str.	C=C str.	C=C str.	C=C str.	C=C str.	C=C str.	C=C str.	C=C str.	C=C str.
[vapor]	1638.7	1637.4	1637.3	1638	1638.8	1638.7	1637.9	1638.7	1638.5
0	1636.8	1636	1635.8	1635.7	1637	1637.2	1637.1	1637.3	1636.6
100	−1.9	−1.4	−1.5	−2.3	−1.8	−1.6	−0.8	−1.4	−1.9
[delta cm^{-1}]									

TABLE 4.7 IR vapor-phase data for trans-alkenes

trans Alkenes	CH=CH str.	a.CH₂ str.	a.CH₃ str.	s.CH₃ str.	s.CH₂ str.	C=C str.	a.CH₃ def.	s.CH₃ def.	HC=CH twist	(A)[a.CH₃ def.]/(A)[HC=CH twist]	(A)[C=C str.]/(A)[CH=CH twist]
2-Butane	3035 (0.310)	2981 (0.600)	2942 (1.250)	2982 (0.325)	2982 (0.325)	inactive	1447 (0.150)	1392 (0.080)	961 (0.455)	0.33	0
2-Pentene	3030 (0283)	2997 (1.235)	2942 (0.440)	2980 (0.568)		?	1460 (0.210)	1390 (0.070)	961 (0.375)	0.56	0
2-Hexene	3021 (0.179)	2965 (1.240)	2940 (1.040)	2890 (0.440)	2880 (0.540)	?	1460 (0.169)	1390 (0.330)	965 (0.330)	0.51	0
3-Hexene	3035 (0.095)	2970 (1.240)	2950 (0.530)	2910 (0.332)	2890 (0.362)	inactive	1461 (0.080)	1385 (0.031)	969 (0.220)	0.36	0
4-Octene	3020 (0.160)	2965 (1.250)	2935 (1.150)	2880 (0.600)	2880 (0.600)	inactive	1460 (0.150)	1385 (0.080)	964 (0.250)	0.61	0
5-Decene	3020 (0.120)	2968 (1.115)	2940 (1.245)	2880 (0.512)	2865 (0.512)	1635 (0.025)?	1466 (0.151)	1385 (0.070)	969 (0.211)	0.72	0.12?
2,2-Dimethyl-3-hexene	3030 (0.090)	2975 (1.245)	2920 (0.315)	2888 (0.275)	2888 (0.275)		1478 (0.120)	1370 (0.100)	971 (0.110)	0.93	0
2,2-Dimethyl-3-heptene	3030 (0.093)	2970 (1.240)	2940 (0.566)	2880 (0.365)	2880 (0.365)		1476 (0.130)	1390 (0.046) 1370 (0.134)	972 (0.158)	0.82	0
2-Butene-1-ol	3030 (0.275) [H-O str.] 3670 (0.130)	2982 (0.630) [C-O str.] 1004 (1.240)	2940 (0.850)	2982 (0.630)	2982 (0.630)	1675 (0.085)	1476 (0.270)	1385 (0.750)	970 (0.975)	2.08	0.09
[Crotonate]											
methyl					[C=O str.] 1754 (1.229) 1750 (1.219) 1741 (1.229)	1663 (0.299)			970 (0.210)	1.42	
ethyl					1745 (1.050) 1740 (1.050)	1665 (0.271)			971 (0.244)	1.1	
1,2-Dichloroethylene	3080 (0.210)				[C.T.] 1661 (0.092) 1654 (0.081) 1641 (0.110)	inactive	[o.p. C-Cl str.] 815 (1.250)		910 (0.255) 897 (0.450) 880 (0.290)		0
1,3-Dichloropropene						1631 (0.510)	[=C-Cl str.] 775 (1.245)	[C-Cl str.] 691 (0.570)	930 (0.356)		1.43
beta-Bromostyrene			[i.p. o.p. 5H Ring def.] 741 (1.240)			1668 (0.060)	[=C-Br str.] 740 (1.245)		938 (0.442)		0.14
Cinnamonitrile	1265 (0.071)	1201 (0.092)		[o.p. Ring def.] 689 (0.640)	[CN str.] 2224 (0.335)	1621 (0.379)			960 (0.862)		0.44
beta-Bromostyrene		1220 (0.271)	732 (0.910)	690 (0.452)		1649 (0.050)					0.06

? not observed.

TABLE 4.8   IR data for the CH=CH twist frequency for trans disubstituted ethylenes

trans disubstituted ethylene XCH=CY X, Y	CH=CH twist cm^{-1}
R,R′	962–966
O, CH_3	959
C_6H_5, C_6H_5	958
CH_3, $(CH_3)CH$-O-	964–967
$CH_2X′$, $CH_2X′$	960–965
C_6H_5, CH_2-O-	964–966
C_6H_5, CN	964–967
CH_3, CN	953
C_6H_5, CN	962
Cl, CH_3	926
Cl, CH_2Cl	931
Br, CH_2Br	935
Cl, CH_2-O-	925–932
Cl, C_6H_5	930–942
Cl, CN	920
Cl, $B(OH)_2$	960
CH_3O, $CH(OCH_3)_2$	929
Cl, Cl	892
Br, Br	896
CH_3, (C=O)H	964
CH_3, (C=O)CCH	967
CH_3, CO_2H	966
CH_3, (C=O)OC_2H_5	968
CH_3, (C=O)$N(CH_3)_2$	966
C_6H_5, (C=O)CH_3	972
C_6H_5, (C=O)C_6H_5	975
C_6H_5, (C=O)H	972
C_6H_5, CO_2H	976
C_6H_5, (C=O)OR	976
RO(C=O), (C=O)OR	976
4-ClC_6H_4, CO_2H	975

TABLE 4.9   IR data for the =CH_2 wag and its overtone for 1,1-disubstituted ethylenes

Cinnamate	C=O str.	C=C str.	[(A)C=C] /[(A)C=O]	HC=CH twist	[(A)HC=CH twist] /[(A)C=C str.]	[(A)C=CH twist] /[(A)C=O str.]
Methyl	1740 (1.141)	1640 (0.431)	0.38	975 (0.181)	0.42	0.16
Ethyl	1735 (1.050)	1640 (0.370)	0.35	975 (0.169)	0.46	0.16
Butyl	1731 (1.141)	1641 (0.379)	0.33	978 (0.205)	0.54	0.18
Isobutyl	1735 (0.806)	1641 (0.310)	0.38	975 (0.130)	0.42	0.16
Isopentyl	1737 (0.830)	1642 (0.282)	0.34	980 (0.171)	0.61	0.21
lsopropyl	1731 (1.250)	1641 (0.500)	0.41	982 (0.310)	0.62	0.25
Tert-butyl	1727 (1.030)	1640 (0.370)	0.36	972 (0.247)	0.67	0.24
Cyclohexyl	1731 (0.654)	1642 (0.214)	0.33	980 (0.129)	0.6	0.2
Benzyl	1739 (0.959)	1640 (0.371)	0.39	980 (0.246)	0.66	0.26

TABLE 4.10   A comparison of the fundamentals for 1,1-dihaloethylenes

Cinnamyl	C=O str.	C=C str.	CCO str.	COC str.	HC=CH twist	[op. Ring def.]	[a.CH_2 str.]	[s. CH_2 str.]
Acetate	1763 (0.740)	1652 (0.035)	1229 (1.240)	1026 (0.295)	962 (0.240)			
Butyrate	1754 (0.950)	1660 (0.011)	1170 (1.240)		968 (0.225)			
Isobutyrate	1753 (1.042)	1655 (0.020)	1155 (1.230)		968 (0.292)			
	[a.CH_3 str.]	[s.CH_3 str.]	[a.CH_3 def.]	[s.CH_3 def.]	[i.p. o.p. 5H def.]	[op. Ring def.]	[a.CH_2 str.]	[s. CH_2 str.]
Acetate	2955 (0.080)	2895 (0.025)	1451 (0.082)	1366 (0.235)	740 (0.140)	689 (0.130)		
Butyrate	2975 (0.210)	2885 (0.085)	1451 (0.080)	1380 (0.075)	740 (0.100)	690 (0.095)	2945 (0.195)	2885 (0.085)
Isobutyrate	2982 (0.330)	2890 (0.080)	1475 (0.130)	1391 (0.100)	742 (0.130)	691 (0.120)	2958 (0.159) [CH_2 bend] 1453 (0.092)	2890 (0.080)
	[(A)C=C str.] /[(A)C=O str.]	[(A)HC=CH twist] /[(A)C=C str.]	[(A)HC=CH twist] /[(A)C=O str.]	[C=O str.]- [CCO str.]	[(A)C=O str.]- /[(A)CCO str.]	[C=C str.]- [HC=CH twist]		
Acetate	0.05	6.9	0.32	534	0.59	690		
Butyrate	0.01	20.5	0.24	584	0.77	692		
Isobutyrate	0.02	14.6	0.28	598	0.85	687		

TABLE 4.11   IR vapor-phase data and assignments for alkyl cinnamates [C=C stretching, CH=CH twisting, and C=O stretching]

1,1-Substituted ethylene $XYC=CH_2$ X, Y	$C=CH_2$ wag CS_2 soln. cm^{-1}	$2(C=CH_2$ wag) CCl_4 soln. cm^{-1}
R,R'	885–890	1785–1795
CH_3, C_6H_5	885–890	1785–1805
CH_3, 2,3-$Cl_2-C_6H_3$	905	1820
CH_3, CH_2Cl	902	1820
C_6H_5, CH_2Cl	907	1820
CH_3, CH_2OH	893	1792
CH_3, $CH(OH)CN$	914	1840
CH_3, Cl	875	1765
C_6H_5, Cl	877	1768
Cl, Cl	867	1744
Br, Br	877	1765
F, F	804 v.p.	1613 v.p.
Cl, CH_2Cl	891	1788
Br, CH_2Br	896	1802
Cl, $CH_2OC_6H_5$	887	
Cl, N=C=N	897	1811
Br, CF_3	929	
CH_3, $O(C=O)CH_3$	869	
CH_3, OCH_3	795	1600
C_2H_5O, OC_2H_5	711	
CH_3, (C=O)OR	939	1882
C_5H_{11}, (C=O)OCH_3	939	1888
CH_3, (C=O)OH	947	1905
C_5H_{11}, (C=O)OH	947	1898
CH_3, (C=O)R	930	1865
C_2H_5, (C=O)CH_3	931	1870
Cl, CO_2H	933	1878
Cl, (C=O)OR	925	1860
CH_3, CN	930	1878
Cl, CN	916	1843
CN, CN	985	1970
$C_2H_5SO_2$, $C_2H_5SO_2$	1004	2020

TABLE 4.11A   IR vapor-phase data and assignments for cinnamyl esters

$CH_2=CCl_2$ cm^{-1}	$CH_2=CBrCl$ cm^{-1}	$CH_2=CBr_2$ cm^{-1}	Assignment cm^{-1}
3130	3140	3112	asym. $CH_2=$ str.
3035	3046	3027	sym. $CH_2=$ str.
1616	1609	1601	$C=C$ str.
1391	1383	1364	$CH_2=$ bend
1088	1074	1070	$CH_2=$ rock
788	765	698	asym. $CX_2=$ str.
601	531	474	sym. $CX_2=$ str.
375	336	324	$CX_2=$ rock
299	240	182	$CX_2=$ bend
874	872	[881] (*1)	$CH_2=$ wag
686	684	[675] (*2)	$CH_2=$ twist
458	427	[404] (*2)	$CX_2=$ wag

*1 Unpublished Dow Chemical Company data.
*2 Reference 5.

TABLE 4.12   Raman data and assignments for vinyl esters of carboxylic acids

Vinyl	C=O str.	RI	C=C str.	RI	C=O str.- C=C str.	RI C=O str. /RI C=C str.
Propionate	1756	1	1648	9	108	0.11
Butyrate	1758	1	1648	4	110	0.25
Deconate	1759	1	1648	6	111	0.17
2-Ethylhexanoate	1753	2	1647	9	108	0.22
Neodecanoate	1748	2	1647	8	101	0.13
tert-Nonanoate	1744	2	1646	9	98	0.22
Pivalate	1749	2	1647	7	102	0.29
Adipate, di	1752	2	1648	9	104	0.22
Sebacate, di	1755	2	1648	9	107	0.22
Acrylate	1740	4	1648	5	92	0.81
2-Butenoate	1734	2	1644	9	90	0.22
2-Furoate	1734	2	1648	1	86	2
Cinnamate	1720	0.5	1636	9	84	0.06
Benzoate	1732	3	1647	2	85	1.5

TABLE 4.13 **Raman data and assignments for styrene monomers**

Compound Styrene	C=C str.	Ring 2	Ring 18		Ring 19	Ring 3		Ring 5	s.C(C)$_3$ str.	
4-Amino	1624 (9)	1610 (8)	1415 (5)		1209 (4)	1179 (6)		843 (4)		
4-Ethyl	1631 (9)	1612 (5)	1425 (1)		1205 (2)	1181 (3)				
4-tert-Butyl	1631 (9)	1612 (5)							789 (4)	
4- Fluoro	1635 (9)	1603 (4)	1404 (2)		1203 (5)			842 (4)		
4-Chloro	1632 (9)	1596 (3)	1423 (1)		1204 (4)	1178 (2)		790 (3)		
4-Cyano	1632 (5)	1608 (9)	1427 (0)		1208 (1)	1178 (6)			415 (1)	
4-Nitro	1631 (4)									
		Ring 4	*Ring 5*	*Ring 15*	*Ring 16*		*Ring 8*	*Ring 10*		*Ring 21*
3-Bromo	1629 (4)			1305 (1)	1200 (3)		997 (9)			309 (4)
3-Chloro	1633 (8)			1307 (3)	1203 (3)		999 (9)	405 (2)		308 (1)
3-Fluoro	1632 (4)	1613 (4)	1415 (1)				1002 (9)	724 (3)		
3-Nitro	1634 (5)	1618 (1)	1410 (1)		1208 (3)					
		Ring 3	*Ring 4*		*Ring 17?*	*Ring 7*	*Ring 18?*			*Ring 21*
2-Fluoro	1633 (9)	1612 (3)	1416 (1)		1230 (3)	1153 (1)				
2-Bromo	1628 (9)						1020 (4)			325 (5)
		Ring 4	*Ring 6*		*Ring 9*				*Ring 17*	
2,4-Dimethyl	1628 (9)	1611 (8)	1420 (3)		1239 (7)				459 (5)	
2,5-Dimethyl	1627 (9)	1612 (9)	1417 (2)		1239 (5)					
alpha-Methyl	1631 (5)									
1,3-Di(alpha-methyl)	1631 (8)									
1-Octene	1642 (8)									

TABLE 4.14   IR group frequency data for styrene and substituted styrenes

Compound	2(C=CH₂ wag) cm⁻¹	C=C str. cm⁻¹	CH=CH twist CS₂ soln. cm⁻¹	C=CH₂ wag CS₂ soln. cm⁻¹	CH=CH twist minus C=CH₂ cm⁻¹	pK$_a$ of corresponding sub. phenol
Styrene	1821	1632	989	907	82	9.9
Styrene sub.						
4-Methyl	1812	1631	987	903	84	10.35
4-Bromo	1826	1631	986	909	77	9.34
4-Chloro	1815	1628	982	904	78	9.2
4-fluoro	1825 [vp]	1638 [vp]	990 [vp]	911 [vp]	79	9.95
4-Cyano	1833 [CS₂]	1627 [CS₂]	982	917	65	
3-Chloro	1831	1631	986	913	73	8.93
3-Hydroxy	1815	1629	984	905	79	9.33
3-Chloromethyl	1832	1637	987	909	78	
3-Acetyl	1828	1637	987	910	77	9.18
3,4-Dichloro	1824	1626	980	910	70	8.4
3,4-Dimethyl	1817	1628	989	903	86	10.17
3,5-Dimethyl	1808	1626	984	903	81	10.1
2,4-Dimethyl	1821	1626	987	907	80	10.49
2-Chloro	1833	1629	986	914	72	8.5
2-Bromo	1835	1629	984	916	68	8.43
2-Methyl	1832 [vp]	1630 [vp]	989 [vp]	914 [vp]	75	10.28
2,3-Dichloro	1842	1626	986	919	67	7.4
2,4-Dichloro	1832	1623	983	917	66	7.8
2,5-Dichloro	1842	1624	986	921	65	7.3
2,6-Dichloro	1872	1637	981	932	49	6.8
2,3,6-Trichloro	1876	1631	978	935	43	6.1
2,4,5-Trichloro	1876	1631	978	933	45	6.4
2,3,4,5-Tetrachloro	1858	1627	974	925	49	7
2,3,4,6-Tetrachloro	1883	1633	978	937	41	5.22
2,3,4,5,6-Pentachloro	1883	1631	975	936	39	4.77
2,4,6-Trimethyl	1843	1632	991	919	72	10.83

TABLE 4.15   IR group frequency data for α-halostyrenes, α-methylstyrenes, and related compounds

Compound	2(CH$_2$ wag) CCl$_4$ soln. cm^{-1}	C=C str. CCl$_4$ soln. cm^{-1}	CH$_2$= wag CS$_2$ soln. cm^{-1}
Styrene			
alpha-(X)			
X			
Chloro	1760	1626	877
Bromo	1775	1617	882
Propyl	1801	1629	897
Phenyl	1809	1616	896
2-Hydroxyethyl	1799	1627	897
α-Methylstyrene	1797	1629	892
Sub.			
4-Chloro	1790	1627	890
4-Bromo	1789	1623	892
4-Hydroxy	1772	1630	882
4-Acetyl	1795	1626	897
4-Methyl	1783	1629	887
3,4-Dichloro	1802	1634	896
3,5-Dichloro	1810	1629	899
4-Chloro, 3-methyl	1790	1631	892
2-Hydroxy	1822	1632	911
2-Chloro	1810	1642	904
2,3-Dichloro	1805	1640	902
2,4-Dichloro	1819	1642	905
2-Chloro, 5-methyl	1814	1642	902
2-Nitro	1822	1643	902
2,4,5-Trimethyl	1803	1642	899
2,3,4,5,6-Pentafluoro	1830	1645	918
1-Isopropenylnaphthalene	1815	1634	904
1-Isopropenylpyridine	1815	1634	904

TABLE 4.16 IR vapor-phase data for butadienes and propadienes

Compound	a.CH₂ str.	CH= str.	a.CH₂= str.	2(CH₂) wag	i.p. (C=C)₂ str.	o.p. (C=C)₂ str.	CH₂= bend
1,3-Butadiene	3102 (0.130)	3030 (0.100)	3000 (0.120)	1830 (0.110) 1822 (0.060) 1810 (0.090)	1648 (0.010)	1604 (0.190) 1594 (0.130) 1585 (0.165)	1390 (0.032) 1379 (0.028) 1369 (0.035)
2-Methyl-1,3-butadiene or Isoprene	3099 (0.380)	3030 (0.185)	2990 (0.430)	1802 (0.120)	1649 (0.060)	1602 (0.420)	1375 (0.100)
Propadiene or Allene	3100 (0.030) perpendicular v8 (e)		3020 (0.090) 3010 (0.060) 2995 (0.070) parallel, vb (b2)	1700 (0.098) 1690 (0.087) 1681 (0.097) parallel	IR inactive a1	1972 (1.250) 1953 (0.590) 1942 (0.860) parallel, v6 (b2)	1409 (0.080) 1397 (0.060) 1378 (0.100) parallel, v7 (b2)
1,2-Butadiene or 1-Methyl-propadiene	3080 (0.231)	3010 (0.690)		1802 (0.120)	1081 (0.091) 1072 (0.082) 1063 (0.100)	1970 (0.590) 1962 (0.540) 1959 (0.550)	
1-Chloro-propadiene*	3079	3079?	3009 [C=C=C bend (a)]	494 [C=C-X bend (a') bend]	1101 [C=C-X str (a)]	1963	1435
1-Bromo-propadiene*	3080	3008?	3005	423*2	1078	1961	1432
1-Iodo-propadiene*	3070 [o.p. cis (HC=CH)₂ str; C=C-X str.]	3070? [C=C=C bend (a')]	3004	387*1	1076	1953	1425
1-Chloro-propadiene	767	592	548 [i.p. (C=C)₂ str]		184*1	cis (CH=CH)₂ o.p. wag	
1-Bromo-propadiene	681	603	519		169*1		
1-Iodo-propadiene	609 o.p. cis (HC=CH)₂ str; C=C-X str.	625 i.p. cis (HC=CH)₂ str	485		154*1 cis (HC=CH)₂ i.p. wag		
2,5-Norbornadiene	3130 (0.115)	3080 (0.285)	1641 (0.020) 1639 (0.010) 1630 (0.010)	1559 (0.100) 1546 (0.160) 1535 (0.090) C=C str.	666 (0.225) cis (HC=CH)₂ i.p. wag	740 (0.370) cis (CH=CH)₂ o.p. wag	
2,5-Norbornadiene-7-yl acetate	3080 (0.111)		1651 (0.040)	1543 (0.154)	654 (0.090)	728 (1.110) 711 (0.330) 735 (0.265)	1759 (0.750) C=O str.
1,3-Cyclohexadiene	3080 (0.790)		1600 (0.030)	1701 (0.040)	748 (0.190)	658 (1.240)	

Compound	i.p. and o.p. CH=CH₂ twist	C=CH₂ wag	RC=CH₂ wag	CH=CH₂ wag	RC=CH₂ wag
1,3-Butadiene	1011 (0.240) 995 (0.200)	908 (1.250)			
2-Methyl-1,3-butadiene	990 (0.371) [CH₂ rock]	907 (1.130)	895 (1.240)	518 (0.060)	758 (0.020)
Propadiene	1040 (0.060) perpendicular v9 (e)	845 (0.690) perpendicular v10 (e)			
1,2-Butadiene	910 (0.256)	859 (1.240) 849 (0.935) 839 (0.949)	−1175 (0.020) [CH= bend]	[CH₃C=C bend]? 560 (0.182) 552 (0.160) 544 (0.200) C=C=C bend (a')	C=C=C bend (a) 548 519 485
1-Chloro-propadiene	999	875	822	767	
1-Bromo-propadiene	1000	862	813	681	
1-Iodo-propadiene	995	854	807	609	

* Reference 18.

TABLE 4.17　IR vapor-phase data for carbon hydrogen stretching vibrations and other vibrations for 1-alkenes

Compound	2(C=CH₂ wag) + s.CH₃ bend	a.CH₂= str.	s.CH₂= str.	a.CH₃ str.	a.CH₂ str.	s.CH₃ str.	s.CH₂ str.	a.CH₃ bend (A)	a.CH₃ bend (A')	CH₂ bend	CH₂= bend	s.CH₃ bend	s.CH₃ bend	C-C str.?	CH₂ wag	CH₂ rock
Propene		3080 (0.170)	3000 (0.320)	2960 (0.580)		2920 (0.300)		1475 (0.110)	1443 (0.220)			1399 (0.079)				
1-Butene	3182 (0.005)	3085 (0.211)		2975 (1.245)		2900 (0.420)	2880 (0.300)	1462 (0.129)	1432 (0.129)			1380 (0.090)			1307 (0.109)	800 (0.020)
3-Methyl-1-butene	3190 (0.010)	3084 (0.180)		2964 (1.245)	2930 (0.300)		2890 (0.300)	14865 (0.111)				1380 (0.096)			1305 (0.050)	
3-Methyl-1-pentene	3180 (0.020)	3084 (0.275)		2970 (1.240)	2930 (0.871)		2885 (0.585)	1461 (0.200)			1421 (0.090)	1387 (0.140)			1290 (0.050)	782 (0.020)
3,4-Dimethyl-1-pentene	3190 (0.010)	3082 (0.150)		2965 (1.245)	2940 (0.370)		2882 (0.464)	1463 (0.140)			1422 (0.040)	1380 (0.131)				
3,7-Dimethyl-1-octene	3199 (0.010)	3090 (0.131)		2970 (1.235)	2940 (0.835)		2882 (0.452)	1469 (0.240)			1423 (0.040)	1385 (0.110)		1170 (0.020)		734 (0.011)
3,3-Dimethyl-1-butene		3100 (0.142)		2970 (1.250)			2890 (0.250)	1472 (0.110)			1421 (0.070)	1380 (0.119)	1370 (0.173)	1214 (0.080)		
3,3-Dimethyl-1-pentene	3200 (0.005)	3095 (0.149)	3009 (0.061)	2980 (1.245)			2892 (0.200)	1470 (0.115)			1420 (0.068)	1381 (0.103)	1369 (0.098)	1185 (0.032)		
3,3-Dimethyl-1-hexene	3205 (0.005)	3095 (0.110)	3008 (0.080)	2970 (1.245)	2940 (0.440)	2905 (0.290)	2885 (0.340)	1470 (0.110)			1420 (0.050)	1380 (0.071)	1370 (0.080)	1200 (0.030)		
Vinylcyclohexane*¹	3082 (0.091)		3000 (0.080)		2935 (1.250)		2862 (0.460)			1454 (0.121)	1415 (0.030)					

*¹ [Ring breathing, 840 (0.012)].

TABLE 4.18 IR data and assignments for the alkyl group of vinyl alkyl ethers

Vinyl alkyl ether	a.CH3 str.	a.CH2 str.	s.CH3 str.	s.CH2 str.	a.CH3 bend	s.CH3 bend	CH3 rock	CH3 rock	CH2 wag and or C–C str.	CH2 twist	CH3 rock and or C–C str.	CH2 rock	C–Cl str.	C–Cl str.
Methyl	2958 (0.310) 2930 (0.240)		2870 (0.100) 2860 (0.095) 2850 (0.110)		1460 (0.152)	1319 (0.225)	1145 (0.260) 1136 (0.190) 1127 (0.210)	1024 (0.186) 1011 (0.175) 997 (0.171)						
Ethyl	2997 (0.434) 2984 (0.370)	2940 (0.270)	2900 (0.260)		1476 (0.087)	1388 (0.200) 1377 (0.210)			1128 (0.481)	1079 (0.226)	1059 (0.271)			
Butyl	2971 (0.790)	2950 (0.690)	2894 (0.410)		1476 (0.135)	1379 (0.190)			1135 (0.249)	1083 (0.270)	1030 (0.177)			
Isobutyl	2965 (0.840)	2925 (0.392)	2897 (0.376)		1475 (0.200)	1388 (0.195)			1145 (0.232)	1080 (0.280)	1019 (0.310)			
2-Ethylhexyl	2970 (1.245)	2938 (1.245)	2882 (0.640)		1470 (0.250)	1380 (0.180)				1079 (0.266)	1015 (0.164)	735 (0.035)		
2-Methoxyethyl	2995 (0.322)	2930 (0.902)	2890 (0.555)	2830 (0.327)	1462 (0.185)	1360 (0.185)			aC–O–C str. 1135 (1.150)	1095 (0.560)	1039 (0.250) 1013 (0.145)			
2-Chloroethyl		2980 (0.105)		2895 (0.065)	1466 (0.060)	1377 (0.100)				1086 (0.240)	1010 (0.138) 1001 (0.142)		764 (0.170)	687 (0.082)? and CH=CH2 wag?
Bis[2-(vinyloxy) ethyl]ether		2938 (0.291)	2880 (0.256)		1460 (0.090)	1359 (0.140)			1140 (0.970)	1092 (0.350)	1010 (0.165) 980 (0.200)			

Alkynes and Compounds Containing C≡C Groups

*Numbers in parentheses indicate in-text page reference.

TERMINAL C≡CH

Table 5.1 lists IR data and assignments for compounds containing the terminal acetylenic group (1). The $v\equiv C-H$ mode occurs near 3300 cm^{-1} with a weak shoulder on the low frequency side of the strong IR band. This weak shoulder has been attributed to a Fermi resonance interaction with $v\equiv C-H$ and the combination tone $vC\equiv C + 2[C\equiv C-H$ bending] (1).

All of these compounds exhibit $vC\equiv C$ in the region 2100–2148 cm^{-1}, and this band is weak in most cases. Compounds where the halogen atom or a carbonyl group is joined to the terminal acetylenic group exhibit strong IR $vc\equiv c$ absorption bands.

A weak IR band in the region 897–961 cm^{-1} is assigned to $vC-C$. The $\equiv C-H$ bending mode is not split in the case of 1-alkynes, and occurs in the region 628–633 cm^{-1}. Substitution of a halogen atom on the 3-carbon atom splits the degeneracy and both in-plane and out-of-plane ≡C—H bending modes are observed in the IR. With the exception of 3-iodopropyne the in-plane bending mode occurs at a higher frequency than the out-of-plane bending modes. For example, 3-iodopropyne exhibits the C—H bending modes at 637 cm^{-1}, while the Cl, Br, and F analogs exhibit the in-plane bending mode at 652, 649, and 674 cm^{-1}, respectively, and the out-of-plane bending mode occurs at 637, 639, and 636 cm^{-1}, respectively (2).

Substitution of deuterium for hydrogen ≡C—H to ≡C—D helps in establishing the fundamental vibrations that result from this portion of the molecule. In the case of 3-chloropropyne-1-d
and 3-bromopropyne-1-d the v≡C—D modes occur at 2618 and 2607 cm^{-1}, respectively (see Table 5.2). In the case of ≡C—D bending, the Cl and Br analogs exhibit the in-plane mode at 516 and 512 cm^{-1}, respectively, while the out-of-plane mode occurs at 502 and 503 cm^{-1}, respectively.

The 3-chloropropyne-1-d, 3-bromopropyne-1-d, and phenylacetylene-1-d are interesting because their v≡C—D and vC≡C modes couple. The vC≡C mode shifts from 2147 to 2000, 2138 to 2006, and 2119 to 1989 cm^{-1} for 3-Cl, Br, and 2-phenyl C≡C—H and D analogs, respectively. On the other hand, the v≡C—H mode and v≡C—D modes are (3325 and 2618), (3315 and 2607), and (3315 and 2596 cm^{-1}) for the 3-Cl, 3-Br and 2-phenyl C≡C—H and D analogs, respectively. The ratio of v≡C—H/v≡C—D is 1.27, 1.28, and 1.28, respectively, and, if v≡C—D were a pure vibration it would be expected to occur at 2340 cm^{-1}. Because vC≡C shifts to lower frequency upon D substitution together with the behavior of vC—D, their frequency behavior is expected when the two modes are coupled. As vC≡C and v≡C—H occur approximately 1170 cm^{-1} apart, the amount of coupling between these two modes is most likely negligible (3–5).

Terminal acetylenic compounds often exhibit the first overtone of ≡C—H bending in the region 1219–1265 cm^{-1}, and for the D analogs near 1000–1043 cm^{-1}. In addition, the C—C≡C bending mode is observed in the region 300–353 cm^{-1} (1).

Table 5.2 lists the 15 vibrational assignments for 3-halopropynes using both IR and Raman data (2,4,5). The vC—H and vC—D and vC≡C modes have been previously discussed. The vibrations most affected by change in the halogen atom for the CH$_2$X group are vC—X skeletal bending, CH$_2$ wagging, CH$_2$ twisting, and CH$_2$ rocking. In most cases these fundamentals decrease in frequency progressing in the series F to I. These molecules have a plane of symmetry, and the 10 vibrations that occur within the plane are designated as a′ fundamentals and the 5 vibrations that occur out-of-the plane are designated as a″ fundamentals. These molecules have C$_s$ symmetry.

Table 5.3 lists IR vapor phase data and assignments for 1-alkynes (3). Most of these assignments are for the alkyl group of these 1-alkynes. The numbers in parenthesis are the measured absorbances, and study of these numbers shows the intensities relative to one another in each spectrum. It is of interest to note that the ratio [(A)CCH bend/(A)$v$asym. CH$_2$] decreases as the number of (CH$_2$)$_n$ increases from 2 to 10. This indicates that the intensity for CCH bending is essentially constant and the intensity for $v$asym. (CH$_2$)$_n$ becomes more intense as n becomes larger. It should be noted that $v$asym. (CH$_2$)$_n$ shifts to lower frequency by 17 cm^{-1} as n is increased from 2 to 10. Most likely the dipole moment change during $v$asym. CH$_2$ changes slightly with this small shift in its frequency.

Figure 5.1 shows a plot of absorbance ratios for the 1-alkynes vs the number of carbon atoms in the 1-alkynes (3). Correlations such as these help in spectra-structure identification of unknown samples.

1-HALOPROPYNES

1-Halopropynes have C_{3v} symmetry, and the 15 fundamentals are distributed: $5a_1$, and $5e$ (The e fundamentals are doubly degenerate). The a_1 modes should yield parallel IR vapor bands and polarized Raman bands while the e modes should yield perpendicular IR vapor bands and depolarized Raman bands. Vibrational assignments in Table 5.4 were made using these criteria (7).

It has already been noted that $v{\equiv}C{-}D$ and $vC{\equiv}C$ coupled considerably in the case of the 1-halopropynes. During a cycle of $vC{\equiv}C$, both the $C{-}C$ and $C{-}X$ bonds must expand and contract.

$$H_3C \leftarrow C{\equiv}C \rightarrow X \leftrightarrow H_3C \rightarrow C{\equiv}C \leftarrow X$$

Therefore, this complex $vC{\equiv}C$ mode must include contribution from both $vC{-}C$ and $vC{-}X$. A comparison of these modes vs those for propyne are presented here:

1-halopropyne	$vC{\equiv}C$ cm^{-1}	$vC{-}C$ cm^{-1}	$vC{-}X$ or $vC{-}H$ cm^{-1}
Br	2239	1037	465
I	2210	1013	403
propyne	2130	930	3320

It is noted that $vC{\equiv}C$ for 1-halopropynes occurs approximately $95\,\text{cm}^{-1}$ higher in frequency than it occurs for propyne. In addition, $vC{-}C$ for 1-halopropynes occurs approximately $90\,\text{cm}^{-1}$ higher in frequency than it occurs for propyne. The $vC{-}X$ mode occurs at lower frequency than the other two vibrations while $vC{-}H$ occurs at higher frequency. Moreover, the force constant for Br is higher than it is for I. All of these facts suggest that the $vC{\equiv}C$ mode for 1-halopropyne is complex and involves a stretching motion of the two adjacent groups.

Table 5.4a lists the Coriolis coupling constants for 1-halopropyne (7,8), propyne (9) and propyne-1-d (10). These Coriolis constant coupling coefficients are included so that the reader has better knowledge about interpreting spectral data. Lord and Merrifield (11) have stated that the physical meaning of the minus sign to the Coriolis constant is that the angular momentum of vibration, which is so related to that of the rotation, produces an increase angular velocity of rotation of the vibrating molecular dipole moment about the molecular axis. The plus sign means that the velocity of rotation of the vibrating dipole moment is less than it would be if no angular momentum of vibration were present. In terms of normal vibrations, this means that there is only a slight decrease in the velocity of rotation during the asym. CH_3 stretching vibration (v_6) in these molecules relative to what the velocity would be for the nonvibrating molecule. A significant increase is noted in the velocity of rotation during the asymmetric CH_3 deformation (v_7), because all Coriolis coupling constants are negative (between -0.33 and -0.39). These results are what is expected if the CH_3 group were considered to be a symmetrical 3-armed flywheel rotating about a fixed axis. The velocity of rotation would be increased if one or more of the arms were bent toward the fixed axis; in the case of v_7 there is an alternating bending of two hydrogen atoms with one hydrogen atom toward and away from the molecular axis. Perhaps this can be more clearly demonstrated in v_{10}(e), which is essentially C—X bend in the case of 1-halopropyne; a shift of such a heavy atom off of the molecular axis would most certainly slow down the velocity of the molecular rotation relative to that of the fixed rigid rotating molecule. The order of magnitude of the Coriolis coupling coefficients has significance in the interpretation of the IR spectra of these compounds. The larger the value the closer the Q peaks are spaced in the subband. Consequently, a value of approximately one might appear to be nearly one broad absorption band with little or no fine structure due to unresolved closely spaced Q peaks in the subband. It is significant to note that the perpendicular (e) modes in solution or in the liquid phases appear to be as broad as they are in the vapor phase. They differ only in that the fine structure (Q peaks) are not observed. This suggests that in the condensed phases the molecules are still rotating, but not as freely as in the vapor phase (7).

PHENYLACETYLENE IN VARIOUS SOLUTIONS

Phenylacetylene in 0 to 100 mol % $CHCl_3/CCl_4$ solutions has been studied utilizing FT—IR spectroscopy (14) (see Table 5.5). Figure 5.2 shows a plot of $v≡C$—H for phenylacetylene vs mole % $CHCl_3/CCl_4$. Figure 5.2 shows that the $v≡C$—H decreases in frequency as the mole % $CHCl_3/CCl_4$ increases. Breaks in the plot near 10 and 45–55 mol % $CHCl_3/CCl_4$ suggest that different complexes are being formed as the mole % $CHCl_3/CCl_4$ changes.

In the case of phenylacetylene in solution, a complex such as ϕ—C≡C—H$\cdots$ClCCl$_3$ is suggested, in intermediate solutions of $CHCl_3/CCl_4$ a complex such as

$$\phi-C≡C\text{-}H\cdots ClCCl_3$$
$$\vdots$$
$$Cl_3CH$$

and in $CHCl_3$ a complex such as

$$\phi-C≡C\text{-}H\cdots ClCHCl_2$$
$$\vdots \qquad \vdots$$
$$Cl_3CH \quad HCCl_3$$

are suggested to explain the changes in both the $v≡C$—H and $vC≡C$ frequency changes with change in the solvent system. Bulk dielectric effects of the solvents also contribute to the group frequency shifts as the mole % change from 0–100.

Figure 5.3 shows a plot of $\nu C \equiv C$ vs mole % $CHCl_3/CCl_4$. This plot shows that $\nu C \equiv C$ decreases in frequency as the mole % $CHCl_3/CCl_4$ is increased. Figure 5.4 shows a plot of the in-plane $\equiv C-H$ bending mode $\delta_{ip} \equiv C-H$ vs mole % $CHCl_3/CCl_4$. The plot stops near 60 mol % $CHCl_3/CCl_4$. The plot stops near 60 mol % CH_3/CCl_4, because absorbance from $CHCl_3$ masks the absorbance of $\delta_{ip} \equiv C-H$. This plot shows that $\delta_{ip} \equiv C-H$ increases in frequency as the mole % $CHCl_3/CCl_4$ is increased. Figure 5.5 shows a plot of the out-of-plane $\equiv C-H$ mode, $\gamma_{op} \equiv C-H$ vs mole % $CHCl_3/CCl_4$ for phenylacetylene, and it shows that it increases in frequency as the mole % $CHCl_3/CCl_4$ is increased. The breaks in Figs. 5.2 and 5.5 indicate that different complexes are forming in these mole % $CHCl_3/CCl_4$ segments of the plots.

Table 5.5 lists IR data for phenylacetylene in 2% wt/vol solutions in mole % $CHCl_3/CCl_4$ solutions (14). Two IR bands are noted near $3300\,cm^{-1}$ in the neat phase and in solution in various solvents. For example, in hexane solution, the strongest band in this set occurs at $3322.91\,cm^{-1}$ and the shoulder occurs at $3311.27\,cm^{-1}$ while in the neat liquid the strongest band in this set occurs at $3291.17\,cm^{-1}$ and the shoulder occurs at $3305.08\,cm^{-1}$. The changes in both frequency and the intensity ratio of this band set is proof that these two bands are in Fermi resonance (F.R.), and these two bands have been corrected for F.R. The corrected frequencies vary between 3297.2 and $3318.1\,cm^{-1}$ for $\nu \equiv C-H$ and between 3299.2 and $3316.1\,cm^{-1}$ for the combination tone. The highest $\nu \equiv C-H$ frequency exhibited by phenylacetylene is when it is in solution with hexane and the lowest when in the neat liquid phase.

A plot of $\gamma_{op} \equiv C-H\,cm^{-1}$ vs $\delta_{ip} \equiv C-H\,cm^{-1}$ for phenylacetylene (recorded in solvents that were not masked by the solvents used in this study and that for the neat liquid phase show a linear relationship). This indicates that both modes are affected equally in a particular solvent, but differently in each solvent in the neat liquid phase (14). These data do not correlate well with the solvent acceptor number (AN), and this may be due to the fact that the AN values do not reflect the intermolecular hydrogen bonding capabilities of solvents such as CH_2Cl_2 and $CHClCl_3$.

1,4-DIPHENYLBUTADIYNE

In the solid state 1,4-diphenylbutadiyne has a monoclinic crystal structure with two molecules in the unit cell. The two molecules in the unit cell have a center of symmetry (12). The space group for 1,4-diphenylbutadiene is $P2_1/C$, which is isomorphous with the C_{2h} point group. Thus, the $\nu_{ip}(C \equiv C)_2$, Ag mode should be only Raman active and the $\nu_{op}(C \equiv C)_2$, Bu mode should be only IR active.

Table 5.6 lists IR data and assignments for 1,4-diphenylbutadiyne in $\sim 10\%$ wt/vol $CHCl_3/CCl_4$ solutions (13). Raman data for this compound in the solid state and in $CHCl_3$ and CCl_4 solutions are also included. A Raman band at 2218.9 (in CCl_4), at 2217.8 (in $CHCl_3$), and at $2214.4\,cm^{-1}$ in the solid phase is assigned to $\nu_{ip}(C \equiv C)_2$. The IR band at $2150\,cm^{-1}$ in the solid state is assigned to $\nu_{op}(C \equiv C)_2$. The $\nu_{ip}(\equiv C)_2$ mode in the solid state occurs $64.4\,cm^{-1}$ higher in frequency than the $\nu_{op}(C \equiv C)_2$ mode in the solid state. There is no evidence for $\nu_{op}(C \equiv C)_2$ in the Raman spectrum in either of the solution or solid phases. The solid phase data and assignments for 1,4-diphenylbutadiyne support the x-ray data (12).

In CCl_4 solution $\nu_{ip}(C \equiv C)_2$ is assigned at $2250.51\,cm^{-1}$ and in $CHCl_3$ solution at $2219.45\,cm^{-1}$, a decrease in frequency of $1.06\,cm^{-1}$. In CCl_4 solution $\nu_{op}(C \equiv C)_2$ is assigned at 2152.23 cm^{-1} and in CHCl$_3$ solution at 2150.22 cm^{-1}, a decrease in frequency of 2.01 cm^{-1} (see Table 5.6). In the IR, the absorbance ratio $[(A)\nu_{ip}(C\equiv C)_2]/[(A)\nu_{op}(C\equiv C)_2]$ generally increases as the mole % CHCl$_3$/CCl$_4$ increases. As $\nu_{ip}(C\equiv C)_2$ is observed in the IR when 1,4-butadiyne is in solution, this indicates that in solution the two phenyl groups are not coplanar as they are in the solid phase (13).

PROPARGYL ALCOHOL VS PROPARGYL FLUORIDE

Table 5.7 list vibrational data and assignments for propargyl alcohol, its −OD and 1-d and O−D analogs and propargyl fluoride (3-fluoropropyne). It is helpful to compare the frequency assignments of a fluoro analog to the corresponding OH analog, because the C−F and C−O modes occur at similar frequency, and the F analog contains 3 less fundamental vibrations. The OH analog has an additional 3 fundamental vibrations. They are νOH, δOH, and τOH (or OH torsion in the vapor phase). The assignments for the propargyl alcohol and its deuterium analogs are included to show the reader the value of using the vibrational assignment for corresponding R−F vs R−OH or R−O analogs. The same will be shown to be of value in the vibrational assignments for Aryl analogs and phosphorus analogs.

The interesting feature in the study of propargyl alcohol is that in the liquid phase it does not have a plane of symmetry because CH$_2$ twist couples with OH bending and CH$_2$ wag, indicating that these fundamentals all belong to the same species. With C$_s$ symmetry, CH$_2$ twisting belongs to the a'''' symmetry species, and δOH and CH$_2$ wag belong to the a' species (15).

1,3-DIHALOPROPYNES

Vapor- and solution-phase infrared spectra of 1,3-dichloropropyne are presented in Figures 5.6 and 5.7, respectively. Complete vibrational assignments have been made for both 1,3-dichloropropyne and its 1,3-Br$_2$ analog (16). Vibrational assignments based on both IR and Raman data for the 15 fundamentals of each analog are presented in Table 5.8. The a' and a'' fundamentals are due to in-plane and out-of-plane vibrations, respectively. The a' modes should yield type A/B IR vapor bands and polarized Raman bands while a'' modes should yield type C IR vapor bands and depolarized Raman bands.

The band at 2261 cm^{-1} is assigned to primarily νC≡C in the case of 1,3-dichloropropyne. The band at 2224 cm^{-1} is assigned to a combination tone ($\nu_5 + \nu_{10}$). The two bands are in Fermi resonance (F.R.) and these two frequencies have not been corrected for F.R. The F.R. correction would lower the unperturbed νC≡C frequency and raise the combination tone frequency.

Assignments in Table 5.8 are simplified because the X−C−C≡C−X stretching modes are expected to be complex as discussed for the 1-halopropynes.

In the case of the 1,3-dihalopropynes it appears as though the νC≡C mode is even more complex than it is in the case of 1-haloalkyne. The C−C≡C−X atoms are all on the molecular axis, and the other X atom is in the plane of molecular symmetry. Therefore, when the C≡C group expands and contracts during a cycle of νC≡C, the C−C of the group, C−C≡, and the ≡C−X groups must compress and expand. In addition the X−C−C≡ bond angle most likely decreases to a small degree in the case of 3-bromopropyne. Comparison of these vibrations for 1,3-dibromopropyne and 1-bromopropyne show that these modes occur at similar frequencies.

Compound	vC≡C cm^{-1}	vC−C≡ cm^{-1}	v≡C−X cm^{-1}
1,3-dibromopropyne	2226	1064	512
1-bromopropyne	2239	1037	464
3-bromopropyne	2138	961	3335

The vC−C≡ and v≡C−X modes for 1,3-dibromopropyne occur at higher frequency than the corresponding modes for 1-bromopropyne. Moreover, the vC≡C and vC−C≡ modes are higher than those exhibited by the 3-halopropynes. These data indicate that vC≡C is a more complex mode than just stretching of the C≡C bond. Nevertheless, vC≡C is considered to be a good group frequency in identifying this group in unknown samples.

Figure 5.8 shows the expected normal vibrations of 1-halopropynes, 3-halopropynes, and 1,3-dihalopropynes (16) and their frequency assignments. These normal modes are most likely oversimplified.

PHENYLACETYLENE AND PHENYLACETYLENE-1D

Table 5.9 lists vibrational data for phenylacetylene and phenylacetylene-1d. The in-plane CC−H and CC−D bending frequencies are assigned at 648 and 482 cm^{-1}, respectively. The out-of-plane CC−H and CC−D bending frequencies are assigned at 612 and 482 cm^{-1}, respectively. Again, the vC≡C mode for the H and D analogs occur at 2119 and 1989 cm^{-1}, respectively, and this is attributed to coupling between vC≡C and vC−D (17).

Assignment of the ring modes will be discussed later.

REFERENCES

1. Nyquist, R. A. and Potts, W. J. (1960). *Spectrochim. Acta*, **16**, 419.

2. Evans, J. C. and Nyquist, R. A. (1963). *Spectrochim. Acta*, **19**, 1153.

3. Nyquist, R. A. (1985). *Appl. Spectrosc.*, **39**, 1088.

4. Nyquist, R. A., Reder, T. L., Ward, G. R., and Kallos, G. J. (1971). *Spectrochim. Acta*, **27A**, 541.

5. Nyquist, R. A., Reder, T. L., Stec, F. F., and Kallos, G. J. (1971). *Spectrochim. Acta*, **27A**, 897.

6. Evans, J. C. and Nyquist, R. A. (1960). *Spectrochim. Acta*, **16**, 918.

7. Nyquist, R. A. (1965). *Spectrochim. Acta*, **7**, 1245.

8. Davidson, D. W., Sundaram, S., and Cleveland, F. F. (1962). *J. Chem. Phys.*, **37**, 1087.

9. Boyd, D. R. J. and Thompson, H. W. (1952). *Trans. Faraday Soc.*, **48**, 493.

10. Grisenthwaite, H. A. J. and Thompson, H. W. (1954). *Trans. Faraday Soc.*, **50**, 212.

11. Lord, R. C. and Merrifield, R. E. (1952). *J. Chem. Phys.*, **20**, 1348.

12. Wiebenga, E. H. (1940). *Z. Kristallogr.*, **102**, 93.

13. Nyquist, R. A. and Putzig, C. L. (1992). *Vib. Spectrosc.*, **4**, 35.

14. Nyquist, R. A. and Fiedler, S. (1994). *Vib. Spectrosc.*, **7**, 149.

15. Nyquist, R. A. (1971). *Spectrochim. Acta*, **27A**, 2513.

16. Nyquist, R. A., Johnson, A. L., and Lo, Y.-S. (1965). *Spectrochim. Acta*, **21**, 77.

17. Evans, J. C. and Nyquist, R. A. (1960). *Spectrochim. Acta*, **16**, 918.

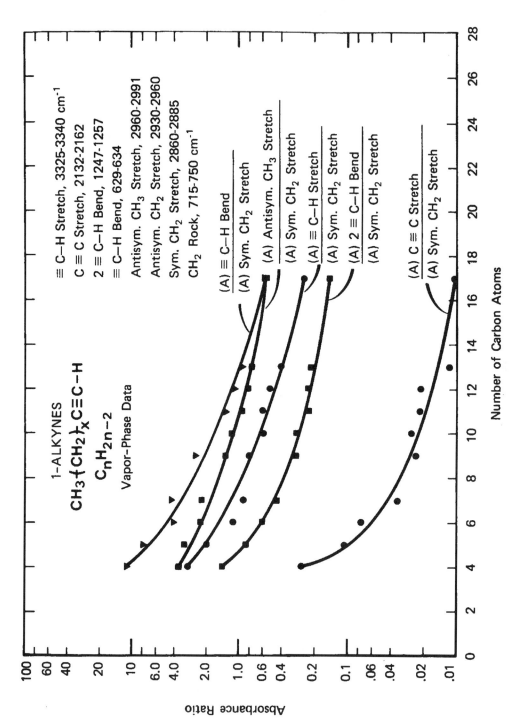

FIGURE 5.1 A plot of absorbance ratios for the 1-alkynes vs the number of carbon atoms in the 1-alkynes.

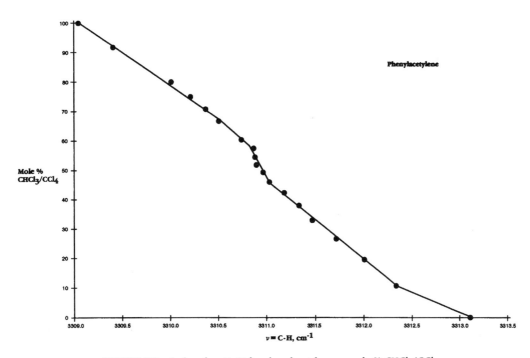

FIGURE 5.2   A plot of $v{\equiv}C{-}H$ for phenylacetylene vs mole % $CHCl_3/CCl_4$.

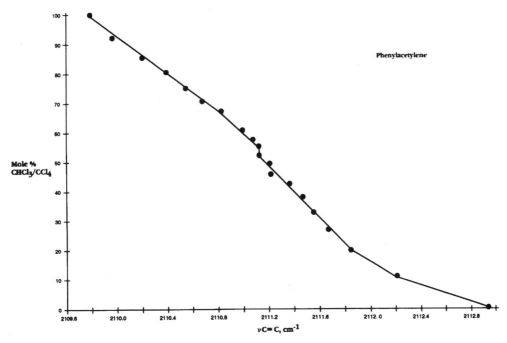

FIGURE 5.3   A plot of $vC{\equiv}C$ for phenyl acetylene vs mole % $CHCl_3/CCl_4$.

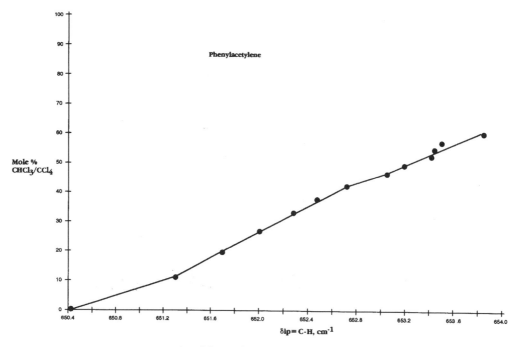

FIGURE 5.4   A plot of the in-plane ≡C−H mode vs mole % CHCl₃/CCl₄.

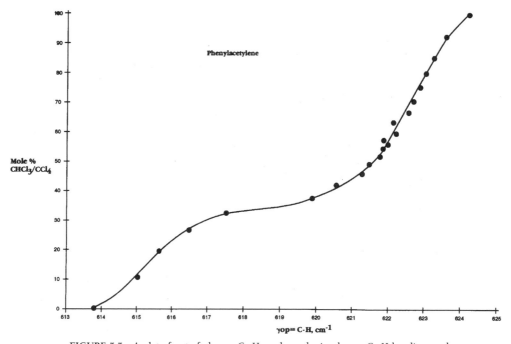

FIGURE 5.5   A plot of out-of-plane ≡C−H mode vs the in-plane ≡C−H bending mode.

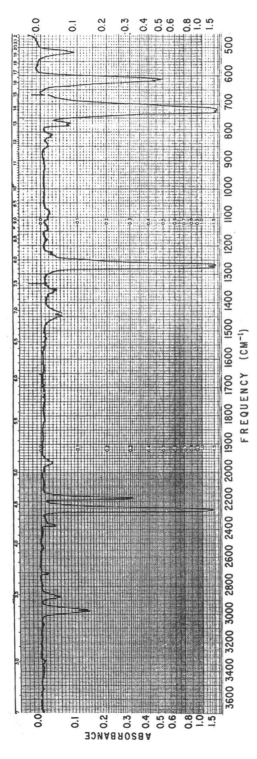

FIGURE 5.6 Vapor- and solution-phase infrared spectra of 1,3-dichloro-propyne.

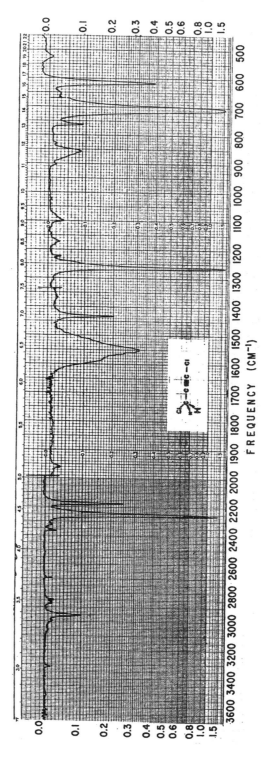

FIGURE 5.7 Vapor- and solution-phase infrared spectra of 1,3-dichloro-propyne.

APPROXIMATE NORMAL MODES
PROPYNE AND 1-HALOPROPYNES

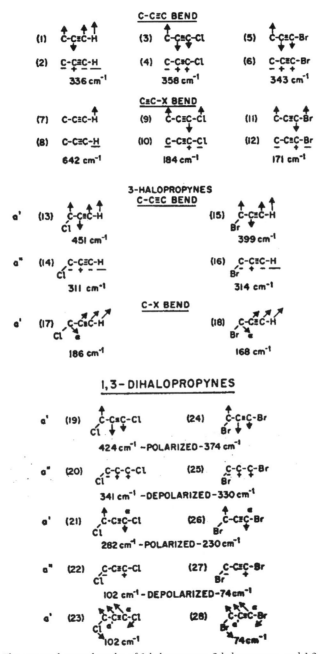

FIGURE 5.8   The expected normal modes of 1-halopropynes, 3-halopropynes, and 1,3-dihalopropynes.

TABLE 5.1   IR data and assignments for terminal acetylenic compounds

Compound type	C—H str.	CC str.	C—C str.	C—H bend	2(C—H bend)	C—CC bend
CH_3—CC—H	3320 3300sh	2130	930	630	1249	
R—CH_2—CC—H	3320 3299sh	2121	924–959	628–633	1242–1249	335
C_6H_5—CC—H [vapor]	3340 3320sh	2120		642 and 613	1219	
C_6H_5—CC—H [soln]	3316 3305sh	2115		648 and 611	1219	353?
ICH_2—CC—H	3315 3298sh	2109sh 2148sh	961	637	1263	
$BrCH_2$—CC—H	3315 3300sh	2126 2121sh	959	649 and 639	1265	313?
$BrCH_2$—CC—D	2599, CD str. 2582sh	1994	943	512 and 503	1000	301
$ClCH_2$—CC—H	3315 3299sh	2131 2126sh	959	652 and 637		310?
$ClCH_2$—CC—D	2618, CD str. 2589sh	2000	943	516 and 502	1043	300
FCH_2—CC—H	3322 3310sh	2148		674 and 636		
$HOCH_2$—CC—H	3316 3296sh	2120	902	650 and 629	1268?	
$HOCH_2C(-R)_2$—CC—H	3314–3316 3294–3300sh	2102–2115		649–655 620–630		
CH_3O—CC—H	3317 3296sh	2119 2104sh	932	663 and 625		
$ROC(-R)_2$—CC—H	3312–3314 3292–3296sh	2103–2111	908–938	651–660 627–630		
$C_6H_5OCH_2$—CC—H	3312–3320 3292–3301sh	2121–2131 2139sh		665–671 628–663		
$C_6H_5CH_2SCH_2$—CC—H	3317 3299sh	2120 2100sh	949	634	1248?	325?
$C_6H_5SCH_2$—CC—H	3318 3300sh	2122 2100sh	950	635		320?
$(R)_2NCH_2$—CC—H	3311–3316 3295–3300sh	~2100	897–925	647–653 622–624		~330?

?tentative assignment.

TABLE 5.2   IR and Raman data and assignments for 3-halopropynes

Species	3-fluoro-propyne	3-chloro-propyne	3-chloro-propyne-1-d	3-bromo-propyne	3-broma-propyne-1-d	3-iodo-propyne	Assignment
a′	3328	3335	2618	3335	2607	3335	CH or CD str.
	2955	2968	2969	2976	2978	2958	s. CH_2 str.
	2150	2147	2000	2138	2006	2130	CC str.
	1465	1441	1442	1431	1436	1423	CH_2 bend
	1381	1271	1265	1218	1215	1160	CH_2 wag
	940	960	943	961	945	959	C−C str.
	1039	725	723	621	634	570	C−X str.
	675	650	516	652	512	640	C−H or C−D bend
	539	451	438	399	386	364	skeletal bend
	211	186	165	168	163	157	skeletal bend
a″	2972	3002	2992	3006	3008	3008	a.CH_2 str.
	1242	1179	1176	1152	1151	1116	CH_2 twist
	1018	908	908	866	866	810	CH_2 rock
	635	637	502	637	501	640	CH or CD bend
	310	311	294	314	296	314	skeletal bend

TABLE 5.3 IR vapor-phase data and assignments for 1-alkynes

Compound	CC–H str.	C.T.	a.CH$_3$ str.	a.CH$_2$ str.	s.CH$_3$ str.	s.CH$_2$ str.	CC str.	CH$_2$ bend	s.CH$_3$ bend	2(CCH bend)	CH$_2$ rock	CCH bend	RI(A) CCH bend/RI(A) a.CH$_2$ str.
1-Pentyne	3330 (0.310)		2978 (0.390)	2952 (0.290)		2890 (0.110)	2135 (0.020)	1460 (0.059)	1390 (0.020)	1250 (0.125)		630 (1.240)	4.28
1-Hexyne	3330 (0.388)		2970 (0.877)	2948 (0.750)	2900 (0.318)	2880 (0.323)	2120 (0.056)	1455 (0.101)	1385 (0.048)	1249 (0.199)	740 (0.020)	630 (1.250)	1.67
1-Heptyne	3330 (0.273)		2970 (0.661)	2945 (0.990)		2980 (0.296)	2125 (0.034)	1460 (0.090)	1380 (0.035)	1250 (0.131)	740 (0.020)	631 (1.230)	1.24
1-Nonyne	3338 (0.305)		2970 (0.491)	2940 (1.240)		2870 (0.385)	2130 (0.030)	1467 (0.084)	1386 (0.025)	1249 (0.120)	725 (0.010)	630 (0.950)	0.77
1-Decyne	3338 (0.259)	3320 (0.106)	2970 (0.540)	2940 (1.143)		2875 (0.490)	2124 (0.024)	1465 (0.100)	1385 (0.049)	1250 (0.139)	724 (0.010)	632 (0.700)	0.61
1-Undecyne	3338 (0.250)	3318 (0.060)	2970 (0.420)	2940 (1.250)		2868 (0.440)	2124 (0.034)	1466 (0.090)	1385 (0.030)	1250 (0.100)	722 (0.030)	630 (0.604)	0.48
1-Dodecyne	3328 (0.270)	3320 (0.070)	2970 (0.440)	2935 (1.240)		2864 (0.420)	2122 (0.025)	1465 (0.100)	1380 (0.022)	1248 (0.119)	719 (0.011)	630 (0.590)	0.47
1-Tridecyne	3330 (0.169)	3310 (0.056)	2970 (0.319)	2935 (1.250)		2860 (0.430)	2121 (0.015)	1465 (0.048)	1385 (0.030)	1246 (0.089)	720 (0.020)	631 (0.409)	0.32

TABLE 5.4   IR data and assignments for 1-chloropropyne and 1-bromopropyne

Species	1-Bromopropyne	1-Iodopropyne	Assignment
a_1	2922	2922	s.CH$_3$ str.
	2239	2210	CC str.
	1368	1364	s.CH$_3$ bend
	1037	1013	C—C str.
	465	403*1	C—X str.
e	2965	2965	a.CH$_3$ str.
	1442	1439	a.CH$_3$ bend
	1027	1021	CH$_3$ rock
	343*1	343	skeletal bend
	171*1	163*1	G—X bend

*1 [liquid].

TABLE 5.4A   Coriolis coupling constants for 1-chloropropyne and 1-bromopropyne

Assignment	1-Chloropropyne	1-Bromopropyne	1-Iodopropyne	Propyne	Propyne-1-d
a.CH$_3$, e [stretch]	0.042	0.051	0.072	0.074	0.071
a.CH$_3$, e [bend]	−0.357	−0.347	−0.336	−0.39	−0.37
a.CH$_3$, e [rock]	0.4	0.393	0.383	0.387	0.4
skeletal bend, e	0.90 less than 1	0.95 less than 1	0.94 less than 1	0.96 less than 1	0.92 less than 1
C—X bend, e	0.90 less than 1	0.95 less than 1	0.94 less than 1	0.96 less than 1	0.92 less than 1

TABLE 5.5   IR data for phenylacetylene in CHCl$_3$/CCl$_4$ solutions

Phenyl-acetylene Mole % CHCl$_3$/CCl$_4$	C—H str. cm^{-1}	CC str. cm^{-1}	i.p. CC—H bend cm^{-1}	o.p. CC—H bend cm^{-1}	v32 (b2) ring def. cm^{-1}	v34 (b2) sk. def. cm^{-1}	v24 (b1) sk. def. cm^{-1}
0	3313.1	2113	650.4	613.8	690.4	530	513.6
26.53	3311.7	2111.7	652	616.5	690.6	530.2	513.5
52	3310.9	2111.1	653.4	621.7	690.9	530.3	513.4
75.06	3310.2	2110.5		622.9	691.2	530.4	513.4
100	3309.1	2109.8		624.2	691.8	530.5	513.4
delta cm^{-1}	−4	−3.2	3	10.4	1.4	0.5	−0.2

TABLE 5.6 IR data and assignments for 1,4-diphenylbutadiyne in CHCl$_3$/CCl$_4$ solutions

1,4-Diphenyl-butadiyne IR Mole % CHCl$_3$/CCl$_4$	i.p. (CC)$_2$ str. cm^{-1}	o.p. (CC)$_2$ str. cm^{-1}	[i.p. (CC)$_2$ str.]-[o.p. (CC)$_2$ str.] cm^{-1}	A[i.p. (CC)$_2$ str.]	A[o p. (CC)$_2$ str.]	A[i.p. (CC)$_2$ str.]/A[o.p. (CC)$_2$ str.] cm^{-1}	IR solid cm^{-1}	Raman i.p. (CC)$_2$ str. cm^{-1} [CCl$_4$]	Raman i.p. (CC)$_2$ str. cm^{-1} [solid]	[i.p. (CC)$_2$ str.]-[o.p. (CC)$_2$ str.]
0	2220.51	2152.23	68.28	0.215	0.32	0.67	2150	2218.9	2214.4	64.4
10.8	2220.34	2151.93	68.41							
10.8	2220.2	2152.15	68.05	0.26	0.36	0.55				
26.7	2220.11	2151.57	68.54							
26.7	2220.13	2151.61	68.52	0.293	0.369	0.79				
37.7	2219.99	2151.29	68.7	0.365	0.422	0.86				
45.9	2119.89	2151.12	68.6	0.414	0.464	0.89				
52.2	2219.83	2150.99	68.84	0.432	0.481	0.9				
63.4	2219.72	2150.79	68.9	0.476	0.518	0.92				
70.8	2219.65	2150.63	69.02	0.516	0.561	0.92				
80.2	2219.55	2150.45	69.1	0.575	0.617	0.93				
92.4	2219.41	2150.25	69.16	0.618	0.65	0.95	[CHCl$_3$]			
100	2219.45	2150.22	69.23	0.649	0.697	0.93	2217.8			67.58
delta cm^{-1}	1.06	2.01	0.95					1.1		

TABLE 5.7 Vibrational assignments for propargyl alcohol, propargyl alcohol-od, propargyl alcohol-1d, od, and propargyl fluoride

H–CC–CH_2–OH cm^{-1}	H–CC–CH_2–OD cm^{-1}	D–CC–CH_2–OD cm^{-1}	H–CC–CH_2–F cm^{-1}	Assignments and approximate descriptions [OH or OD analogs are all assigned to the a species]
3319	3319	2597	3320	C–H str., a′ or C–D str. coupled with CC str.
2950	2950	2960	2980	a.CH_2 str., a″
2925	2930	2937	2957	s.CH_2 str. a′
2138	2133	1987	2146	CC str. or CC str. coupled with C–D str.
1452	1456	1457	1457	CH_2 bend, a″
1382	1362	1366	1375	CH_2 wag, a′
1227 [H-bonded]	1220	1165	1240	CH_2 twist, a″
1197 [unassociated]				
1032	1030	1035	1020	C–F str., a′ or C–O str.
972	971	981	1018	CH_2 rock, a″
907	907	938	943	C–C str., a′
955	656	519 or 493	675	C–H bend, a′ or C–D bend
629	629	493 or 519	637	C–H bend, a″ or C–D bend
551	553	562	544	C–C–F bend, a′ or C–C–O bend
312	313	304	310	C–CC bend, a″
235	~230	211	211	C–CC bend, a′
3663 [vapor] and 3625 [soln.]		2706 [vapor]		O–H or O–D str. [unassociated]
~3500bd [soln.]	2678 [soln.]	2640 [liquid]		O–H or O–D str. [bonded]
1298 [vapor]				C–O–H or C–O–D bend [unassociated]
1420 [liquid]				C–O–H or C–O–D bend [bonded]
192 [soln.]		~150 [soln.]		C–O–H or C–O–D torsion

TABLE 5.8 Vibrational data for 1,3-dichloropropyne and 1,3-dibromopropyne

1,3-Dichloropropyne cm^{-1}	1,3-Dibromopropyne cm^{-1}	Assignment
		a′ species
2957	2959	CH_2 str.
2261	2226	CC str.
1433	1423	CH_2 bend
1264	1205	CH_2 wag
1098	1064	C−C str.
709	613	C−X str.
617	512	CC−X str.
424	374	C−CC bend
282	230	skeletal bend
102	74	CC−X bend
		a″ species
2994	3004	CH_2 str.
1172	1140	CH_2 twist
904	857	CH_2 rock
341	330	C−CC bend
102	74	CC−X bend

TABLE 5.9 Vibrational data and assignments for phenylacetylene and phenylacetylene-1d

Phenylacetylene cm^{-1}	Phenylacetylene-d1 cm^{-1}	Assignment	Description
		a1	
3315		v1	CC−H str.
	2596	v1	CC−D str.
3065	3068	v2	
3058	3059	v3	
3035	3037	v4	
2119		v5	CC str.
	1989	v5	CC str.
1597	1596	v6	
1490	1489	v7	
1192	1189	v8	
1178	1174	v9	
1028	1023	v10	
1000	998	v11	
763	759	v12	
467	462	v13	
		b1	
3101	3101	v14	
3083	3085	v15	
1573	1572	v16	
1446	1443	v17	
1332	1328	v18	
1285	1279	v19	
1158	1157	v20	
1071	1069	v21	
648		v22	CC−H bend
	482	v22	CC−D bend
610	623	v23	
515	529	v24	
351	344	v25	
		a2	
967	968	v26	
842	840	v27	
418	418	v28	
		b2	
983	982	v29	
917	914	v30	
754	754	v31	
688	688	v32	
612		v33	o.p. CC−H bend
	482	v33	op. CC−D bend
529	529	v34	
351	344	v35	
165	152	v36	

1-Halopropadienes and 1-Bromopropadiene-1d

*Numbers in parentheses indicate in-text page reference.

Propadienes were discussed in Chapter 4, and assignments were presented in Table 4.16. Chapter 6 is presented separately from previous discussion of propadienes because these compounds are formed from the rearrangement of 3-halopropynes. By study of the rearrangement of 3-bromopropyne-1d, it was possible to determine the rearrangement mechanism.

Assignments of these propadienes were based upon both IR and Raman data. Table 6.1 lists the frequencies and assignments for these 1-halopropadienes and 1-bromopropadiene-1d (1,2). Infrared solution and vapor-phase spectra of 3-bromopropyne-1d impure with 3-bromopropyne are shown in Fig. 6.1 and the Raman spectra are shown in Fig. 6.2. Figure 6.3 shows IR solution and vapor-phase spectra and Fig. 6.3a shows the Raman spectrum of 1-bromopropadiene-1d impure with 1-bromopropadiene (2).

The exchange between C≡C−H and O−D to form C≡C−D is known to occur under basic conditions, and it was used to prepare phenylacetylene-1d (3). The method for deuterium change utilizes a basic column packing. The sample of 3-bromopropyne-1d was passed through a preparative gas-liquid chromatography column packed with chromosorb w(30-60 mesh) coated with 15% E20M and 10% KOH that had been pretreated with D_2O (4). The sample of 1-bromopropadiene-1d was prepared using a method developed at The Dow Chemical Company (5). Using 2 ml of the ~90% sample of 3-bromopropyne-1d as starting material, in solution in 20 ml of dimethylformamide in the presence of 50 mg of NaBr under atmospheric pressure and 100 °C for 5 h, the sample of 1-bromopropadiene-1d was synthesized. The solution was then passed through a preparative liquid-gas chromatographic column, and the sample collected in a trap. The sample of 1-bromopropadiene-1d contained about 10% 1-bromopropadiene as an impurity.

The halopropadienes have C_s symmetry, and the 15 normal modes are distributed as: 10a′ and 5a″. The a′ modes are symmetrical and the a″ modes antisymmetrical with respect to the plane of symmetry. The a′ modes should yield type A/B IR vapor bands and polarized Raman bands, while a″ modes should yield type C IR vapor bands and depolarized Raman bands. The question to be answered is whether the product formed is HDC=C=CHBr or CH_2=C=CDBr. In other words, did the deuterium atom or the bromine atom move in the chemical rearrangement?

IR vapor-phase spectra for 1-chloro-, 1-bromo-, and 1-iodopropadiene are shown in Figs. 6.4, 6.5, and 6.6, respectively. IR vapor-phase type A/B bands are noted at (888/868 cm^{-1}), (873/852 cm^{-1}), and (860/848 cm^{-1}) for the Cl, Br, and I analogs, respectively. In addition, a type A/B band is noted at (872/865 cm^{-1}) in the case of the deuterated propadiene analog. Each of these compounds exhibits a type A/B band at (1758/1745 cm^{-1}), (1743/1732 cm^{-1}), (1717/1707 cm^{-1}), and (1735/1725 cm^{-1}) for the Cl, Br, I, and BrD analogs, respectively. The calculated first overtone of the type A/B bands reported here are: (1776/1736 cm^{-1}), (1746, 1704 cm^{-1}), (1720/1696 cm^{-1}), and 1744/1730 cm^{-1} for the Cl, Br, I, and BrD analogs, respectively. Assignment of the type A/B bands in the region 848–888 cm^{-1} to CH_2 = wag, and the type A/B bands in the region 1725–1758 cm^{-1} as 2(CH_2 = wag) is correct, because the CH_2 = wag mode always exhibits negative anharmonicity (occurs at higher frequency than the calculated 2(CH_2 = wag) (6).

A vapor-phase type C band is observed at 822, 812, and 807 cm^{-1} in the IR spectra of 1-chloro-, 1-bromo-, and 1-iodopropadiene, and a type C band is observed at 681 cm^{-1} in the case of the BrD analog. The ratio 812 cm^{-1}/681 cm^{-1} is equal to 1.19. The bands in the region 807–882 cm^{-1} are assigned to the out-of-plane =C−H bending vibration for the Cl, Br, and I propadiene analogs, respectively. In the case of the BrD analog the out-of-plane C−D bending mode is assigned at 681 cm^{-1}. The out-of-plane C=C=C bending mode for 1-bromopropadiene is assigned at 519 cm^{-1} and for the BrD analog it is assigned at 501 cm^{-1}. It is most likely that the out-of-plane bending mode and the out-of-plane C−D bending modes are coupled, as the out-of-plane mode occurs at higher frequency than predicted (~574 cm^{-1}) and the out-of-plane C=C=C bending mode occurs 18 cm^{-1} lower than it does for 1-bromopropadiene. Thus, this rearrangement involves migration of the halogen atom rather than the proton or deuteron atoms in a bimolecular reaction (2).

Plots of the fundamental vibrations and some combination and overtones for the 1-halopropadienes vs Pauling electronegativity values have been published (7). The dashed portion of these plots is where these vibrations would be predicted in the case of 1-fluoropropadiene. In our studies of 1-halopropynes, no evidence was observed for the presence of 1-fluoropropadiene. This may not seem surprising, as the strength of the C−X bond decreases in the order F to I. As already noted, it is the halogen atom that is involved in the bi-molecular rearrangement, and the activation energy required to break the C−F bond in this rearrangement is apparently larger than it is for the Cl, Br, and I analogs.

REFERENCES

1. Nyquist, R. A., Lo, Y-S, and Evans, J. C. (1964). *Spectrochim. Acta*, **20**, 619.
2. Nyquist, R. A., Reder, T. L., Stec, F. F., and Kallos, G. J. (1991). *Spectrochim. Acta*, **27A**, 897.

3. Evans, J. C. and Nyquist, R. A. (1960). *Spectrochim. Acta*, **16**.

4. Kallos, G. J. and Westover, L. B. (1967). *Tetrahedron Lett.*, 1223.

5. Pawloski, C. E. and Stewart, R. L. (1960). Method for the synthesis of 1-bromopropadiene. The Dow Chemical Company, Midland, MI.

6. Potts, W. J. and Nyquist, R. A. (1959). *Spectrochim. Acta*, **15**, 679.

7. Pauling, L. (1948). *The Nature of the Chemical Bond*, Ithaca, New York: Cornell University Press.

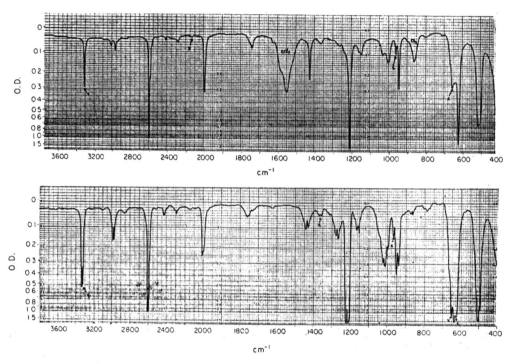

FIGURE 6.1   Infrared solution and vapor-phase spectra of 3-bromopropyne-1d impure with 3-bromopropyne.

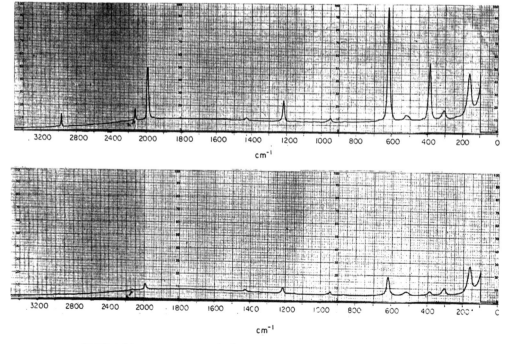

FIGURE 6.2   Raman spectra of 3-bromopropyne-1d impure with 3-bromopropyne.

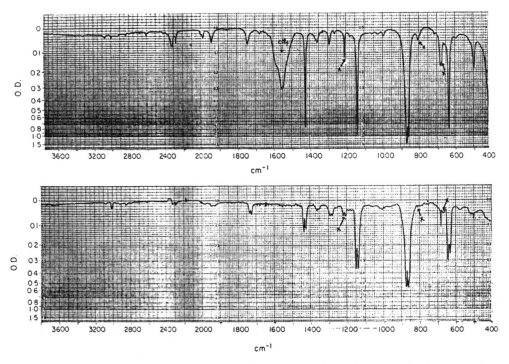

FIGURE 6.3   IR solution and vapor-phase spectra of 1-bromopropadiene-1d impure with 1-bromopropadiene.

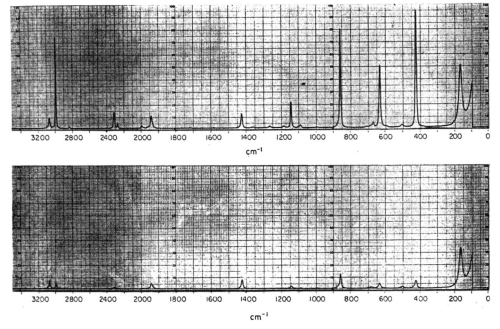

FIGURE 6.3A   Raman spectrum of 1-bromopropadiene-1d impure with 1-bromopropadiene.

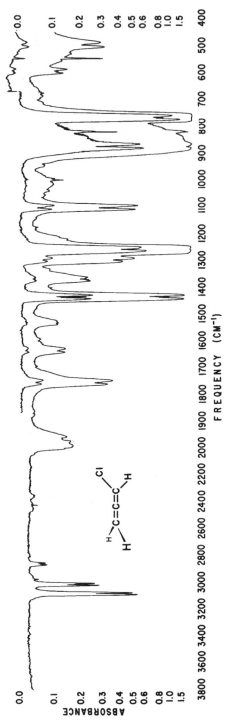

FIGURE 6.4   IR vapor-phase spectra of 1-chloropropadiene.

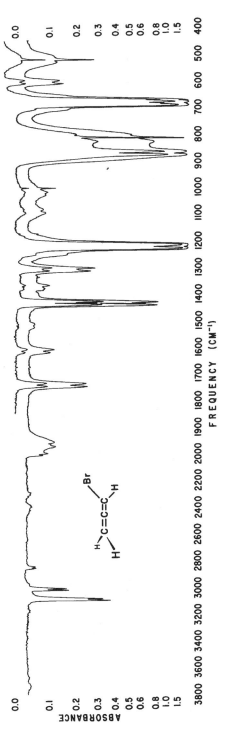

FIGURE 6.5   IR vapor-phase spectra of 1-bromopropadiene.

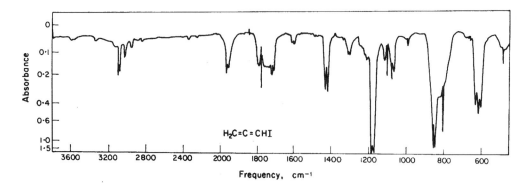

FIGURE 6.6   IR vapor-phase spectra of 1-iodopropadiene.

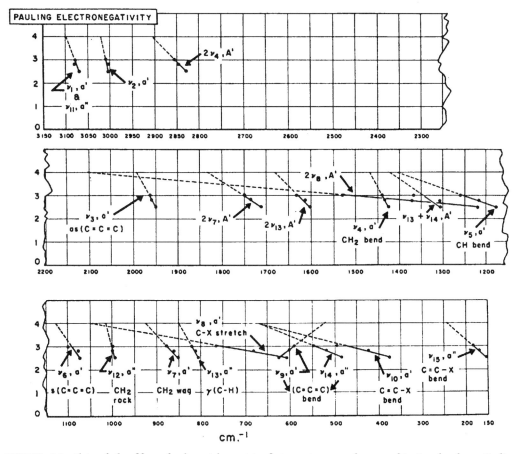

FIGURE 6.7   Plots of the fifteen fundamentals, certain first overtones, andone combination band vs. Pauling electronegativity. The solid lines represent observed data, and the dashed extension is the extrapolation used to predict the partial spectrum of fluoropropadiene (see text).

TABLE 6.1   IR and Raman data and assignments for 1-halopropadienes and 1-bromopropadiene-1d

Species	Chloro-propadiene	Bromo-propadiene	1-Bromo-propadiene-1d	Iodo-propadiene	Assignment
a′	3079	3080	2316	3070	CH str. or CD str.
	3009	3005	3005	3004	CH_2 str.
	1963	1961	1936	1953	a.C=C=C str.
	1435	1432	1426	1425	CH_2 bend
	1256	1217	858	1178	CH bend or CD bend
	1101	1078	1141	1076	s.C=C=C str.
	875	862	867	854	CH_2 wag
	767	681	636	609	C−X str.
	592	603	576	625	C=C=C bend
	494	423*[1]	426	387*[2]	C=C−X
a″	3079	2080	3075	3070	CH_2 str.
	999	1000	994	995	CH_2 rock
	822	812	681	807	CH bend or CD bend
	548	519	501	485	C=C=C bend
	184*[2]	169*[2]	170	154*[2]	C=C−X bend

*[1] [liquid]
*[2] [CS2 soln.]

Alcohols and Phenols

*Numbers in parentheses indicate in-text page reference.

Both alcohols and phenols are widely used for their unique properties, or as intermediates in the manufacture of other chemicals.

OH STRETCHING FOR ALCOHOLS

Table 7.1 lists IR vapor-phase data for alcohols (1). Primary alcohols, R—CH$_2$-OH, exhibit the OH stretching frequency in the region 3670–3680 cm^{-1}, secondary alcohols, (R-)$_2$CH-OH, 3650–3660 cm^{-1}, and tertiary alcohols, (R-)$_3$C—OH, 3640–3648 cm^{-1}. This decrease in vOH frequency progressing in the series primary, secondary, tertiary alcohols is attributed to the increasing inductive effect of the alkyl groups, which weaken the OH bond (1). In dilute CCl$_4$ solution, the primary, secondary, and tertiary alcohols occur in the regions 3630–3634, 3620–3635, and 3600–3620 cm^{-1}, respectively (2). The vOH frequencies occur at lower frequency in CCl$_4$ solution as a result of intermolecular hydrogen bonding between the OH proton and the Cl atom of CCl$_4$ (e.g., O—H$\cdots$Cl—C Cl$_3$)(3). In the vapor phase and in dilute solution the vOH band is sharp, and has relatively weak intensity. This is in contrast to intermolecular hydrogen-bonded OH:OH frequencies, which occur at lower frequency, 3200–3400 cm^{-1}, and have very strong broad band intensities.

Cycloalkanols are also secondary alcohols and their νOH frequencies in dilute CCl_4 solution occur in the region 3621.4–3627.7 cm^{-1}. These νOH frequencies decrease progressively in the order C_4 through C_8 (3). It is interesting to compare the νOH frequencies in the vapor and in dilute CCl_4 solution.

	Vapor cm^{-1}(1)	CCl_4 solution cm^{-1}(3)	Vapor-CCl_4 cm^{-1}
cyclopentanol	3660	3625.5	34.5
cyclohexanol	3659	3623.9	35.1
cycloheptanol	3655	3621.7	34.3
cyclooctanol	3650	3621.4	28.6
Δ cm^{-1}	10	4.1	5.9

In the vapor phase and in CCl_4 solution νOH decreases 10 and 4.1 cm^{-1} progressing in the series C_5 to C_8 for these cyclic secondary alcohols. In CCl_4 solution the OH is intermolecularly bonded to a Cl atom such as $OH\cdots ClCCl_3$. This intermolecular hydrogen bond weakens the OH bond; consequently, νOH occurs at lower frequency in CCl_4 solution than it occurs in the vapor phase. In addition, the increasing inductive contribution of $(CH_2)_3$ to $(CH_2)_7$ to CHOH is the reason the νOH frequency shifts to lower frequency as the series progresses from C_5 to C_8. Perhaps the reason that νOH does not shift as much to lower frequency in CCl_4 solution as it does in the vapor phase is that the α-bond

angle increases as n increases from 2 to 5. With the increasing α-bond angle, the α-carbon atoms prevent the OH and CCl_4 Cl atoms from coming as close in space to form the $OH\cdots Cl$ bond. In other words the intermolecular $OH\cdots Cl$ bond distance increases as $(CH_2)_n$ increases due to the increase in the α-bond angle, which increases the steric factor of the two adjacent CH_2 groups.

All alcohols form intermolecular hydrogen bonds in the condensed phase, if steric factors are not present. These intermolecular hydrogen bonds are formed between $(OH:OH)_n$ groups. In the case of the alcohols, the OH group is the most basic site as indicated. In the vapor phase, OH group for ordinary alcohols do not form intermolecular hydrogen bonds at elevated temperature. However, in the vapor phase at elevated temperature alcohols do form intramolecular hydrogen bonds with other available basic sites within the molecule to form 5-, 6-, 7-, or 8-membered intramolecular $OH\cdots X$ bonds. Examples are presented in what follows (1).

The compound 2-methoxyethanol in the vapor phase exhibits a weak shoulder at 3680 cm^{-1} assigned as νOH, and the 3640 cm^{-1} band is assigned to the intramolecular hydrogen bonded $OH\cdots O$ group forming a 5-membered group (1). In the case of 2-(2-methoxyethoxy) ethanol, the unassociated νOH is assigned at 3678 cm^{-1}, and the IR bands at 3639 and 3540 cm^{-1} are assigned to intramolecular hydrogen bond OH:O groups forming 5- and 8-membered rings, respectively. In both the 5- and 8-membered rings the OH group is bonded to an ether oxygen atom and both have comparable basicity. The 8-membered $\nu OH\cdots O$ group occurs at lower frequency than that for the 5-membered ring, because the $OH\cdots O$ groups are closer in space and form a stronger intramolecular hydrogen bond. Numerous examples of intramolecular hydrogen bonds for primary, secondary, and tertiary alcohols with OH, R−O−R, S, halogen, C=C, phenyl, and C=O groups are presented in Reference (1). Those engaged in GC/FT-IR experiments would benefit from information available in this text.

ALCOHOL C−O STRETCHING

The C−O stretching frequency, νC−O, in the vapor phase occurs in the region 1031–$1060\,cm^{-1}$, 1135–$1147\,cm^{-1}$, and 1141–$1180\,cm^{-1}$ for the primary, secondary, and tertiary alcohols, respectively (1). The absorbance (A) for νC−O for the primary alcohols generally decreases in intensity progressing in the series ethanol through 1-decanol, and it is attributed to an increase in the number of (CH_2) groups. The νC−O frequency increase progressing in the series primary, secondary, tertiary alcohol is attributed to increased branching on the C−O−H carbon atom, and νC−O in these cases includes some stretching of the C−C bonds.

PRIMARY ALCOHOLS AND CYCLOALKANOLS (CARBON-HYDROGEN VIBRATIONS)

The νasym. CH_3 and νsym. CH_3 modes occur in the region 2950–$2980\,cm^{-1}$ and 2899–$2920\,cm^{-1}$, respectively. The νasym. CH_2 and νsym. CH_2 modes occur in the region 2930–$2940\,cm^{-1}$ and 2865–$2910\,cm^{-1}$, respectively. The CH_2 bending and sym. CH_3 bending modes occur in the region 1452–$1469\,cm^{-1}$ and 1384–$1392\,cm^{-1}$, respectively. The CH_2 rocking mode occurs in the region 720–$778\,cm^{-1}$.

PHENOLS

In the solid phase (Nujol mull) phenols, which are not intramolecular, exhibit intermolecular hydrogen-bonded hydroxyl groups, $\nu(OH\cdots OH)_n$ in the range 3180–$3400\,cm^{-1}$. The 2-alkylphenols exhibit $\nu(OH\cdots OH)_n$ in the range 3438–$3535\,cm^{-1}$, and the frequency increases with the increasing steric factor of the alkyl group [CH_3 to $C(CH_3)_3$] (12). The increasing steric factor increases the bond distance between $(OH\cdots OH)_n$ group. Table 7.2 lists IR vapor-phase data for νOH, νOH$\cdots$X, in-plane OH bending, and phenyl oxygen stretching.

OH STRETCHING

Unassociated νOH for these phenols is assigned in the region 3642–$3660\,cm^{-1}$. In cases where the OH group is in the cis spatial configuration to a tert.-butyl group, the νOH frequency occurs at higher frequency in the region 3670–$3680\,cm^{-1}$. The higher frequency is attributed to repulsion between the OH proton and the protons on the tert.-butyl group. In cases where only one 2-tert.-butyl group is present, a νOH band will occur in each of these regions of the spectrum, and when 2,6-di-tert.-butyl groups are present only one band will occur in the lower region. The phenol νOH:X frequencies occur in the region 3278–$3630\,cm^{-1}$ depending upon which atom or groups are in the 2- or 6-positions. The frequency is dependent upon both the basicity of the X group, the acidity of the OH proton, and the spatial distance between OH and X in the cyclic intramolecular hydrogen bond.

PHENYL-OXYGEN STRETCHING AND OH BENDING

Phenols show a strong band in the region $1209-1295 \, \text{cm}^{-1}$ assigned to a complex in-plane ring mode, which includes stretching of the phenyl-O bond, $v\phi-\text{O}$ (1). A band in the region $1152-1224 \, \text{cm}^{-1}$ is attributed to in-plane OH bending.

INTRAMOLECULAR vOH$\cdots$X FREQUENCIES IN THE VAPOR AND CCL$_4$ SOLUTION PHASES

Table 7.3 compares the vOH, trans vOH, and vOH$\cdots$X frequencies in the vapor and CCl$_4$ solution phases.

In the case of phenol, vOH ($3650 \, \text{cm}^{-1}$) in the vapor decreases $40 \, \text{cm}^{-1}$ in CCl$_4$ solution ($3610 \, \text{cm}^{-1}$). The decrease in the vOH frequency is attributed to the formation of an intermolecular hydrogen bond of form OH$\cdots$Cl with CCl$_4$ molecules when phenol is in dilute solution in CCl$_4$. The vOH:X frequencies listed occur in region ($3202-3595 \, \text{cm}^{-1}$) in the vapor and occur at lower frequency by $17-35 \, \text{cm}^{-1}$ in CCl$_4$ solution ($3185-3560 \, \text{cm}^{-1}$). This decrease in frequency we attribute to the formation of X$\cdots$H—O which further weakens the OH bond.

$$\overset{\displaystyle\cdot\atop\displaystyle\cdot}{\text{Cl}}\text{CCl}_3$$

Table 7.4 lists the vOH frequencies for 1% wt./vol. 4-hydroxybenzaldehyde in 0–100 mol % CHCl$_3$/CCl$_4$ (4). In this case vOH for 4-hydroxybenzaldehyde decreases $12.4 \, \text{cm}^{-1}$ in going from solution in CCl$_4$ to solution in CHCl$_3$. Moreover, vOH decreases continually as the mole % CHCl$_3$/CCl$_4$ is increased (see Fig. 7.1). This decrease in frequency results from an increase in the Solvent Field effect as the mole % CHCl$_3$/CCl$_4$ is increased.

One would expect that the Cl atoms of CCl$_4$ would be more basic than the Cl atoms of CHCl$_3$ and that the vOH:Cl in the case of CCl$_4$ would occur at a lower frequency than in CHCl$_3$, and the opposite is observed in the case of 4-hydroxybenzaldehyde. Apparently, the situation is much more complex in the case of CHCl$_3$/Cl$_4$ solutions. There is, of course, intermolecular bonding such as (Cl$_3$CH$\cdots$ClCCl$_3$)$_n$ in the mixed solvent system. The Cl$_3$CH proton would also bond intermolecularly with the C=O group (C=O$\cdots$HCCl$_3$), both sides of the π system of the phenyl group, and to the OH group such as

All of these intermolecular hydrogen bonding sites filled with CHCl$_3$ would weaken the OH bond because it would be more acidic. Consequently, the vOH:X mode would occur at lower frequency than that exhibited by the much simpler case of CCl$_4$ solution. Apparently, the hydrogen bonding equilibrium shifts as the mole % CHCl$_3$/CCl$_4$ changes.

Table 7.5 lists IR data for phenol and intramolecular hydrogen-bonded phenols 10% wt./vol. in CCl$_4$ solution in the region $3800-1333 \, \text{cm}^{-1}$ and in CS$_2$ solution in the region 1333–

$400 \, \text{cm}^{-1}$ using 0.1-mn cells and 10 and 2% wt./vol. in 2,2,4-trimethylpentane in the region $450\text{–}280 \, \text{cm}^{-1}$ using a 0.4-mn cell unless otherwise indicated (5).

In the case of phenol, vOH occurs at $3610 \, \text{cm}^{-1}$ and OH torsion occurs at $300 \, \text{cm}^{-1}$ (6). In this case the OH group turns about the phenyl-O bond in a circle in and out of the plane of the phenyl group. However, in the case of the intramolecular hydrogen-bonded phenols the OH proton is in the plane of the phenyl ring and the 2-X group. Thus, its out-of-plane OH deformation is best described as γOH or as γOH$\cdots$X. The maximum peak of absorption for vOH$\cdots$X occurs in the region $3598\text{–}3180 \, \text{cm}^{-1}$ for most of the compounds listed in Table 7.5. However, compounds considered to exist in resonance forms such as

exhibit very broad vOH bands (2500 to $3500 \, \text{cm}^{-1}$ with a maximum near $3050 \, \text{cm}^{-1}$ in the case of the 2-hydroxyacetophenones), for 2,4-dibenzoyl resorcinol (very broad in this region 2200–$3600 \, \text{cm}^{-1}$ with the maximum near $3000 \, \text{cm}^{-1}$), and for 2-hydroxybenzophenone (very broad in the region $2200\text{–}3600 \, \text{cm}^{-1}$ with the maximum near $3100 \, \text{cm}^{-1}$). The γOH or γOH$\cdots$X frequencies occur in the region $366\text{–}858 \, \text{cm}^{-1}$. The γOH$\cdots$X frequencies increase in frequency as the vOH$\cdots$X frequencies decrease in frequency as demonstrated in Fig. 7.2. It is usually easy to detect γOH$\cdots$X due to the following: (a) its absorption band is always uniquely broad, and it is easily distinguished from other fundamental absorption bands occurring in this region of the spectrum; (b) the -OD analogs exhibit vOD$\cdots$X lower in frequency than γOH$\cdots$X by a factor of ~1.35, and this indicates that γOH$\cdots$X is essentially motion of the proton (or deuteron) alone; and (c) as the vOH$\cdots$X bond becomes stronger its frequency decreases, and with a stronger hydrogen bond it becomes more difficult to twist the OH proton out of the plane causing γOH$\cdots$X to shift to higher frequency.

One can perform a simple experiment utilizing IR to determine whether a phenol is inter- or intra-molecularly hydrogen bonded.

When IR spectra of phenolic compounds are recorded in the solid, condensed, or say, 10% wt./vol. in CCl_4 and CS_2, the OH group forms polymeric $(\text{OH}\cdots\text{OH})_n$ bonds which appear as broad bands centered near $3200\text{–}3400 \, \text{cm}^{-1}$. This happens even in cases when the OH is intramolecularly hydrogen bonded if the OH oxygen atom is more basic than the atom or group substituted in the 2-position. Examination of the sample in dilute CCl_4 or CS_2 solution using 01.0–mm cell in all cases shows only an unassociated phenolic vOH band or an intramolecular vOH$\cdots$X band for compounds containing only 2-x groups (which is not very basic). When the 2-x group is very basic or in the case of compounds such as 2-hydroxybenzophenone, the intramolecular hydrogen bond does not shift significantly or change in intensity. In all classes of OH containing compounds, the inter- or intra-molecularly hydrogen-bonded OH group has much more intensity than the same OH group existing in an unassociated state, or say, intermolecularly hydrogen bonded to CCl_4 or CS_2.

WEAK INTRAMOLECULARLY HYDROGEN-BONDED PHENOLS

The weak intramolecularly hydrogen-bonded phenols exhibit γOH$\cdots$X frequencies in the range 366–400 cm^{-1} and νOH$\cdots$X frequencies above 3500 cm^{-1}. The γOH$\cdots$X frequencies for 2-F, 2-I, 2-Br, and 2-Cl phenol occur at 366, 379, 394, and 359 cm^{-1}, respectively, and their pK$_a$ values (7) are 9.37, 9.04, 9.01, and 8.99, respectively. The halogen atoms increase in size in the order F, Cl, Br, and I, and the data show that the size of the halogen atom does not correlate with γOH$\cdots$X, but it does correlate with the pK$_a$ values.

Compounds such as 2-(methylthio) phenol and 2-(methoxy) phenol exhibit their νOH$\cdots$X and γOH$\cdots$X frequencies at (3415 and 537 cm^{-1}) and (3560 and 428 cm^{-1}), respectively. In this case, the sulfur atom is larger than the oxygen atom. Consequently, the S$\cdots$HO spatial distance of the intramolecular hydrogen bond is shorter than the O$\cdots$HO spatial distance of the intramolecular hydrogen bond. Thus, a stronger intramolecular hydrogen bond is formed in the case of 2-(methylthio) phenol than in the case of 2-(methoxy) phenol.

In the case of 2-phenylphenol the two rings are not coplanar, and the OH group intramolecularly hydrogen bonds to the π electron system of the 2-phenyl group (8). In this case, νOH$\cdots\pi\phi$ and γOH$\cdots\pi\phi$ occur at 3570 and 386 cm^{-1}, respectively.

The 2-hydroxy acetophenones form very strong intramolecular hydrogen bonds, and as already noted here, produce a very broad multipeaked absorption band extending over the range 2500–3500 cm^{-1} with a maximum near 3050 cm^{-1}. This is caused by the resonance forms I and II shown on page 127 and this affects both the νOH$\cdots$O=C and γOH$\cdots$O=C frequencies as discussed here. Enhancement of structure I would elongate the OH bond, thus making the proton more acidic. Enhancement of structure II would tend to increase the carbonyl group. Both of these effects would lower νOH$\cdots$O=C, and raise γOH$\cdots$O=C frequencies, because the intramolecular hydrogen bond would be stronger. In addition, these resonance effects would induce more double bond character into the C—O bond, which would also contribute to an increase in the γOH$\cdots$O=C frequency, as the νOH$\cdots$O=C bond would be stronger. In addition, these resonance effects would induce more double bond character into the C—O bond, which would also contribute to an increase in γOH$\cdots$O=C. Because the νOH$\cdots$O=C band is so broad, it is difficult to correlate νOH$\cdots$O=C with Tafts $(\sigma_\rho - \sigma') + (\sigma_m - \sigma')$ values. Tafts $(\sigma_\rho - \sigma')$ and $\sigma_m - \sigma'$ are measures of the resonance effects of the 5-substituent upon the OH and COCH$_3$ groups (9,10). Figure 7.3 shows a plot of $(\sigma_\rho - \sigma') + (\sigma_m - \sigma')$ vs the γOH$\cdots$O=C frequencies. The sum of the Taft values is expected to yield a parameter with which to compare the γOH$\cdots$O=C frequencies for 2-hydroxy-5-X- acetophenones. Similar curves are obtained by plotting γOH$\cdots$O=C vs $(\sigma_\rho - \sigma')$ or $(\sigma_m - \sigma')$ alone. In Fig. 7.3, the plot is continued with the dashed line representation through the two solid points that one would predict from the $(\sigma_\rho - \sigma')$ and $(\sigma_m - \sigma')$ values for acetyl and nitro, respectively. The x points are values previously obtained for phenols (10), and the agreement is quite good. A similar plot of γOH$\cdots$O=C vs $(\sigma_\rho - \sigma')$ for these compounds when extropolated as a straight line to higher values allows one to predict $(\sigma_\rho - \sigma')$ values for the acetyl group (+0.48) and the nitro group (+0.62), and the values obtained are +0.60 and +0.64 for acetyl and nitro, respectively (10). In the case of 2-methylsulfonylphenol, νOH$\cdots$O$_2$S occurs at 3330 and νOD$\cdots$O$_2$S at 2475 cm^{-1}. The νOH/νOD ratio is ~1.35, and the γOH$\cdots$O$_2$S/γOD$\cdots$O$_2$S frequency ratio

645 cm^{-1}/529 cm^{-1} is equal to ~1.22. The 529 cm^{-1} band is the only broad band in this region of the spectrum. An SO$_2$ deformation has been assigned in this region of the spectrum (11), and it is suggested that the reason γOH$\cdots$O$_2$S occurs at higher frequency is because it is coupled with an SO$_2$ deformation.

TEMPERATURE EFFECTS

Figure 7.4 shows plots of the intramolecular hydrogen-bonded OH : Cl stretching frequencies for 2-chlorophenol, 2,4,5-trichlorophenol and 2,6-dichlorophenol in CS$_2$ solution vs temperature in °C. These plots show that the νOH : Cl frequencies decrease in frequency in a linear manner as temperature is decreased (13). As the temperature is decreased the CS$_2$ volume contracts, and the Field effect of CS$_2$ increases. This then causes the νOH : Cl frequencies to decrease as the temperature is lowered.

REFERENCES

1. Nyquist, R. A. (1984). *The Interpretation of Vapor-phase Infrared Spectra: Group Frequency Data*, Philadelphia: Sadtler Research Laboratories, Div. of Bio-Rad.

2. Bellamy, L. J. (1975). *The Infrared Spectra of Complex Molecules*, vol. I, 3rd ed., London: Chapman and Hall; New York: John Wiley & Sons, Inc. p. 108.

3. van der Maas, J. H. and Lutz, E. T. G. (1974). *Spectrochim. Acta*, 30A, 2005.

4. Nyquist, R. A., Settineri, S. E., and Luoma, D. A. (1992). *Appl. Spectrosc.*, **46**, 293.

5. Nyquist, R. A. (1963). *Spectrochim. Acta*, **19**, 1655.

6. Evans, J. C. (1960). *Spectrochim. Acta*, **16**, 1382.

7. Bennett, G. M., Brooks, G. L., and Glasstone, S. (1935). *J. Chem. Soc.*, 1821.

8. Baker, A. W. and Shulgin, A. T. (1958). *J. Am. Chem. Soc.*, **80**, 5358.

9. Taft, R. W. Jr. (1956). *Steric Effects in Organic Chemistry*, Chap. 13, New York: J. Wiley.

10. Taft, R. W. Jr., Deno, N. C., and Skell, P. S. (1958). *Ann. Rev. Phys. Chem.*, **9**, 287.

11. Simon, A., Kriegsman, H., and Dutz, H. (1956). *Chem. Ber.*, **89**, 2378.

12. Lin-Vien, D., Colthup, N. B., Fateley, W. G., and Grasselli, J. G. (1991). *The Handbook of Infrared and Raman Characteristic Frequencies of Organic Molecule*, San Diego: Academic Press.

13. Nyquist, R. A. (1986). *Appl. Spectrosc.*, **40**, 79.

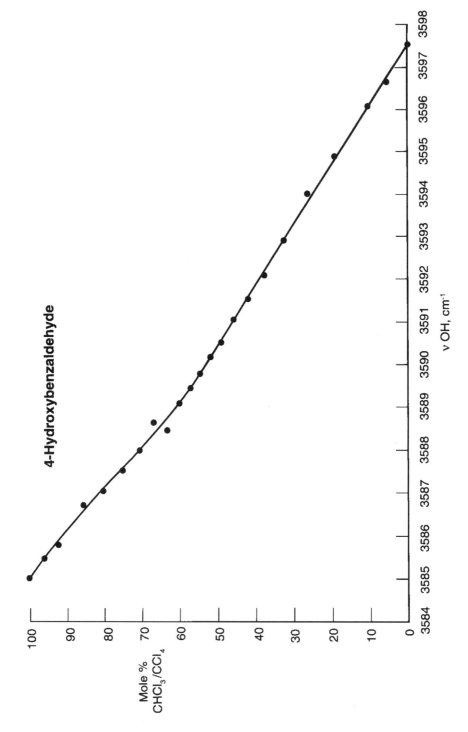

FIGURE 7.1   A plot of νOH for 4-hydroxybenzaldehyde vs mole % CHCl₃/CCl₄.

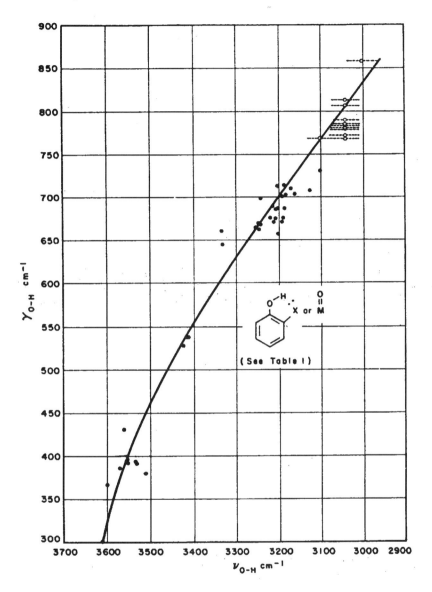

FIGURE 7.2 A plot of the $\gamma OH \cdots X$ frequencies vs the $\nu OH \cdots X$ frequencies for 2-X-substituted phenols.

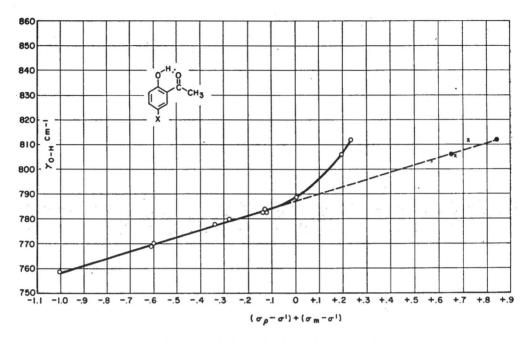

FIGURE 7.3   A plot of Tafts $(\sigma_p - \sigma') + (\sigma_m - \sigma')$ vs γOH$\cdots$OC frequencies.

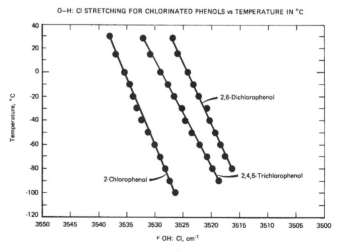

FIGURE 7.4   Plots of intramolecular hydrogen-bonded OH:Cl stretching frequencies for chlorinated phenols in CS_2 solution vs temperature in $^\circ$C.

TABLE 7.1 IR vapor-phase data and assignments for alcohols

Compound	OH str.	C-O str.	a.CH₃ str.	a.CH₂ str.	s.CH₃ str.	2 (a.CH₃ bend) in F.R. s.CH₃ str.	s.CH₂ str.	a.CH₃ bend	CH₂ bend	s.CH₃ bend	CH₂ rock
Methanol	3680 (0.150)	1031 (1.150)	2980 (0.620)		2920 (0.600)	2820 (0.370)	2910 (1.150)	1458 (0.129)			
Ethanol	3679 (0.240)	1060 (1.230)	2975 (1.150)				2890 (1.150)		1452 (0.150)	1392 (0.460)	800 (0.030)
Propanol	3675 (0.160)	1065 (0.840)	2965 (1.230)	2940 (1.150)	2900 (0.700)		2890 (0.495)		1461 (0.140)	1390 (0.190)	740 (0.020)
1-Butanol	3670 (0.161)	1041 (0.572)	2950 (1.230)				2890 (0.820)		1466 (0.151)	1389 (0.1510)	749 (0.035)
1-Hexanol	3675 (0.111)	1053 (0.400)	2970 (0.850)	2940 (1.230)	2899 (0.590)		2880 (0.600)		1467 (0.133)	1389 (0.130)	728 (0.020)
1-Heptanol	3670 (0.111)	1050 (0.341)	2970 (0.750)	2935 (1.250)			2878 (0.651)		1464 (0.130)	1389 (0.130)	722 (0.022)
1-Octanol	3675 (0.081)	1050 (0.320)	2970 (0.700)	2935 (1.250)			2870 (0.680)		1464 (0.141)	1388 (0.122)	723 (0.020)
1-Nonanol	3670 (0.070)	1052 (0.251)	2970 (0.640)	2935 (1.250)			2865 (0.475)		1464 (0.130)	1388 (0.130)	720 (0.005)
1-Decanol	3679 (0.055)	1052 (0.188)	2970 (0.500)	2935 (1250)			2865 (0.551)		1464 (0.102)	1385 (0.079)	
1-Dodecanol	3670 (0.060)	1050 (0.211)	2970 (0.555)	2935 (1.252)			2865 (0.770)		1460 (0.150)	1384 (0.090)	725 (0.022)
Cyclopentanol	3660 (0.160)	1004 (0.380)		2970 (1.250)			2900 (0.540)		1452 (0.100)		
Cyclohexanol	3659 (0.050)	1070 (0.250)		2940 (1.250)			2864 (0.330)		1458 (0.100)		
Cycloheptanol	3655 (0.053)	1034 (0.245)		2940 (1.230)			2870 (0.330)		1461 (0.110)		
Cyclooctanol	3650 (0.040)	1054 (0.135)		2930 (1.230)			2864 (0.290)		1456 (0.109)		

(continued)

TABLE 7.1 (*continued*)

Compound	OH str.	C-O str.	a.CH$_3$ str.	a.CH$_2$ str.	s.CH$_3$ str.	s.CH$_2$ str.	in F.R. s.CH$_3$ str.	2 (a.CH$_3$ bend)	a.CH$_3$ bend	CH$_2$ bend	s.CH$_3$ bend	CH$_2$ rock
2-Propanol	3660	1147										
2-Butanol	3360	1135										
2-Pentanol	3659	1140										
2-Hexanol	3360	1135										
2-Heptanol	3660	1140										
2-Octanol	3660	1135										
2-Decanol	3658	1135										
2,2-Dimethylethanol	3644	1141										
2,3-Dimethyl-2-butanol	3645	1180										
3-Methyl-3-pentanol	3642	1169										
3-Ethyl-3-pentanol	3645	1162										
2-Methyl-2-hexanol	3644	1170										
3-Methyl-3-hexanol	3642	1165										
2-Methyl-2-octanol	3642	1160										
2-Methyl-2-nonanol	3642	1159										

TABLE 7.2   IR vapor-phase data for substituted phenols

X-Phenol 4-X	OH str. cm^{-1}	Phenyl-O str. cm^{-1}	delta OH cm^{-1}	OH str. cis[t.-C$_4$H$_9$] cm^{-1}	OH:X str. cm^{-1}
H	3650	1260	1182		
CH$_3$	3655	1255	1171		
isoC$_3$H$_7$	3655	1257	1172		
tert.-C$_4$H$_9$	3658	1260	1172		
F	3660	1230	1178		
Cl	3658	1255	1172		
Br	3655	1256	1171		
OH	3658	1243	1170		
CH$_3$O	3655	1239	1174		
C$_6$H$_5$O	3655	1255	1171		
NO$_2$	3645	1270	1190		
CN	33645	1270	1178		
3-X					
CH$_3$	3650	1278	1158		
tert.-C$_4$H$_9$	3655	1285	1160		
Cl	3658	1289	1180		
Br	3655	1291	1181		
I	3642	1288	1179		
CH$_3$O	3659	1295	1150		
CN	3650	1285	1152		
2,4-X$_2$					
CH$_3$,CH$_3$	3658	1262	1191		
2-t-C$_4$H$_9$,4-CH$_3$	3642	1241	1177	3670	
2,4-Di-t.-C$_4$H$_9$	3646	1248	1178	3678	
2-NO$_2$,4-CH$_3$		1250	1185		3278
2,4-Di-Cl$_2$	3660	1275	1192		3580
2,4-Di-Br$_2$	3655	1278	1188		3562
2-Br,4-t.-C$_4$H$_9$	3655	1281	1183		3560
2,5-X$_2$					
2,5-Di-t.-C$_4$H$_9$	3645	1200	1149	3678	
2-Br,5-OH	3655	12114	1149		3559
2,6-Di-X$_2$					
2,6-Di-CH$_3$	3655	1269	1192		
2-CH$_3$,6-t.C$_4$H$_9$	3644	1270	1200	3670	
2,6-Di-isoC$_3$H$_7$	3650	1263	1201		
2-t.-C$_4$H$_9$,6-isoC$_3$H$_7$	3645	1263	1201	3670	
2,6-Di-CH$_2$CH=CH$_2$	3650	1258	1200		3585
2,6-Di-CH$_3$O		1285	1220		3580
2-CH$_3$,6-NO$_2$		1250	1159		3222
2,6-Di-NO$_2$		1271	1170		3230
2,6-Di-Cl$_2$		1240	1175		3570
2,6-Di-Br$_2$		1231	1172		3545
2,4,6-Tri-X$_3$					
2,4,6-Tri-CH$_3$	3658	1233	1198		
2,4-Di-CH$_3$,6-t.-C$_4$H$_9$	3650	1224	1178	3680	
4-CH$_3$,2,6-Di-t.-C$_4$H$_9$		1230	1161	3675	
2,4-Di-Cl,6-NO$_2$		1245	1152		3560[Cl] 3240[NO$_2$]

(continued)

TABLE 7.2 (*continued*)

X-Phenol 4-X	OH str. cm^{-1}	Phenyl-O str. cm^{-1}	delta OH cm^{-1}	OH str. cis[t.-C$_4$H$_9$] cm^{-1}	OH:X str. cm^{-1}
2,4-Di-Br,6-NO$_2$		1248	1160		3545[Br] 3230[NO$_2$]
2,4,6-Tri-Cl		1221	1166		3578
2,4,6-Tri-Br		1228	1160		3544
2,3,6-Tri-Cl		1298	1167		3562
2,3,4,6-Tetra-Cl		1285	1200		3564
2,3,5,6-Tetra-Cl		1290	1218		3558
2,3,4,5,6-Penta-F		1209	1224		3630

TABLE 7.3 A comparison of OH:X stretching frequencies in the vapor and CCl$_4$ solution phases

2-X-phenol X	OH:X str. vapor cm^{-1}	OH:X str. CCl$_4$ soln. cm^{-1}	[vapor]-CCl$_4$ soln.] cm^{-1}	OH str, or trans vapor cm^{-1}	OH str. CCl$_4$ soln. cm^{-1}
H				3650	3610
Cl	3580	3553	27	3650	
Br,4-Br	3562	3534	28	3655	
CH$_3$O	3595	3560	35	3655	
CH$_3$S	3445	3415	30	3650	
NO$_2$	3270	3245	25	3520	
CH(=O)	3202	3185	17		
NO$_2$,4-NO$_2$	3238	3215	23		
Cl,4-Cl,6-NO$_2$	3560(Cl) 3240(NO$_2$)	3209	31		

TABLE 7.4 The OH stretching frequency for 4-hydroxybenzaldehyde in 0 to 100 mol % $CHCl_3/CCl_4$ solutions

4-Hydroxy-benzaldehyde 1%(wt./vol.) Mole % CHCl3/CCl4	O-H str.
0	3597.5
5.68	3596.7
10.74	3596.1
19.4	3594.9
26.53	3594.1
32.5	3592.9
37.57	3592.1
41.93	3591.5
45.73	3591.1
49.06	3590.5
52	3590.2
54.62	3589.8
57.22	3589.4
60.07	3589.1
63.28	3588.5
66.74	3588.6
70.65	3587.9
75.06	3587.5
80.05	3587.1
85.75	3586.7
92.33	3585.8
96.01	3585.4
100	3585.1
delta O-H str.	[−12.4]

TABLE 7.5 The OH stretching frequencies for phenol and intramolecular hydrogen-bonded phenols and the OH torsion frequency for phenol and the out-of-plane OH : X deformation frequencies for intramolecular hydrogen-bonded phenols

Compound Phenol[1]	O-H str. 3610	O-H torsion 300 gamma O-H	O-D str.	gamma O-D	see text
2-phenyl	3570	386[2]			
2-fluoro	3598[1]	366[1]			
	3587				
2-iodo	3511[1]	379[1]			
	3507				
2-bromo	3534[1]	394[1]			
2,4-dibromo	3533[1]	392[1]			
	3534				
2-chloro	3553[1]	395[1]	2622[2]	298[2]	
	3553	400[2]			
2,4,5-trichloro	3553[1]	395[1]			
	3535				
2-methoxy	3560	433[2]			
		428[3]			
2-methylthio	3415	537	2540	not recorded	0
2-ethylthio	3413	537			
4-methyl-2-methylthio	3422	528			−0.3
2-methylsulfonyl	3330	645	2475	529	
4-chloro-isopropyl-6-nitro	3200	687			−0.24
4-chloro-2-s.-butyl-6-nitro	3205	686			−0.24
2-t-butyl-4,6-dinitro	~3100	731			0.15
2,4-di-t-butyl-6-nitro	~3125	698			−0.13
4,6-dinitro-2methyl	3170	710			0.15
4-chloro-6-nitro-2-methyl	3212	679			−0.24
2-bromo-4,6-dinitro	3159	700 or 706			0.15
2-bromo-3,4-dichloro-6-nitro	3185	786			−0.34
2-chloro-4,6-dinitro	3159	708			0.15
2-chloro-4-cyclohexyl-6-nitro	3219	676			−0.13
2-chloro-4-bromo-6-nitro	3205	675			−0.22
2,4-dichloro-6-nitro	3209	671			−0.24
4-t-butyl-2,6-dinitro	3186	678			−0.13
4-chloro-2,6-dinitro	3190	672			−0.24
4-fluoro-2,6-dinitro	3200	657			−0.44
2,4-dinitro	3215	680			0.15
2-nitro	3245	669			0.15
5-bromo-2-nitro	3220	668			−0.06
4-chloro-2-nitro	3252	664			−0.24
4,5-dichloro-2-nitro	3245	662			−0.34
Salicylate					
p-chlorophenylthio	3190	701			
methyl	3200	713			
phenyl	3240	698			
chloride	3331	660			
Aldehyde					
salicyl	3185	713	2360[4]	520[4]	0

(continued)

TABLE 7.5   (*continued*)

Compound Phenol[1] Phenol	O-H str. 3610	O-H torsion 300 gamma O-H	O-D str.	gamma O-D	see text
5-chlorosalicyl	3180	700			−0.34
2-Hydroxyacetophenone	[5]	789	2361sh 2294 2138sh		0
5-nitro		812			0.23
5-acetyl		806			0.19
4-phenyl		784			−0.13
5-iodo		783			−0.13
5-methyl		783			−0.14
5-bromo		780			−0.28
5-chloro		778			−0.34
5-fluoro		770			−0.6
5-methoxy		769			−0.61
5-amino		758			−1.01
2,4-dibenzoyl resorcinol			∼2255 broad	635	
2-hydroxybenzophenone		∼768			

[1] [2,2,4-trimethyl-pentane soln.]
[2] [CCl$_4$ soln.]
[3] & [4] [CS$_2$ soln.]

Aliphatic Amines

*Numbers in parentheses indicate in-text page reference.

In the discussion of aliphatic amines it is necessary to coin symbols for each type of amine, and this was established in Reference (1).

The symbols P, S, and T are used <u>first</u> to denote primary, secondary, and tertiary amines (NH$_2$, NH, and N), respectively. The symbols P′, S′, and T′ are used <u>second</u> to denote the structure of the alkyl portion of the amine [RCH$_2$-, (R)$_2$CH, and (R)$_3$C] for the primary, secondary, and tertiary, respectively. For example, dimethylamine would be denoted as SP′P′, disopropylamine as SS′S′, methylamine as PP′, and tert-butylamine as PT′.

NH$_2$ STRETCHING FREQUENCIES FOR ALIPHATIC AMINES

In the vapor phase, $v$asym. NH$_2$ and $v$sym. NH$_2$ have weak IR band intensity, and often the $v$asym. NH$_2$ mode is not observed. The $v$asym. NH$_2$ mode is assigned in the region 3404–3422 cm^{-1} and $v$sym. NH$_2$ in the region 3340–3361 cm^{-1}. In most cases, $v$sym. NH$_2$ has more intensity than $v$asym. NH$_2$ (1).

However, the situation is reversed in the 2-alkoxyethylamine series, but is normal in the 3-alkoxypropylamine series. Thus, this intensity reversal results from weak intramolecular hydrogen bonding between an N—H proton and the free pair of electrons on the 2-alkoxy oxygen atom.

Primary aliphatic amines with PS′ structure exhibit $v$sym. NH_2 in the region ~3330–3340 cm^{-1}, and this IR band is weak. When observed, $v$asym. NH_2 is assigned in the region ~3400–3422 cm^{-1}.

Primary aliphatic amines with PT′ structure exhibit $v$sym. NH_2 in the region 3322–3335 cm^{-1}, and $v$asym. NH_2 when observed in the region 3395–3400 cm^{-1}. Both IR bands are weak. In summary, the vNH$_2$ frequency progressing in the order PP′, PS′, and PT′ is shown here (1):

Structure	$v$asym. NH_2 cm^{-1}	$v$sym. NH_2 cm^{-1}	Type
$R-CH_2-NH_2$	3404–3422	3340–3361	PP′
$(R-)CH-NH_2$	~3400–3422	3330–3340	PS′
$(R-)C(-NH_2)$	3395–3400	3322–3335	PT′
$HO-CH_2-CH_2-NH_2$	3465	3342	PP′
$HO-CH_2-CHCH_3-NH_2$	3414	3348	PS′
$HO-CH_2-C(CH_3)_2-NH_2$	3400	3335	PT′

NH$_2$ WAG FREQUENCIES FOR ALIPHATIC AMINES

Table 8.1 lists IR vapor-phase data and assignments for primary alkylamines. The compounds with PP′ structure exhibit a strong relatively broad band in the region 764–780 cm^{-1}. The absorbance (A) ratios: (A) [NH$_2$ wag]/(A) [$v$sym. CH$_2$] and (A) [NH$_2$ wag]/(A)[CH$_2$ bend] show that the values decrease progressing in the order C$_2$ to C$_{19}$. This presumably indicates that (A) for NH$_2$ wag is relatively constant, and that (A) for $v$sym. CH$_2$ and (A) for CH$_2$ bend increase as the number of (CH$_2$)$_n$ increases. Correlations such as these are valuable in spectra-structure identification of unknown aliphatic amines.

The cycloalkyamines with the structure PS′ exhibit NH$_2$ wag in the region 755–785 cm^{-1}, and with the exception of cyclohexylamine increase in frequency as the cycloalkyl ring increases in size from C$_3$ to C$_8$. The absorbance (A) ratios for NH$_2$ wag/$v$sym. CH$_2$ and NH$_2$ wag/CH$_2$ bend also decrease in value as the number of CH$_2$ groups increase from 3 to 8.

The NH$_2$ wag for alkylamines is also affected by the structure of the alkyl group. For example, PP′, PS′, and PT′ exhibit NH$_2$ wag in the regions 760–780 cm^{-1}, 779–799 cm^{-1}, and 800–813 cm^{-1}, respectively. In addition, a weak band assigned to vC−N is also affected by the nature of the alkyl group. For example, vC−N for PP′, PS′, and PT′ occur in the regions 1043–1085 cm^{-1}, 1111–1170 cm^{-1}, and 1185–1265 cm^{-1}, respectively. Both NH$_2$ wag and vC−N increase in frequency with increased branching on the α-carbon atom of the alkyl group. As already noted here, $v$sym. NH$_2$ decreased in frequency with increased branching on the alkyl α-carbon atom. The apparent reason for the decrease in frequency of $v$sym. NH$_2$ is that with increased branching on the α-carbon atom more electrons are released to the C−N group (inductive affect). This causes the N atom to become more basic; consequently, the NH$_2$ bonds are weakened causing the vNH$_2$ modes to vibrate at lower frequencies. Also, the inductive effect increases in the order P′, S′, and T′, and this causes the C−N bond strength to increase in the same order. Consequently, vC−N increases in frequency as the strength of the hydrogen bond increases. In addition, the NH$_2$ wag increases in frequency in the order P′, S′, and T′. This is because it takes more energy for the two relatively charged NH$_2$ protons to wag about the relatively negative free pair of electrons in the C−N plane as the protons and the nitrogen atom have increasingly relative opposite electrical charges. A similar explanation was presented for the vOH, vC−O, OH torsion for alkanols (see Chapter 6).

NH$_2$ BENDING FOR ALKYLAMINES

The NH$_2$ bending mode for the alkylamines with PP′ structure occurs in the region 1599–1629 cm^{-1}, with PS′ in the region 1612–1621 cm^{-1}, and with PT′ in the region 1610–1616 cm^{-1}, and has weak to weak-medium IR band intensity in the vapor phase. Therefore, the NH$_2$ bending mode frequencies are not useful for distinguishing between alkylamines with PP′, PS′, and PT′ structures. In the liquid it is assigned in the region 1590–1627 cm^{-1} with medium intensity (3). In contrast, NH$_2$ bend for arylamines (anilines) occurs at higher frequency with strong IR band intensity (3). (See the next chapter.)

RAMAN DATA FOR PRIMARY AMINES

Table 8.2 lists Raman data and assignments for the neat phase for primary amines.

In the neat phase, νasym. NH$_2$ and νsym. NH$_2$ occur in the region 3367–3379 cm^{-1} and 3307–3322 cm^{-1}, respectively. In all cases, the νsym. NH$_2$ mode has more relative band intensity than the νasym. NH$_2$ modes as indicated by the number in parenthesis (2).

It is interesting to compare the νasym. NH$_2$ and νsym. NH$_2$ frequencies obtained for butylamine (1-aminobutane) in the vapor and liquid phases: νasym. NH$_2$ (3411 cm^{-1}, vap. and 3376 cm^{-1}, liq.) and νsym. NH$_2$ (3345 cm^{-1}, vap. and 3322 cm^{-1}, liquid). This comparison shows that νasym. NH$_2$ and νsym. NH$_2$ occur at lower frequency in the liquid phase by 35 and 23 cm^{-1}, respectively.

Two of the examples given in Table 8.2 contain a 4-aminocyclohexyl group, and both exhibit a very strong Raman band at 784 cm^{-1}, which is assigned to the cyclohexyl breathing mode.

CHEMICAL REACTIONS OF ALKYLAMINES CONTAINING NH$_2$ OR NH GROUPS

This section is brought to the reader's attention for matters of safety and because it can cause confusion when interpreting spectral data.

From experience it has been noted that heat generated from the chemical reaction of aliphatic primary or secondary amines with CS$_2$ can cause the entire content of the volumetric flask to blow out of its mouth. This can cause an injury, or possibly a fire, because CS$_2$ has a low flashpoint.

These chemical reactions occur as follows:

$$2\text{R-NH}_2 + \text{CS}_2 \rightarrow \underset{\underset{\text{H}}{|}}{\overset{\overset{\text{S}}{\|}}{\text{R-N-C-S}^-}} \quad \text{N}^+\text{H}_3\text{-R}$$

$$2(\text{R-})_2\text{NH} + \text{CS}_2 \rightarrow \overset{\overset{\text{S}}{\|}}{(\text{R-})_2\text{N-C-S}^-} \quad {}^+\text{NH(-R)}_2$$

As well, if exposed to air for a period of time these same alkylamines can undergo comparable reactions with CO_2 (replace CS_2 with CO_2). In addition, alkylamines can react slowly with solvents such as CCl_4, $CHCl_3$, and CH_2Cl_2, and this can cause problems in specifically identifying the original sample.

SECONDARY ALIPHATIC AMINES

The N—H stretching band for compounds of form $(R-)_2NH$ is weak, and is not readily detected in compounds whose IR spectra have been recorded in the vapor phase (1). However, N—H wag has strong IR band intensity with half bandwidths varying between 50 and 100 cm^{-1} and occurs in the region 686–750 cm^{-1} in the vapor phase. Thus, it occurs at lower frequency than NH_2 wag. In the liquid, the weak IR band occurs in the region 3320–3280 cm^{-1} and in dilute solution in the region 3310–3360 cm^{-1} (3).

Dialkylamines with SP'P' structure exhibit NH wag in the region 699–715 cm^{-1}, with SP'S' structure near ~686 cm^{-1}. These data suggest that NH wag decreases in frequency in the order SS'S', SP'S', and SP'P', which is the opposite order for the primary alkylamines (see the preceding materials here).

In the dialkyl amine series with SP'P' structure a band assigned as $\nu N(-C)_2$ occurs in the region 1132–1151 cm^{-1} (1).

REFERENCES

1. Nyquist, R. A. (1984). *The Interpretation of Vapor-Phase Infrared Spectra: Group Frequency Data*, Philadelphia: Sadtler Research Laboratories, Division of Bio-Rad Laboratories, Inc.

2. *Raman Data from the Sadtler Research Laboratories*, Philadelphia: Division of Bio-Rad Laboratories, Inc.

3. Lin-Vien, D., Colthup, N. B., Fateley, W. G., and Grasselli, J. G. (1991). *The Handbook of Infrared and Raman Characteristic Frequencies of Organic Molecules*, San Diego: Academic Press.

TABLE 8.1   IR vapor-phase data and assignments for primary amines

Amine	a.CH$_3$ str.	a.CH$_2$ str.	s.CH$_3$ str.	s.CH$_2$ str.	CH$_2$ bend	s.CH$_3$ bend	NH$_2$ wag	(A)[NH$_2$ wag] (A)[s.CH$_2$ str.]	(A)[NH$_2$ wag] (A)[CH$_2$ bend]
Methyl	2960 (0.640)		2898 (0.600)				778 (0.784)		
Ethyl		2965 (1.240)	2922 (0.940)	2870 (0.640)	1456 (0.164)	1398 (0.280)	772 (1.087)	1.69	6.63
Propyl		2978 (1.230)		2880 (0.679)	1465 (0.135)	1387 (0.129)	764 (0.630)	0.93	4.67
Butyl		2940 (1.250)		2880 (0.621)	1465 (0.120)	1387 (0.111)	779 (0.664)	1.07	5.53
Heptyl	2970 (0.700)	2940 (1.250)		2868 (0.510)	1468 (0.130)	1377 (0.080)	770 (0.248)	0.49	1.9
Octyl	2970 (0.462)	2930 (1.250)		2964 (0.510)	1465 (0.091)	1385 (0.050)	771 (0.190)	0.37	2.09
Nonyl	2970 (0.460)	2930 (1.250)		2864 (0.530)	1466 (0.100)	1388 (0.066)	775 (0.189)	0.36	1.89
Decyl	2970 (0.290)	2935 (1.250)		2864 (0.380)	1466 (0.080)	1385 (0.035)	780 (0.120)	0.32	1.51
Undecyl	2970 (0.410)	2935 (1.210)		2864 (0.570)	1465 (0.100)	1385 (0.050)	775 (0.131)	0.23	1.11
Tridecyl		2935 (1.250)		2864 (0.310)	1460 (0.050)	1389 (0.020)	775 (0.050)	0.16	1.01
Tetradecyl		2938 (1.250)		2864 (0.310)	1460 (0.058)	1385 (0.050)	770 (0.060)	0.19	1.11
Pentadecyl		2930 (1.250)		2860 (0.310)	1460 (0.059)	1385 (0.011)	770 (0.040)	0.13	0.68
Octadecyl	2970 (0.040)	2935 (1.250)		2962 (0.310)	1465 (0.045)	1385 (0.005)	775 (0.035)	0.11	0.77
Nonadecyl		2930 (1.250)		2860 (0.400)	1460 (0.089)	1385 (0.005)	775 (0.044)	0.11	0.49
Cyclopropyl		3100 (0.220)		3010 (0.250)	1455 (0.140)		755 (1.230)	4.03	8.79
Cyclopentyl		2968 (1.250)		2884 (0.420)	1459 (0.061)		779 (0.223)	0.53	3.66
		2935 (1.250)		2864 (0.420)	1458 (0.130)		770 (0.170)	0.41	1.42
Cycloheptyl		2930 (1.250)		2860 (0.320)	1460 (0.100)		784 (0.151)	0.47	1.51
Cyclooctyl		2932 (1.150)		2868 (0.199)	1459 (0.069)		785 (0.060)	0.31	0.87

TABLE 8.2 Raman data and assignments for primary amines

1-Amino-butane	bis-(4-Amino-cyclohexyl) methane	Ethoxyethyl amine	Hexa-methylene diamine	2-Aminoethyl methyl ether	3-Methoxy-propyl amine	4,4′-bis(p-Amino-cyclo-hexyl) methane	Assignment
3376(1)		3379(1)	3367(1)	3379(1)	3377(2)	3367(0)	a.NH$_2$
3322(3)	3311(1)	3322(3)	3307(3)	3322(3)	3319(5)	3312(1)	s.NH$_2$
2963(5)		2977(5)		2985(5)			a.CH$_3$ str.
2938(7)		2934(9)		2950(6)	2929(7)	2931(2)	a.CH$_2$ str.
			2899(1)				a.CH$_2$ str.
2875(4)		2870(5)			2868(4)		s.CH$_2$ str.
			2853(2)	2823(4)	2826(6)	2846(3)	s.CH$_2$ str.
					2811(5)		C.T.
		2797(2)		2723(2)	2755(3)		C.T.
			1640(1)	1597(1)			NH$_2$ bend
1444(7)	1441(5)	1459(6)	1440(9)	1453(9)	1450(9)	1441(5)	CH$_2$ bend
	784(9)					784(9)	Ring breathing

Arylamines (Anilines), Azines, and Oximes

Table 9.1 lists IR data for the $v$asym. NH$_2$ and $v$sym. NH$_2$ frequencies for 3-X and 4-X-anilines in the vapor phase, and in (0.5% wt./vol. or less) n-hexane, CCl$_4$, and CHCl$_3$ solutions (1). The IR vapor-phase data for arylamines are also given in Reference (2).

Califano and Moccia (3) have shown that the $v$asym. NH$_2$ and $v$sym. NH$_2$ frequencies and intensities recorded in CCl$_4$ solution correlate with Hammett σ values. The ranges for $v$asym. NH$_2$ and $v$sym. NH$_2$ in each of the solvents are given in Table 9.1. In general, the $v$asym. NH$_2$ and $v$sym. NH$_2$ IR vapor-phase frequencies correlate with Hammet σ values as shown previously (3) (see Fig. 9.1 and 9.2 for plots of the CCl$_4$ solution data). In all cases these modes occur at the highest frequency in the vapor phase.

Figure 9.3 shows a plot of $v$asym. NH$_2$ vs $v$sym. NH$_2$ frequencies for 3-X and 4-X anilines in each solvent. Three of the plots are essentially linear, and the plot of the lower frequency bands observed in the case of the CHCl$_3$ solutions is nonlinear. In each case where the $v$asym. NH$_2$ and $v$sym. NH$_2$ frequencies have been recorded in both hexane and CCl$_4$ solutions, the $v$asym. NH$_2$ mode occurs from 0.36 to 7.51 cm^{-1} lower in frequency in CCl$_4$ than in hexane, while the $v$sym. NH$_2$ mode occurs at higher frequency by 2.05 cm^{-1} to 3.45 cm^{-1} in CCl$_4$ than in hexane (1).

The $v$asym. NH$_2$ mode for 4-X and 3-X anilines occurs from 1.04 to 9.19 cm^{-1} higher in frequency in CHCl$_3$ solution than in CCl$_4$, while $v$sym. NH$_2$ occurs from 0.33 to 7.34 cm^{-1} higher in frequency in CHCl$_3$ than in CCl$_4$.

The $v$asym. NH$_2$ and $v$sym. NH$_2$ frequency decreases in going from solution in hexane to solution in CCl$_4$ because the NH$_2$ protons intermolecularly bond to the Cl electron pairs of CCl$_4$

(1). The same explanation can be used to support the analogous behavior of one plot in $CHCl_3$ solution. The second plot in the case of $CHCl_3$ solution most likely is the result of:

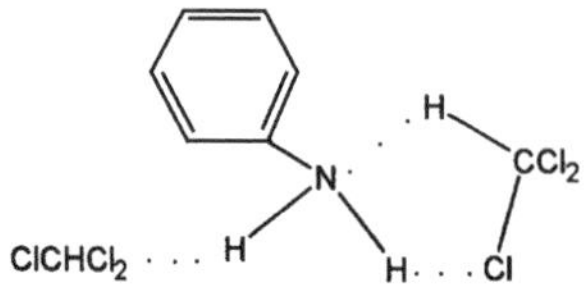

Interaction of the CCl_3H proton with the N atom would cause the NH_2 modes to weaken further, and both $v$asym. NH_2 and $v$sym. NH_2 would shift to even lower frequency as observed. Of course, the $CHCl_3$ proton also intermolecularly bonds with other basic sites such as the phenyl ring, C=C, and C=O groups (see the chapters that follow).

Linnett (4) has developed equations to calculate the NH_2 bond angles of 3-X and 4-X anilines, and there are:

$$4\pi^2 v\text{asym. } NH_2 = k[1/^mH + (1 + \cos^\theta)/^mN]$$
$$4\pi^2 v\text{sym. } NH_2 = k[1/^mH + (1 + \cos^\theta)/^mN]$$

In these equations ^mH is the mass of hydrogen, and ^mN is the mass of nitrogen used to determine the bond angles of the NH_2 group. The $v$asym. NH_2 and $v$sym. NH_2 frequencies recorded in each of the solvents were utilized in the calculations. Figure 9.4 shows a plot of σ_m and σ_p vs the calculated NH_2 bond angles (Table 9.2). This plot shows that the NH_2 bond angles for these anilines increase as the Hammett σ values increase, and this result is in agreement with Krueger's conclusion (5). However, this does not exclude some change in the NH_2 bond lengths with change in the inductive effect as discussed for alkylamines in Chapter 8. Both of these factors would lower the vNH_2 frequencies.

The frequency separation between $v$asym. NH_2 and $v$sym. NH_2 changes in each solvent system (see Table 9.3 and 9.3a). In solution, these separations are larger in the cases of CCl_4 and $CHCl_3$ solution than in hexane solution, and it is less in the other complex existing in $CHCl_3$ solution. In addition, these separations generally appear to increase in CCl_4 and $CHCl_3$ solutions as the σ values increase.

NH_2 STRETCHING FREQUENCIES VS TEMPERATURE

Table 9.4 lists $v$asym. NH_2 and $v$sym. NH_2 frequencies recorded in CS_2 solution over the range +27 to $-60\,^\circ C$ for both aniline and 4-chloroaniline. In both compounds the $v$asym. NH_2 and $v$sym. NH_2 decrease in frequency with a decrease in temperature. In addition, because the $v$sym. NH_2 mode decreases more in frequency than the $v$asym. NH_2 mode, there is a small increase in the frequency separation as the temperature decreases. Figure 9.5 shows plots of both vNH_2 modes for aniline and 4-chloroaniline vs $^\circ C$, and within experimental error the plots are linear. These data are included to demonstrate to the reader that vibrational modes are also temperature dependent. These vNH_2 modes most likely decrease in solution with a decrease in temperature, because the CS_2 volume contracts with a decrease in temperature forcing the NH_2 bonds closer to CS_2 molecules. Thus, the weak hydrogen bonds formed between $S=C=S$ bonds are somewhat stronger, causing the νNH_2 modes to decrease in frequency (6).

AZINES

Table 9.5 lists infrared and Raman data and assignments for azines. Azines are formed by the reaction of hydrazine with either an aldehyde or a ketone. These compounds have the following empirical structures:

A. $(H_2C=N-)_2$

B. $(RCH=N-)$

C. $\phi CH=N-)_2$

D. $[(R)_2C=N-]_2$

E. $[(\phi)_2C=N-]_2$

These compounds are reported to exist in either s-trans, s-cis, and/or gauche isomers depending upon their physical state. In the solid state, formaldazine, empirical structure A, is reported to exist in the s-trans configuration, and in the liquid phase, a small amount of the gauche isomer is also present. In the s-trans configuration $\nu asym.$ $(C=N-)_2$ is IR active and only $\nu sym.$ $(C=N-)_2$ is Raman active because these molecules have a center of symmetry located between the $N-N$ bond. Thus, for formaldazine the medium IR band at $1637\,cm^{-1}$ and the strong Raman band at $1612\,cm^{-1}$ are assigned to the $\nu asym.$ $(C=N-)_2$ and $\nu sym.$ $(C=N-)_2$ modes, respectively (8).

In studies of aldazines and ketazines, $\nu asym.$ $(C=N-)_2$ was assigned to a strong IR band in the region $1636-1663\,cm^{-1}$ and $\nu sym.$ $(C=N-)_2$ was assigned to a strong Raman band in the region $1608-1625\,cm^{-1}$ (9–12). King et al. (13) have assigned $\nu asym.$ $(C=N-)_2$ to the strong IR band at $1747\,cm^{-1}$ and the strong Raman band at $1758\,cm^{-1}$ to $\nu sym.$ $(C=N-)_2$, for $(CF_2=N-)_2$. In this latter case, the $(C=N-)_2$ modes are reversed when compared to those already reported here.

The arylaldehyde azines reported in Table 9.5 were assigned on the same basis as reported in the foregoing. These compounds were all recorded in the solid phase. In making these assignments both their IR and Raman spectra were recorded, and the IR data were also compared to the IR data for the correspondingly substituted benzaldehydes. The $\nu asym.$ $(C=N-)_2$ mode was assigned to the IR band in the region $1606-1632\,cm^{-1}$ and $\nu sym.$ $(C=N-)_2$ was assigned to the Raman band in the region $1539-1563\,cm^{-1}$.

In all the aldehyde azines studied, $\nu sym.$ $(C=N-)_2$ was assigned to the most intense Raman band in the region $1539-1587\,cm^{-1}$. In the case of nitrobenzenes, the $\nu sym.$ NO_2 mode has very strong intensity. The argument for assigning $\nu sym.$ $(C=N-)_2$ to the region $1539-1587\,cm^{-1}$ is that they exhibit the most intense bands in the Raman spectra, and that the $1560\,cm^{-1}$ Raman band for 2-nitrobenzaldehyde azine is twice as strong as the $1348\,cm^{-1}$ Raman band assigned as $\nu sym.$ NO_2 (7). In the case of the 2,6-dichloro isomer, it is not possible for the phenyl group and the $C=N$ groups to be coplanar. As $\nu sym.$ $(C=N-)_2$ for the 2,6-dichlorophenyl isomer occurs $24-48\,cm^{-1}$ higher in frequency than the other benzalded azines, this indicates that there is no conjugation between the phenyl and $C=N$ groups ($\phi - C=N-$) groups. In the case of the 2,6-

dichlorophenyl isomer the reason why $v$sym. $(C=N-)_2$ occur at lower frequency for other benzaldehyde azines is that the phenyl and $C=N$ groups are conjugated. Further support for this conclusion is that $v$sym. $(C=N-)_2$ occurs at $1587\,cm^{-1}$ for the $2,6-Cl_2$ analog, which is intermediate between those assigned to the other benzaldehyde azines ($1539-1563\,cm^{-1}$) and those for the alkylaldehyde azines ($1608-1625\,cm^{-1}$) (7). In conclusion, the benzaldehyde azines exist in an s-trans configuration in the solid state (7).

OXIMES

In the vapor phase aliphatic aldehyde and ketone oximes exhibit vC$=$N in the range 1650–1665 cm^{-1}, and in the liquid and solid phase in the range 1649–1670 cm^{-1} (2, 18). In the vapor phase vOH occurs in the range 3642–3654 (2), and in CCl_4 solution in the range 3580–3600 cm^{-1} (18). In CCl_4 solution, the decrease in frequency in going from the vapor to CCl_4 solution is attributed to intermolecular hydrogen bonding of form $(OH\cdots ClCCl_3)$. In dilute CCl_4 solution, the vOH frequencies form a linear plot when plotted vs Hammett σ values (19). In the neat or solid phase $v(OH\cdots HO)_n$ occurs in the range 3100–3330 cm^{-1}, and is comparable to that which is exhibited by alcohols and phenols. In the vapor phase vN$-$O for both aliphatic and aromatic oximes are reported to occur in the range 910–980 cm^{-1} (2), and in the liquid or solid phase for a larger number of oximes in the range 870–1030 cm^{-1} (19).

Table 9.6 lists IR and Raman data for glyoximes in the solid state. Glyoximes have the following empirical structure: $HO-N=CX-CY=N-OH$. These compounds contain two $C=N$ groups and two hydroxyl groups, and it would be expected that the hydroxyl groups would be hydrogen bonded either inter- or intramolecularly. An intramolecular hydrogen bond between the OH group and Cl $(OH\cdots Cl)_2$ is possible in the case of the dichloro analog, and as shown here:

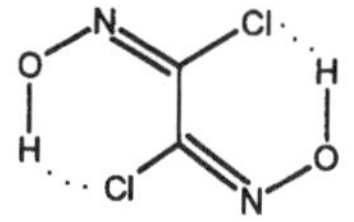

this structure has a center of symmetry at the midpoint of the C$-$C bond.

With a center of symmetry, a mode such as $v$sym. $(N=C-)_2$ would be only Raman active and $v$asym. $(N=C-H)_2$ would be only IR active. Dichloroglyoxime is reported to have a very strong Raman band at $1588\,cm^{-1}$, and a very weak IR band at $\sim1620\,cm^{-1}$. Moreover, the $\sim1620\,cm^{-1}$ IR band is not observed in the Raman spectrum, nor is the $1588\,cm^{-1}$ band observed in the IR spectrum. These data support that dichloroglyoxime has a center of symmetry. In the case of the CN, CN analog, there is also the possibility that intramolecular hydrogen bonding can take place between OH and CN. If so, the two $v(N=C-)_2$ modes are reversed from those exhibited by the Cl, Cl analog because the very strong Raman band ($1593\,cm^{-1}$) occurs at a higher frequency than the weak IR band ($1582\,cm^{-1}$). Thus, $v$sym. $(N=C-)_2$ would be assigned at $1593\,cm^{-1}$ and $v$sym. $(N=C-)_2$ at $1582\,cm^{-1}$.

In the case of glyoxime the intramolecular molecular hydrogen-bonded structure would not exist. However, $v$sym. $(N=C-)_2$ is observed as a strong band in the Raman at $1636\,cm^{-1}$ and $v$asym. $(N=C-)_2$ is observed as a weak band in the IR at $1610\,cm^{-1}$. Thus, glyoxime apparently also has a center of symmetry.

The glyoxime analogs, such as (H and CH_3) and (H and NO_2), do not have a center of symmetry. Therefore, both vasym. $(N=C-)_2$ and vsym. $(N=C-)_2$ are allowed in both the IR and Raman spectra. They are also allowed in cis and gauche structures. In the Raman, the symmetric modes usually have stronger intensity than the antisymmetric modes, and vice versa in the IR. In the case of the (H and CH_3) analog, strong Raman bands are reported at 1630 and 1516 cm^{-1}, and in the case of the (H and NO_2) analog, strong Raman bands are noted at 1650 and 1608 cm^{-1}. A medium IR band is observed at 1650 and 1608 cm^{-1} for the (H and NO_2) analog. Based on intensity arguments, the medium IR band at 1650 cm^{-1} and the weak IR band at 1608 cm^{-1} would be assigned as vasym. $(N=C-)_2$ and vsym. $(N=C-)_2$, respectively. On the other hand, the assignments would be reversed in the case of the (CH_3 and CH_3) analog: vsym. $(N=C-H)_2$ at 1650 cm^{-1} and vasym. $(N=C)_2$ at $\sim$1512 cm^{-1}, because the 1650 cm^{-1} Raman band is stronger than the $\sim$1512 Raman band. However, the vasym. $(N=C-)_2$ mode is forbidden in the case of the (CH_3 and CH_3) analog with a center of symmetry. Therefore, the (CH_3 and CH_3) analog does not have a plane of symmetry.

In Table 9.6 the higher $v(N=C-)_2$ frequencies are listed in the third (IR) and fourth (Raman) columns, and the fifth (IR) and sixth (Raman) columns list the lower frequencies. It has been suggested that the frequencies in columns 3 and 4 are in all cases attributed to vsym. $(N=C-)_2$, and the frequencies in columns 5 and 6 are assigned to vasym. $(N=C-)_2$ (18), but the foregoing discussion does not support all of these assignments. Raman bands in the regions 1627–1650 cm^{-1} and $\sim$1495–1608 cm^{-1} do appear to be characteristic for these glyoximes. Comparison of CH_3, CH_3 glyoxime $v(CN=C-)_2$ frequencies (Raman: 1650, $\sim$1512 cm^{-1}) vs C_6H_5, C_6H_5 glyoxime (Raman: 1627, 1495 cm^{-1}) suggests that conjugation plays a role in decreasing $v(N=C-)_2$ frequencies just as in the case of azines.

REFERENCES

1. Nyquist, R. A., Luoma, D. A., and Puehl, C. W. (1992). *Appl. Spectrosc.*, **46**, 1273.

2. Nyquist, R. A. (1984). *The Interpretation of Vapor-Phase Infrared Spectra: Group Frequency Data*, Philadelphia: Sadtler Research Laboratories, Division of Bio-Rad Laboratories.

3. Califano, S. and Moccia, R. (1956). *Gazz. Chem.*, **86**, 1014.

4. Linnett, I. W. (1945). *Trans. Farad. Soc.*, **41**, 223.

5. Krueger, P. J. (1962). *Nature*, **194**, 1077.

6. Nyquist, R. A. (1986). *Appl. Spectrosc.*, **40**, 79.

7. Nyquist, R. A., Peters, T. L., and Budde, P. B. (1978). *Spectrochim. Acta*, **34A**, 503.

8. Bondybey, V. E. and Nibbler, J. W. (1973). *Spectrochim. Acta*, **29A**, 645.

9. Kirrman, A. (1943). *Compt. Rend.*, **217**, 148.

10. West, W. and Killingsworth, R. B. (1938). *J. Chem. Phys.*, **6**, 1.

11. Kitaev, Yu. P., Nivorozhkin, L. E., Plegontov, S. A., Raevskii, O. A., and Titova, S. Z. (1968). *Dokl. Acad. Sci. USSR*, **178**, 1328.

12. Ogilvie, J. F. and Cole, K. C. (1971). *Spectrochim. Acta*, **27A**, 877.

13. King, S. T., Overend, J., Mitsch, R. A., and Ogden, P. H. (1970). *Spectrochim. Acta*, **26A**, 2253.

14. Bardet, L. and Alain, M. (1970). *C. R. Acad. Sc.*, Paris, **217A**, 710.

15. Cherskaya, N. O., Raktin, O. A., and Shlyapochnikov, V. A. (1987). *Acad. Sci., USSR, Div. Chem. Sci.*, **53**, 2150.

16. (1976). *Sadtler Standard Raman Spectra*, Philadelphia: Sadtler Research Laboratories, Division of Bio-Rad Laboratories.

17. Kahovec, L. and Kohlrausch, K. W. (1952). *Monatsch. Chem.*, 83, 614.

18. Lin-Vien, D., Colthup, N. B., Fateley, W. G., and Grasselli, J. G. (1991). *The Handbook of Infrared and Raman Characteristic Frequencies of Organic Molecules*, San Diego: Academic Press, Inc.

19. Bellamy, L. J. (1968). *Advances in Infrared Group Frequencies*, Bungay, Suffolk, England: Chaucer Press.

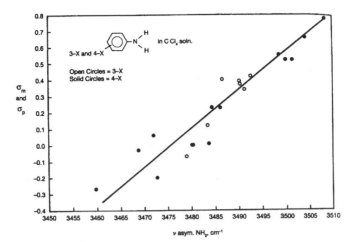

FIGURE 9.1   A plot of *v*asym. NH$_2$ for 3-X and 4-X anilines in CCl$_4$ solutions vs Hammett σ_m and σ_p values for the 3-X and 4-X atoms or groups. The open circles are for 3-X anilines and the closed circles are for 4-X anilines.

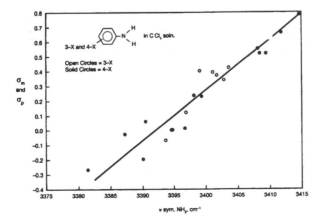

FIGURE 9.2   A plot of *v*sym. NH$_2$ for 3-X and 4-X anilines in CCl$_4$ solutions vs Hammett σ_m and σ_p values for the 3-X and 4-X atoms or groups. The open circles are for 3-X anilines and the closed circles are for 4-X anilines.

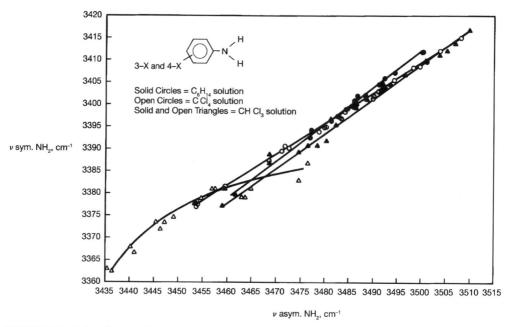

FIGURE 9.3   A plot of νasym. NH_2 vs νsym. NH_2 for 3-X and 4-X anilines in solution with C_6H_{14}, CCl_4, and $CHCl_3$.

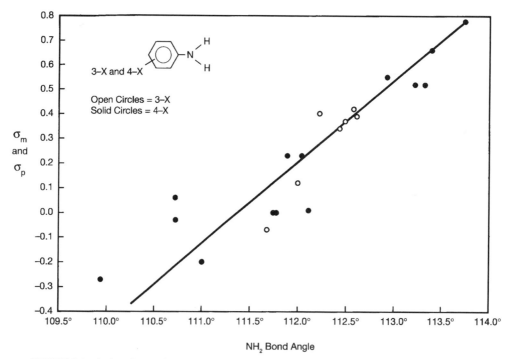

FIGURE 9.4   A plot of σ_m and σ_p vs the calculated NH_2 bond angles for the m- and p-substituted anilines.

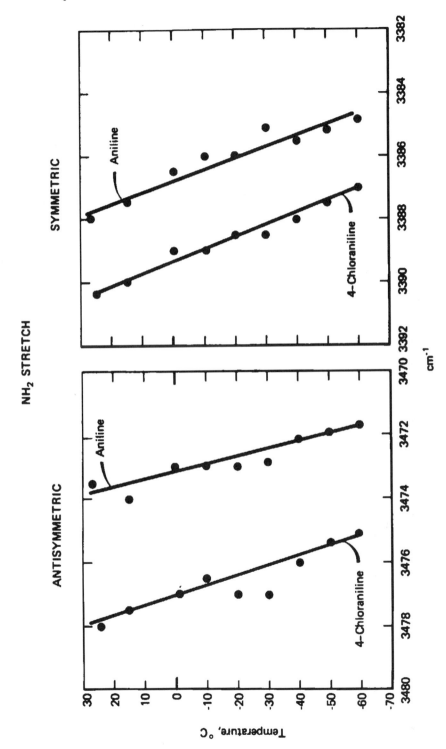

FIGURE 9.5 A plot of both $v\text{NH}_2$ frequencies for aniline and 4-chloroaniline in CS_2 solution vs °C.

TABLE 9.1   IR data for the a. and s. NH_2 stretching frequencies of 3-X and 4-X-anilines in the vapor-phase, and in n-C_6H_{14}, CCl_4, and $CHCl_3$ solutions

4-X-aniline X	vapor phase a.NH_2 str. cm^{-1}	Hexane a.NH_2 str. cm^{-1}	CCl_4 a.NH_2 str. cm^{-1}	$CHCl_3$ a.NH_2 str. cm^{-1}	$CHCl_3$ a.NH_2:$HCCl_3$ str. cm^{-1}	vapor phase s.NH_2 str. cm^{-1}	Hexane s.NH_2 str. cm^{-1}	CCl_4 s.NH_2 str. cm^{-1}	$CHCl_3$ s.NH2 str.	$CHCl_3$ s.NH_2:$HCCl_3$	σ p
OH	3481			3468.9	3440.31	3402			3386.8	3367.89	−0.36
OCH_3	3480		3459.71	3468.9	3441.09	3400		3381.41	3388.75	3366.62	−0.27
$N(CH_3)_2$		3461.63	3454.12	3459.13	3435.4			3377.35	3377.02	3362.96	
NH_2	3480	3453.81			3436.41	3400	3379.54	3376.79	3388.75	3362.43	
OC_6H_5			3468.78	3474.77	3445.62			3387.3	3389.28	3373.49	−0.03
F	3498	3477.02	3472.02	3476.72	3449.2	3415	3394.23	3390.56	3390.71	3374.57	0.06
Cl	3500	3486.41	3484.52	3486.89	3457.74	3418	3400.84	3398.35	3400.11	3380.93	0.23
Br	3502	3488.72	3486.23	3488.96	3459.32	3422	3402.06	3399.48	3401.15	3380.93	0.23
H	3500	3482.98	3480.29	3481.75	3453.95	3400	3397.36	3394.77	3396.43	3378.03	0
H	3500	3482.99	3480.53	3481.57	3453.78	3400	3397.35	3394.88	3396.75	3377.76	0
C_2H_5	3500	3477.22	3471.32	3478.68	3446.47	3418	3392.48	3389.44	3390.77	3371.91	−0.2
$C(CH_3)_3$	3500	3477.18	3472.72	3480.63	3447.29	3417	3392.52	3390.13	3391.79	3373.36	0.01
C_6H_5	3500	3484.73	3483.8	3486.75	3454.79		3398.89	3396.84	3399.33	3378.79	0.52
$C(=O)OC_2H_5$		3500.82	3500.08	3504.25	3478.68	3422	3411.85	3408.59	3411.06		0.66
CN	3510		3504.31	3507.43		3420	3414.42	3412.05	3413.73		0.55
CF_3	3510	3500.65	3498.73	3501.31	3476.7	3421	3411.8	3408.35	3409.82	3386.8	0.52
$C(=O)CH_3$	3502		3501.51	3505.77		3430		3409.6	3412.09		0.52
NO_2	3518		3508.61	3510.27				3414.96	3416.69		0.78
3-X-aniline X											
OCH_3	3500	3485.02	3483.54	3485.08	3457.05	3420	3399.2	3397.02	3399.07	3380.93	0.12
F	3501	3492.7	3491.35	3492.65	3464.99	3420	3405.86	3402.99	3404.1	3380.93	0.34
Cl	3502	3492.17	3490.49	3492.02	3464.99	3420	3404.82	3401.94	3403.23	3380.93	0.37
Br	3500	3491.55	3490.31	3491.94	3463.79	3420	3404.5	3401.31	3403.05	3378.98	0.39
CH_3	3500	3481.7	3479.05	3482.59	3453.48	3420	3396.27	3393.8	3395.42	3377.62	−0.07
$C(=O)OC_2H_5$	3491	3487.08	3486.72	3488.89	3459.56	3410	3401.9	3399.26	3401.39	3380.93	0.4
CN			3494.7	3496.96	3474.77			3405.49	3406.75	3382.89	
CF_3	3510	3494.9	3492.74	3493.9	3463.04	3424	3407.14	3403.81	3404.53	3378.98	0.42
Range	3480–3518	3461.63–3500.82	3453.81–3508.61	3459.13–3510.27	3435.40–3478.72		3379.54–3414.42	3376.79–3414.96	3377.02–3416.69	3362.43–3386.8	

TABLE 9.2 The calculated NH$_2$ bond angles for 3-X and 4-X anilines in n-C$_6$H$_{14}$, CHCl$_3$, and CCl$_4$ solutions

4-X-aniline 4-X	degree NH$_2$ C$_6$H$_{14}$	degree NH$_2$ CCl$_4$	degree NH$_2$ CHCl$_3$	degree NH$_2$ NH$_2$: HCCl$_3$
OH			110.9	108.48
OCH$_3$		109.94	110.36	109.02
N(CH$_3$)$_2$	110.94	109.56	110.97	108.51
NH$_2$		109.63		108.91
OC$_6$H$_5$		110.73	111.78	108.37
F	111.15	110.73	111.91	109.01
Cl	111.72	111.9	112.05	109.55
Br	111	112.04	112.31	109.96
H	111.76	111.75	111.69	109.33
H	111.76	111.78	111.56	109.36
C$_2$H$_5$	111.56	110.83	112.41	109.01
C(CH$_3$)$_3$	111.54	111.01	112.65	108.84
C$_6$H$_5$	111.81	112.12	113.86	109.35
C(=O)OC$_2$H$_5$	112.54	113.23	113.67	
CN		113.41	113.78	
CF$_3$	112.51	112.94	113.23	112.97
C(=O)CH$_3$		113.34	113.79	
NO$_2$		113.76	113.73	
3-X-aniline				
3-X				
OCH$_3$	111.8	112	111.85	109.37
F	112.02	112.44	112.49	111.46
Cl	112.17	112.5	112.56	111.46
Br	112.09	112.63	112.58	111.67
CH$_3$	111.72	111.69	112.18	109.32
C(=O)OC$_2$H$_5$	111.61	112.23	112.23	110.03
CN		112.65	112.91	113.52
CF$_3$	112.26	112.59	112.7	111.47

TABLE 9.3   IR data for the frequency separation between a.NH$_2$ and s.NH$_2$ stretching in the vapor, and in n-C$_6$H$_{14}$, Cl and CHCl$_3$ solutions for 3-X and 4-X-anilines

4-X-aniline X	[a.NH$_2$ str.]-[s.NH$_2$ str.] Vapor phase delta cm^{-1}	[a.NH$_2$ str.]-[s.NH$_2$ str.] Hexane delta cm^{-1}	[a.NH$_2$ str.]-[s.NH$_2$ str.] CCl$_4$ delta cm^{-1}	[a.NH$_2$ str.]-[s.NH$_2$ str.] CHCl$_3$ delta cm^{-1}	[a.NH$_2$:HCCl$_3$ str.]-[s.NH$_2$:HCCl$_3$ str.] CHCl$_3$ delta cm^{-1}
OH	79			82.11	72.42
OCH$_3$	80		78.3	80.15	74.47
N(CH$_3$)$_2$		82.09	76.77	82.08	72.44
NH$_2$	80		77.02		73.98
OC$_6$H$_5$			81.48	85.49	72.13
F	83	83.23	81.31	86.01	
Cl	82	85.57	86.17	86.78	76.81
Br	80	86.68	86.75	87.81	78.39
H	100	85.62	85.52	85.32	75.92
H	100	85.64	85.65	84.82	76.02
C$_2$H$_5$		84.74	81.88	80.55	74.56
C(CH$_3$)$_3$	82	84.96	82.59	88.84	73.93
C$_6$H$_5$	83	85.84	86.96	87.42	76
C(=O)OC$_2$H$_5$		88.97	91.49	93.19	
CN	88		92.26	93.7	
CF$_3$	90	88.85	90.38	92.11	89.92
C(=O)CH$_3$	81		91.91	93.68	
NO$_2$	88		93.65	93.58	
3-X-aniline X					
OCH$_3$	80	85.82	86.52	86.01	76.12
F	81	86.84	88.36	88.55	84.06
Cl	82	87.35	88.55	88.79	84.06
Br	80	87.05	89	88.89	84.81
CH$_3$	80	85.43	85.25	87.17	75.86
C(=O)OC$_2$H$_5$	81	85.18	87.46	87.5	78.63
CN			89.21	90.21	91.88
CF$_3$	86	87.76	88.93	89.37	84.06
Range		82.08–88.85	76.77–93.65	80.15–93.70	72.42–91.88

TABLE 9.3A. IR data for the frequency separation between a. and s.NH2 bending for 3-X and 4R-X-anilines in the vapor-phase and in solutions with n-C_6H_{14}, CCl_4, and $CHCl_3$

4-X-aniline 4-X	[Vapor]-[Hexane] delta a.NH$_2$ str. cm^{-1}	[Vapor]-[Hexane] delta s.NH$_2$ str. cm^{-1}	[Hexane]-[CCl$_4$] delta a.NH$_2$ str. cm^{-1}	[Hexane]-[CCl$_4$] delta s.NH$_2$ str. cm^{-1}	[CHCl$_3$]-[CCl$_4$] delta a.NH$_2$ str. cm^{-1}	[CHCl$_3$]-[CCl$_4$] delta s.NH$_2$ str. cm^{-1}
OH						
OCH$_3$					9.19	7.34
N(CH$_3$)$_2$			7.51	2.19	5.01	[−0.33]
NH$_2$	26.14	20.46				
OC$_6$H$_5$					5.99	1.98
F	20.98	20.77	5	3.67	4.7	0.15
Cl	13.59	17.16	1.89	2.49*	2.37	1.76
Br	13.28	19.94	2.49	2.58*	2.73	1.67
H	17.02	2.64	2.69	2.59	1.46	1.66*
H	17.01		2.46	2.47*	1.104	1.87*
C$_2$H$_5$			5.9	3.04	7.63	1.33
C(CH$_3$)$_3$	22.82	25.48	4.46	2.39	7.91	1.66
C$_6$H$_5$	15.27	18.11	0.93	2.05*	2.95	2.49
C(=O)OC$_2$H$_5$			0.74	3.26*	4.17	2.47
CN				2.37	3.12	1.68
CF$_3$	9.35	8.2	1.92	3.45*	2.58	1.47
C(=O)CH$_3$					4.26	2.49
NO$_2$					1.66	1.73*
3-X						
OCH$_3$	14.98	20.8	1.48	2.18*	1.54	2.05*
F	8.3	15.14	1.35	2.87*	1.3	1.11
Cl	9.83	15.18	1.68	2.88*	1.53	1.29
Br	8.45	15.5	1.24	3.19*	1.64	1.73*
CH$_3$	18.3	23.73	2.65	2.47	3.54	1.62
C(=O)OC$_2$H$_5$	3.92	8.1	0.36	2.64*	2.17	2.13
CN					2.26	1.26
CF$_3$	15.1	16.86	2.16	3.33*	1.16	0.72

*delta s.NH$_2$ str is larger than delta a.NH$_2$ str.

TABLE 9.4 IR data for the NH_2 stretching frequencies of aniline and 4-chloroaniline in CS_2 solution in the temperature between 27 and 60°C

°C	Aniline a.NH_2 str. [CS_2] cm^{-1}	Aniline s.NH_2 str. [CS_2] cm^{-1}	[a.NH_2 str.]-s.NH_2 str.] cm^{-1}	°C	4-Chloroaniline a.NH_2 str. [CCS_2] cm^{-1}	4-Chloroaniline s.NH_2 str. [CS_2] cm^{-1}	[a.NH_2 str.]-[s.NH_2 str. cm^{-1}
27	3473.5	3388	85.5	25	3478	3390.8	87.2
15	3474	3387.5	86.5	15	3477.5	3390	87.5
0	3473	3386.5	86.5	0	3477	3389	88
−10	3473	3386	87	−10	3476.5	3389	87.5
−20	3473	3386	87	−20	3477	3388.5	88.5
−30	3472.8	3385.1	87.7	−30	3477	3388.5	88.5
−40	3472.1	3385.5	86.6	−40	3476	3388	88
−50	3471.9	3385.2	86.7	−50	3475.8	3387.5	88.3
−60	3471.7	3384.8	86.9	−60	3475.3	3387	88.3
delta C [−87]	delta a.NH_2 str. [−1.8]	delta s.NH_2 str. [−3.2]	delta cm^{-1} [1.4]	delta C [−85]	delta a.NH_2 str. [−2.7]	delta s.NH_2 str. [−3.8]	delta cm^{-1} [1.1]

TABLE 9.5 IR and Raman data and assignments for azines

Compound Benzaldehyde Azine	IR a.(C=N)$_2$ str. solid cm^{-1}	Raman s.(C=N−)$_2$ str. solid cm^{-1}	IR a.(C=N−)$_2$ str. vapor cm^{-1}(A)	IR vapor cm^{-1}(A)	IR vapor cm^{-1}(A)
4-dimethylamino	1608	1539			
4-methoxy	1606	1553			
4-methyl	1623	1553			
4-hydrogen	1628	1556	1630(1.240)		
4-fluoro	1633	1561			
4-chloro	1627	1547			
4-acetoxy	1631	1562			
4-trifluoromethyl	1631	1561			
4-cyano	1624	1541			
3-nitro	1629	1551			
2-methoxy	1619	1552			
2-chloro	1618	*^{1}			
2-hydroxy	1630	1555			
2-nitro	1627	1560			
2,4-dihydroxy	1632	1563			
2,4-dimethoxy	1617	1546			
3,4-dimethoxy	1626	1557			
2-hydroxy-4-methoxy	1630	1554			
3-methoxy-4-hydroxy	1629	1558			
2,6-dichloro	1629	1587*^{2}		o.p.Ring	
Range	1606–1632	1539–1563 1587*^{2}			
2-furaldehyde Azine			1630(0.940)	745(1.221)	
Acetophenone azine			1630(1.240)	o.p.Ring 756(0.730)	o.p.Ring 690(1.130)
Azine Cyclooctanone Azine			1627(0.120)	a.CH2 str. 2935(1.200)	s.CH2 str. 2864(0.210)
(CF2=N−)$_2$*^{3}	1747	1758			
(RCH=N−)$_2$ and (RR'C=N−)$_2$ Range	1636–1663	1608–1625			

*^{1} not recorded.
*^{2} reference 7, (C=N−)$_2$ not planar of 2,6−Cl$_2$C$_6$H$_3$ rings.
*^{3} reference 13.

TABLE 9.6   IR and Raman data for glyoximes in the solid state [$(NC-)_2$ stretching vibrations]

Atom or Group X	Atom or Group X	$(N=C-)_2$ str. IR cm^{-1}	Raman cm^{-1}	$(N=C-)_2$ str. IR cm^{-1}	Raman cm^{-1}	References
H	H	[- - - - -]	1636 stg.	1610 vwk	[- - - - -]	14,15
CH$_3$	CH$_3$	[- - - - -]	1650 stg.	[- - - - -]	~1512 m	15–17
C$_6$H$_5$	C$_6$H$_5$	[- - - - -]	1627 stg.	1495 m ?	1495 m ?	15,16
Cl	Cl	~1620 vw	[- - - - -]	[- - - - -]	1588 vstg	
CN	CN	[- - - - -]	1593 vstg	1582 vw	[- - - - -]	15
H	CH$_3$	[- - - - -]	1630 stg.	[- - - - -]	1516 stg	17
H	NO$_2$	1650 m	1650 stg.	1608 w	1608 stg	15

?tentative assignment.

CHAPTER 10

Carboxylic Acids

*Numbers in parentheses indicate in-text page reference.

In the condensed phase or in concentrated solution carboxylic acids exist as a double hydrogen bonded dimer as illustrated here:

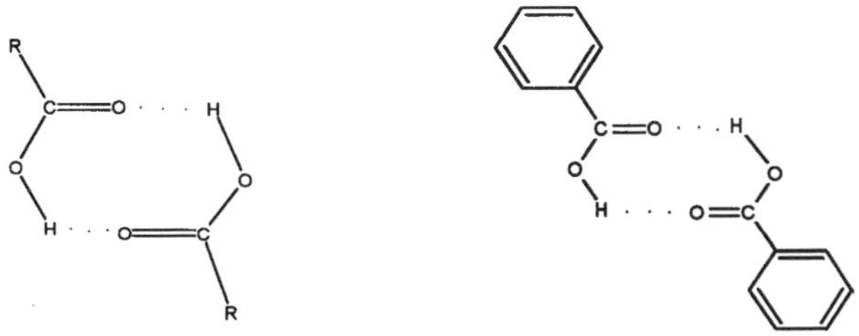

Structure A

Thus, these dimers have a center of symmetry, and only the out-of-phase $(C=O)_2$ stretching mode is IR active, and only the in-phase $(C=O)_2$ stretching mode is Raman active. However, in the vapor phase at elevated temperatures carboxylic acids exist in the monomeric form ($R-CO_2H$ or $\phi-CO_2H$).

The νasym. $(OH\cdots O=C)_2$ mode for structure A is very broad over the range 2500–3000 with subsidiary maxima, which are due to combination and overtones in Fermi resonance with νasym. $(OH\cdots O=C)_2$. The $\nu_{ip}(C=O)_2$ occurs in the region 1625–1687 cm^{-1} in the Raman, in-plane $(C-OH)_2$ bend in the region 1395–1445 cm^{-1}, and $\gamma(OH)_2$ in the range 875–960 cm^{-1}.

Table 10.1 lists IR vapor-phase data, and CCl_4 and $CHCl_3$ solution-phase data for aliphatic carboxylic acids (1,2). The monomeric νOH frequency for the aliphatic carboxylic acid occurs near 3580, 3534, and 3519 cm^{-1} in the vapor, CCl_4, and $CHCl_3$ phases, respectively. In CCl_4 solution the decrease in the νOH frequency is due to intermolecular hydrogen bonding between the acid proton and the Cl atom of CCl_4 ($CO_2H\cdots ClCCl_3$). The Cl atoms are less basic in $CHCl_3$ than they are in CCl_4; however, the νOH still occurs at lower frequency in $CHCl_3$ than in CCl_4 solution. This is attributed to the following doubly hydrogen-bonded complex in $CHCl_3$, which causes νOH to shift to a lower frequency than in CCl_4 solution (2).

Structure B

The $\nu C=O$ frequencies for these aliphatic carboxylic acids decrease in frequency in the order for vapor (1789–1769 cm^{-1}), CCl_4 (1767.3–1750.3 cm^{-1}), and $CHCl_3$ (1756.3–1769 cm^{-1}), and in the order of increased branching on the α-carbon atom [CH_3 to $C(CH_3)_3$] within each physical phase.

Figure 10.1 shows plots of $\nu C=O$ for aliphatic carboxylic acids in CCl_4 solution and in $CHCl_3$ solution vs the number of protons on the acid α-carbon atom (2). The numbers 1–5 on each curve in Figs. 1–4 are for acetic acid, propionic acid, butyric acid, isobutyric acid, and trimethylacetic acid, respectively. Similar curves were obtained plotting $\nu C=O$ vs σ^* and $\nu C=O$ vs E_s. The σ^* values are a measure of the inductive contribution of the alkyl group and E_s is a measure of the steric factor of the alkyl group (3). Thus, it appears that both inductive and steric factors affect $\nu C=O$ frequencies as well as the intermolecular hydrogen-bonded complex formed, as discussed previously.

The out-of-phase $(C=O)_2$ stretching, $\nu_{op}(C=O)_2$, modes occur in the range 1702.2–1714.6 cm^{-1} and 1699.6–1712.5 cm^{-1} in CCl_4 and $CHCl_3$ solutions, respectively, and within each solvent the frequencies decrease as σ^* and E_s increase. In the case of $\nu_{ip}(C=O)_2$, it is not IR active.

Figures 10.2 and 10.3 show, respectively, plots of $\nu C=O$ vs mole % $CHCl_3/CCl_4$ and $\nu_{op}(C=O)_2$ vs mole % $CHCl_3/CCl_4$ for aliphatic carboxylic acids. In CCl_4 solution $\nu C=O$ occur at the highest frequency, and then decrease in frequency as the mole % $CHCl_3/CCl_4$ is increased from $\sim$10 to 100. On the other hand, $\nu_{op}(C=O)_2$ decreases linearly as the mole % $CHCl_3/CCl_4$ is increased from 0 to 100. The essentially linear decrease in both $\nu C=O$ and $\nu_{op}(C=O)_2$ after the initial intermolecular hydrogen bond formation in the case of $vC=O$ is attributed to the reaction field of the solvent system. The reaction field is defined as follows: $[R] = (\in -1)/(2 \in +n^2)$ where $\in$ is the dielectric constant and n is the refracture index of the solvent system (4). It has been shown that there is a linear relationship between the mole % $CHCl_3/CCl_4$ and the reaction field (3). The reaction field increases as the mole % $CHCl_3/CCl_4$ increases and this is a result of an increasing dielectric contribution of the solvent system accompanied by a decrease in the refractive index of the solvent system. The decrease in the $v_{op}(C=O)_2$ frequency in going from solution in CCl_4 to solution in $CHCl_3$ is attributed to the following intermolecularly hydrogen-bonded structure (2):

Structure C

The gradual decrease in frequency is attributed to the increase in the reaction field as the mole % $CHCl_3/CCl_4$ is increased.

Figure 10.4 show plots of $A(vC=O)/A[v_{op} (C=O)_2$ vs mole % $CHCl_3/CCl_4$. These plots indicate that as this ratio increases, as the mole % $CHCl_3/CCl_4$ increases, the concentration of molecules existing as Structure B increases while compounds existing as Structure A decrease with a subsequent increase in Structure C.

IR vapor-phase bands in the range $1105-1178\,cm^{-1}$ are assigned to $vC-C-O$ and IR vapor-phase bands in the range $571-580\,cm^{-1}$ are assigned to $\gamma C=O$ (1). The frequency separation between these two modes in the vapor phase varies between 520 and $598\,cm^{-1}$.

Table 10.2 lists Raman data and assignments for $v(C=O)_2$ and $vC=C$ for carboxylic acids (6).

Four of the acids listed in Table 10.2 exhibit a Raman band, which can be assigned as $v_{op}(C=O)_2$ in the range $1712-1730\,cm^{-1}$. Apparently these four hydrogen-bonded carboxylic acids do not have a center of symmetry in the neat phase. The $v_{ip}(C=O)_2$ has weak to strong relative Raman band intensity and occurs in the range $1630-1694\,cm^{-1}$. The lowest frequency is exhibited by 3-(4-hydroxy-3-methoxyphenyl)-2-propionic acid, and the highest frequency is exhibited by polymethacrylic acid.

The $vC=C$ mode occurs as low as $1557\,cm^{-1}$ in the case of trichloroacrylic acid to as high as $1690\,cm^{-1}$ in the case of itaconic acid (6).

Table 10.3 lists IR group frequency data for acetic acid and its derivatives in the vapor and solution phases (1,7). The $vOH\cdots ClCCl_3$ frequency occurs in the region $3534.3-3500.5\,cm^{-1}$ in CCl_4 solution, and for $vOH\cdots HClCCl_2$ in the range $3521.0-3479.6\,cm^{-1}$. The lower frequencies in $CHCl_3$ solution are attributed to the double hydrogen bond complexed as previously discussed here. The lowest vOH frequency is exhibited by trifluoroacetic acid and the highest vOH frequency is exhibited by trimethylacetic acid. The lowest Taft σ^* value is for trimethyl-acetic acid $[-0.300]$ and the highest Taft σ^* value is for trifluoroacetic acid $[2.778]$ (3). Moreover, the highest pK_a value is for trimethylacetic acid $[5.03]$ and the lowest pK_a value is for trifluoroacetic acid [0.05] (8). These data show that as the OH proton becomes more acidic the νOH mode decreases in frequency in both CCl$_4$ and CHCl$_3$ solutions. Moreover, in general the frequency separation between CCl$_4$ and CHCl$_3$ solution becomes larger as the OH proton becomes more acidic due to the formation of stronger OH···Cl bonds.

Figure 10.5 is a schematic of the number of possible rotational conformers for the tri-, di-, and haloacetic acid monomers (14). X is one halogen atom, 2X is two halogen atoms, 3X is three halogen atoms. The following symbols are used: E(X) indicates that the carbonyl group is eclipsed by X; E(H) indicates that the carbonyl group is eclipsed by H; A(X) indicates that X is anti to the carbonyl group; and A(H) indicates that H is anti to the carbonyl group.

Taft σ values have been correlated with the νC=O and absorbance values for acetic acid and its derivatives (9). There does not appear to be a linear relationship between νC=O and Taft σ^* values. For example, Cl acetic (νC=O, 1791 vs 1.05), Cl$_2$ acetic (νC=O, 1784 vs 1.94), and Cl$_3$ acetic (νC=O, 1789 vs 2.65) in CCl$_4$ solution. On the other hand the ν_{op} (C=O)$_2$ frequencies for the mono-, di-, and trichloroacetic acids increase in frequency as the Taft σ^* constants increase in value (Cl acetic acid, 1737 cm^{-1} vs 1.05; Cl$_2$ acetic acid, 1744 vs 1.94; and Cl$_3$ acetic acid, 1752 cm^{-1} vs 2.65). Therefore, factors other than Taft's inductive effect must affect νC=O and ν_{op} (C=O)$_2$. Intermolecular hydrogen bonding between C=O and (C=O)$_2$, as discussed previously, is one factor that affects these stretching modes. The other factor has been reported to be the existence of rotational conformers (7).

The α-halogenated acetic acids existing in rotational forms 1X—A(X), 2X—A(X), and 3X—A(X)X can form intramolecular hydrogen bonds with the α-halo atom as depicted in Structure D. The intramolecular hydrogen bond would be expected to lower the νC=O frequency, because the C=O bond would be weakened due to the induced contribution for Structure E.

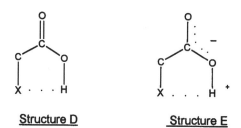

Therefore, the polar A(X) forms would be expected to exhibit the lowest νC=O frequencies within each halo series. Compounds such as 2-methoxyacetic acid and pyruvic acid in the vapor phase also exist in structures such as (D) and (E). In these cases the acid proton is intramolecularly bonding to the oxygen atom of the CH$_3$O or CH$_3$C=O group, respectively (1,13).

Figures 10.6–10.9 show that in either CCl$_4$ or CHCl$_3$ solution, the νC=O mode decreases in frequency within each series (X$_3$CO$_2$H, X$_2$CHCOO$_2$H, XCH$_2$CO$_2$H, and RCO$_2$H) as the pK$_a$ increases or as the acid becomes less acidic (e.g., see F, Cl, Br, and I). Figures 10.10–10.13 (14) show that in either CCl$_4$ or CHCl$_3$ solution the νC=O or ν_{op} (C=O)$_2$ mode increases in frequency within each of the four series as σ^* values increase (e.g., see Br$_2$, Cl$_2$, and F$_2$). Thus, as more sigma (σ) electrons are donated to the acid carbonyl group, the νC=O or ν_{op}(C=O) mode decreases in frequency due to σ electrons being donated to the carbonyl group, and as more σ electrons are withdrawn from the acid C=O or (C=O)$_2$ groups, the νC=O or ν_{op}(C=O)$_2$ increase in frequency. This agrees with what was discussed previously here. However, the opposite or an erratic behavior is noted in Fig. 10.6, Br$_3$ to Br, Cl$_3$ to Cl, F$_3$ to F; Fig. 10.7, Br$_3$ to Br, Cl$_3$ to Cl; Fig. 10.8, Br to Br$_3$, Cl to Cl$_3$, F to F$_3$; Figs. 10.9 and 10.10, Br to Br$_3$, Cl to Cl$_3$, F to F$_3$; Fig. 10.11, Br to Br$_3$, Cl to Cl$_3$; and Fig. 10.12, Br to Br$_3$. In these cases the νC=O or ν_{op}(C=O)$_2$ mode decreases in frequency as the pK_a value is increased or that the νC=O or ν_{op}(C=O)$_2$ does not increase in frequency as the Taft σ^* value increases. This erratic behavior is attributed to the existence of rotational conformers, the result of rotation of the XCH$_2$, X$_2$CH, and CX$_3$ groups about the C—C=O bond as depicted in Fig. 10.5.

Figure 10.13 shows plots of νC=O for the haloacetic acids vs mole % CHCl$_3$/CCl$_4$ solutions. In Fig. 10.13, two plots are noted for iodoacetic acid (14). The IR band near 1769 cm^{-1} is assigned to the rotational conformer 1I—E(I) and the IR band near 1736 cm^{-1} is assigned to rotational conformer 1I—E(H). In the case of chloroacetic acid, two rotational conformers are noted at high mole % CHCl$_3$/CCl$_4$. The lower frequency νC=O band is assigned to the rotational conformer 1Cl—E(H) and the higher frequency IR band to rotational conformer 1Cl—E(Cl). In the case of fluoroacetic acid at low mole % CHCl$_3$/CCl$_4$ solutions, the low frequency IR band is assigned to rotational conformer 1F—E(H), the higher frequency IR band is assigned to rotational conformer 1F—E(F). The other 1-haloacetic acids exist in the form of rotational conformer 1X—E(X). The dihalogenated acid νC=O frequencies are assigned to the 2X—A(X) rotational conformers. Dibromoacetic acid exhibits a band in the region 1797.4–1799.3 cm^{-1} in going from CCl$_4$ solution to CHCl$_3$ solution, and this IR band is assigned to rotational conformer 2Br—E(Br). The trihaloacetic acid ν(C=O)IR band frequencies (14) are assigned to rotational conformer 3X—A(X).

Figure 10.14 shows plots of ν_{op}(C=O)$_2$ vs mole % CHCl$_3$/CCl$_4$ solutions. In Fig. 10.14, the plot for trifluoroacetic acid shows that ν_{op}(C=O)$_2$ increases in frequency as the mole % CHCl$_3$/CCl$_4$ increases (14). This noted exception in this series of plots is attributed to the following—the two CF$_3$ groups in the intermolecularly hydrogen-bonded dimer are rotating to the more polar form as the mole % CHCl$_3$/CCl$_4$ is increased. The most polar form is where the CF$_3$ groups are in rotational conformer 3F—E(F). The field effect of the eclipsed F atom apparently overrides the field effect of the solvent system, and the intermolecular hydrogen bonding between CHCl$_3$ protons and the carbonyl groups, because ν_{op}(C=O)$_2$ increases rather than decreases in frequency as the mole % CHCl$_3$/CCl$_4$ is increased. The 3F—E(F) rotational conformer for ν_{op}(C=O)$_2$ is reasonable, because the OH protons in this case are intramolecularly hydrogen-bonded with the two C=O groups, whereas in the case of the monomer, the 3X—A rotational conformer is stabilized by intramolecular hydrogen bonding via (X⋯HO).

In conclusion, these halogenated acetic acids exist as rotational isomers. Within a series the highest νC=O frequency results from the conformer where the halogen atom eclipses the carbonyl oxygen atom denoted as E(X). The lowest νC=O frequency is where the halogen atom is anti with the carbonyl oxygen atom denoted as A(X). In the anti configuration the acid proton bonds intramolecularly with the anti halogen atom, causing a weakening of the C=O bond. The inductive effect of the halogen atom(s) is (are) independent of molecular geometry, but the field effect is dependent upon molecular geometry. Thus, the field effect upon the C=O group is responsible for the relatively high νC=O frequency with rotational conformers with E(X) structure, but not with A(X) structure (14).

It has been found that the type of carboxylic acid of form RCO_2H had to be identified before its approximate pK_a value could be calculated from $vC=O$ and $v_{op}(C=O)_2$ frequencies recorded in CCl_4 solution (10). The $vC=O$ and $v_{op}(C=O)_2$ frequencies recorded in CCl_4 solution have been reported to correlate in a linear manner with pK_a values (11). It has been reported that $v_{op}(C=O)_2$ band intensities for aromatic carboxylic acids are higher than those for aliphatic carboxylic acids (12). In the vapor-phase at temperatures above 150 °C, carboxylic acids exist only in the monomeric state. At lower temperature both the monomer and dimer carboxylic acid can exist in equilibrium. With increase in temperature the equilibrium shifts toward the monomeric species (1,13).

4-X-BENZOIC ACIDS

Table 10.4 lists IR data and assignments for 4-X-benzoic acids in the vapor, CCl_4 and/or $CHCl_3$ solution phases. The OH, C=O, and (C=O) frequency ranges and assignments are compared here:

	vapor, cm^{-1} (1)	CCl$_4$, cm^{-1} (14)	CHCl$_3$, cm^{-1} (14)
	vOH	$vOH\cdots ClCCl_3$	$vOH\cdots ClHCCl_2$
range	3582–3595	3529–3544	3519–3528
	$vC=O$	$vC=O$	$vC=O\cdots HCCl_3$
range	1758–1768	1735–1751	1719–1744
	$v_{op}(C=O)_2$	$v_{op}(C=O)_2$	$v_{op}(C=O)_2\cdots(HCCl_3)_2$
range	- - - - -	1689–1707	1687–1707

This comparison shows that the vOH, $vC=O$, and $v$asym. $(C=O)_2$ decrease in frequency progressing in the order vapor, CCl_4, and $CHCl_3$ solutions. In the case of $v_{op}(C=O)_2$, the frequency change in going from CCl_4 and $CHCl_3$ is small due to the fact that in the dimer form each O—H group is already hydrogen bonded to each of the C=O groups, and the O—H oxygen atom is much less basic; therefore, the $CHCCl_3$ proton will not form a strong intermolecular hydrogen bond as in the case of $vC=O\cdots HCCl_3$. See Table 10.4a for the factors that affect the CO_2H and $(CO_2)_2$ groups for 4X-benzoic acids in $CHCl_3$ and CCl_4 solutions.

Figure 10.15 shows plots of vOH for 4-X-benzoic acids vs Hammett's σ_p values. These plots show a relationship with the Hammett σ_p values, but many of the points fall off the linear line.

Figure 10.16 shows plots of $vC=O$ for 4-X-benzoic acids vs Hammetts σ_p values. Linear relationships are apparent except for the 4-tert-butyl and 4-methoxy analogs for $CHCl_3$ solution (14).

Figure 10.17 shows plots of $v_{op}(C=O)_2$ for 4-X-benzoic acids vs Hammetts σ_p values. The plots for both CCl_4 and $CHCl_3$ solutions are linear in three different segments (14).

Figure 10.18 shows plots of $v_{op}(C=O)_2$ for 4-X-benzoic acids vs mole % $CHCl_3/CCl_4$. The Cl, Br, H, CH_3, tert-butyl, and methoxy analogs all decrease essentially linearly in going from CCl_4 to $CHCl_3$ solutions. In addition, the frequency separation between these linear plots decreases as Hammetts σ_p values increase in value. However, in the case of the 4-nitro and 4-cyano analogs the $v_{op}(C=O)_2$ frequencies actually increase in going from CCl_4 to $CHCl_3$ solution.

In CCl_4 solution, the 4-nitro analog would have the most acidic OH proton and the most basic C—O oxygen atoms. Therefore, a complex such as (F) would cause the C=O bond to be

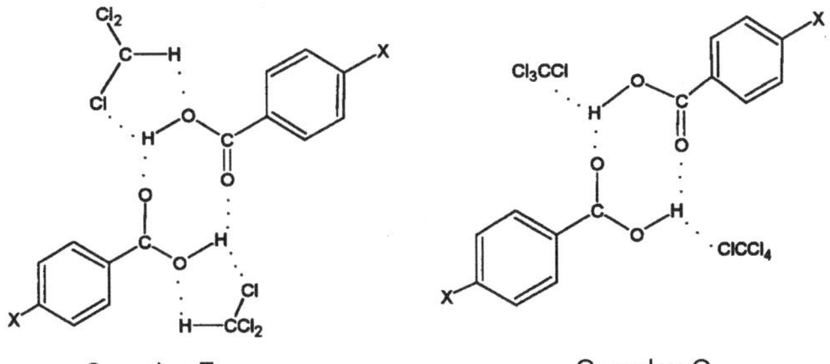

Complex F   Complex G

strengthened. Therefore, the $v_{op}(C=O)_2$ mode would vibrate at a higher frequency than in a case such as Complex G. In addition, the NO_2 group of aryl NO_2 has been shown to form intermolecular hydrogen bonds with $CHCl_3$ (15), and between $Cl_3CH\cdots NC$ for benzonitrile (16). Intermolecular hydrogen bonding of the $CHCl_3$ proton with the 4-NO_2 or 4-cyano group would also cause an increase of $v_{op}(C=O)_2$ in these $(Cl_3C-HClCCl_3)_n$ complex solutions. Moreover, intermolecular hydrogen complexes involving the CH_3Cl proton with the π system on one or both sides of the planar phenyl group would also cause $vC=O$ to increase in frequency. These $v_{op}(C=O)_2$ shifts are 0.27 and 0.65 cm^{-1} for the NO_2 and CH_3O analogs, respectively. For the Cl, Br, H, CH_3, $(CH_3)C$, and CH_3O these shifts are −1.01, −1.10, −1.33, −2.34, −2.27, and −2.52, respectively.

Figure 10.18 shows plots of $vC=O$ for 4-X-benzoic acids vs mole % $CHCl_3/CCl_4$. In CCl_4 solution, the Br analog occurs at a higher frequency than the Cl analog, and in $CHCl_3$ solution the CN analog occurs at a higher frequency than the NO_2 analog. These cases are exceptions to the correlation of $vC=O$ vs Hammett σ_p values.

The 4-tert.-butylbenzoic exhibits two $vC=O$ bands. The lower frequency band frequencies listed by the computer are in the range 1721.55–1718.0 cm^{-1}. The $vC=O$ for the 1721.55–1718.0 cm^{-1} IR band is in good agreement with the $vC=O$ vs Hammett σ_p values. The higher $vC=O$ frequency band for 4-tert-butylbenzoic increases in frequency as the mole % $CHCl_3/CCl_4$ is increased while none of the other 4-X-benzoic acids show this opposite trend. A possible explanation for this $vC=O$ frequency behavior is that the tert-butyl group is hyperconjugated, and that the Cl_3C-H proton is hydrogen bonded to the phenyl π system while the positively charged tertiary carbon atom is associated with the Cl_3CH chlorine atom. Such an interaction would cause the $vC=O$ mode to increase in frequency as observed. Because 4-tert.-butylbenzoic acid exhibits two $vC=O$ frequencies it is apparent that it exists as clusters in $CHCl_3$ solutions. All of the 4-X-benzoic acids show breaks in the plots in the range 10–30 mol % $CHCl_3/CCl_4$ solution, which indicates that other $CHCl_3/CCl_4$ complexes of $(Cl_3C-H\cdots ClCCl_3)_n$ with the 4-X-benzoic are also present.

Table 10.5 lists IR vapor-phase data and assignments for anthranilic acids. These acids have the following basic structure:

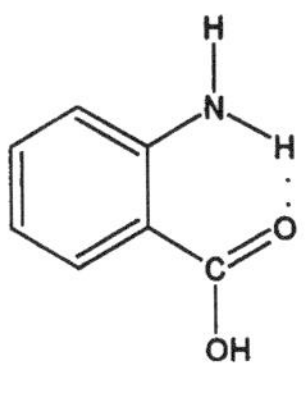

The νO$-$H stretching frequencies occur in the range 3484–3494 cm^{-1}. The νasym. NH$_2$ mode occurs in the range 3518–3525 cm^{-1} and νsym. NH$_2$ in the range 3382–3395 cm^{-1}. The frequency separation of 123–141 cm^{-1} indicates that the N$-$H proton is intramolecularly hydrogen bonded to the C=O group as was illustrated here. A weak band in the region 3430–3436 cm^{-1} is assigned to νOH$\cdots$N. The νC=O mode is in the range 1724–1732 cm^{-1}. The NH$_2$ bending mode is exhibited in the range 1613–1625 cm^{-1}. A medium-strong band in the range 1155–1180 cm^{-1} is attributed to O$-$C= stretch.

N-methyl anthranilic acid and N-phenyl anthranilic acid exhibit νN$-$H$\cdots$O=C at 3392 and 3348 cm^{-1}, respectively. The N$-$H$\cdots$O=C intramolecular hydrogen bond is stronger in the case of the N-phenyl analog due to the inductive effect of the phenyl group compared to that for the N-methyl group.

The νC=O$\cdots$H$-$N frequencies occur near 1719 cm^{-1} for N-methyl and N-phenyl anthranilic acid compared to 1724–1732 cm^{-1} for the other anthranilic acids studied.

Table 10.6 lists infrared data for acrylic acid and methacrylic acid in 0 to 100 mol % CHCl$_3$/CCl$_4$ solutions. The sym. CH$_3$ bending mode for methacrylic acid increases steadily from 1375.5 to 1377.8 cm^{-1} as the mole % CHCl$_3$/CCl$_4$ is increased from 0 to 100, while the CH2= wag mode for methacrylic acid increases wag only 0.5 cm^{-1}, from 948.7 to 949.2 cm^{-1}. In the case of acrylic acid, the CH$_2$= bending mode decreases steadily from 1433.1 to 1429.9 cm^{-1} as the mole % CHCl$_3$/CCl$_4$ increases from 0 to 100, while the vinyl twist mode varies from 983.9 to 983.6 to 984 cm^{-1}.

Table 10.7 lists Raman data for carboxylic acid salts (6). In some cases νasym. CO$_2$ is observed in the Raman, and its intensity is always less than the Raman intensity for νsym. CO$_2$. The Raman bands in the region 1383–1468 cm^{-1} are assigned to νsym. CO$_2$.

Frequency assignments for νasym. CO$_2$ and νsym. CO$_2$ are summarized in Table 10.8. Study of this table suggests that the inductive effect causes νasym. CO$_2$ for the dichloroacetate ion to occur at higher frequency than the other carboxylate ions.

Table 10.9 lists IR data for carboxylic acid salts. This table demonstrates as well that the inductive effect also increases the νasym. CO$_2$ frequency as it increases from 1585 cm^{-1} for sodium acetate to 1640 cm^{-1} for sodium difluoroacetate. It appears that a divalent cation such as calcium (Ca) lowers the νasym. CO$_2$ frequency compared to a monovalent cation such as lithium (Li) or sodium (Na). Compare lithium and calcium formate (1604 vs 1596 cm^{-1}) and sodium and calcium 2-ethylhexanoate (1555 vs 1545 cm^{-1}) while the opposite effect is noted for νsym. CO$_2$ (1382 vs 1400 cm^{-1}) and (1415 and 1424 cm^{-1}) for the formate and 2-ethyl-hexanoates, respectively.

HALF SALTS OF CARBOXYLIC ACIDS

Sodium hydrogen diacetate is an example of a transmission anomaly observed within broad absorption bands of solids. This anomaly indicates that there are perturbations between overlapping energy levels in the solid state.

Sodium hydrogen diacetate is classified as a type A acid salt with a centrosymmetric anion:

Sodium hydrogen diacetate has 24 molecules per unit cell in its crystal structure (18). The broad intense absorption is determined by coupling between two or more vibrating ($-$O$\cdots$H$\cdots$O$-$)$_n$ groups (19).

There are two C—C stretch modes in the centrosymmetric anion, a νsym. $(C-C)_2$ and νasym., $(C-C)_2$. The νasym. $(C-C)_2$ mode yields a strong Raman band at $920\,cm^{-1}$, and it is this mode that interacts, causing the transmission window in the broadband at $920\,cm^{-1}$ (see Fig. 10.19a). Figure 10.19b is that for the $(CD_3)_2$ analog, and in this case the νasym. C—C mode has shifted to $874\,cm^{-1}$ exhibiting the transmission anomaly at this frequency. Another anomaly near $840\,cm^{-1}$ results from the CD_3 rocking mode at this frequency (20).

Application of the perturbation theory redistributes both energy levels and absorption intensity. Near E_1^o the interaction is greatest and the resulting shift of the levels leads to a fall in energy-level density. The loss in absorption intensity in this region appears as a gain in the other regions and the perturbation involves all the E_1^o and E_2^o levels. The intensity may be redistributed over a relatively large frequency range. This yields within the broadband a narrow region of increased transmission, with some regions of increased absorption nearby; the latter might not be easily noted in the spectrum (20).

Because Evans was the first to explain these peculiar effects that appear sometimes in the solid state IR spectra, it is now referred to as the "Evans Transmission Hole" or "Evans Hole."

REFERENCES

1. Nyquist, R. A. (1984). *The Interpretation of Vapor-phase Infrared Spectra: Group Frequency Data*. Philadelphia: Sadtler Research Laboratories, Division of Bio-Rad.

2. Nyquist, R. A., Clark, T. D., and Streck, R. (1994). *Vib. Spectrosc.*, 7, 275.

3. Taft, R. W. Jr. (1956). In *Steric Effects in Organic Chemistry*, M. S. Newman (ed.), Chap. 3, New York: Wiley.

4. Buckingham, A. D. (1960). *Can. J. Chem.*, **308**, 300.

5. Nyquist, R. A., Putzig, C. L., and Hasha, D. L. (1989). *Appl. Spectrosc.*, 1049.

6. (1984). *Sadtler Standard Raman Spectra*, Philadelphia: Sadtler Research Laboratories, Division of Bio-Rad Laboratories.

7. Nyquist, R. A. and Clark, T. D. (1996). *Vib. Spectrosc.*, **10**, 203.

8. Christensen, J. J., Hansen, L. D., and Isatt, R. M. (1976). *Handbook of Proton Ionization and Related-Thermodynamic Quantities*, New York: Wiley Interscience.

9. Bellanato, J. and Baraceló, J. R. (1960). *Spectrochim. Acta*, **16**, 1333.

10. Golden, J. D. S. (1954). *Spectrochim. Acta*, **6**, 129.

11. St. Fleet, M. (1962). *Spectrochim. Acta*, **18**, 1537.

12. Brooks, C. J. W., Eglinton, G., and Moran, J. F. (1961). *J. Chem. Soc.*, **106**.

13. Welti, D. (1970). *Infrared Vapour Spectra*, London: Heyden, in cooperation with Sadtler Research Laboratories, Philadelphia, PA.

14. Nyquist, R. A. and Clark, T. D. (1995). *Vib. Spectrosc.*, **8**, 387.

15. Nyquist, R. A. and Settineri, S. E. (1990). *Appl. Spectrosc.*, **44**, 1552.

16. Nyquist, R. A. (1990). *Appl. Spectrosc.*, **44**, 1405.

17. Nyquist, R. A., Putzig, C. L., and Leugers, M. Anne (1997). *Infrared and Raman Spectral Atlas of Inorganic Compounds and Organic Salts*, vol. 1, p. 72. Academic Press, Boston.

18. Splakman, J. C. and Mills, H. H. (1961). *J. Chem. Soc.*, 1164.

19. Albert, N. and Badger, R. M. (1958). *J. Chem. Phys.*, **29**, 1193.

20. Evans, J. C. (1962). *Spectrochim. Acta*, **18**, 507.

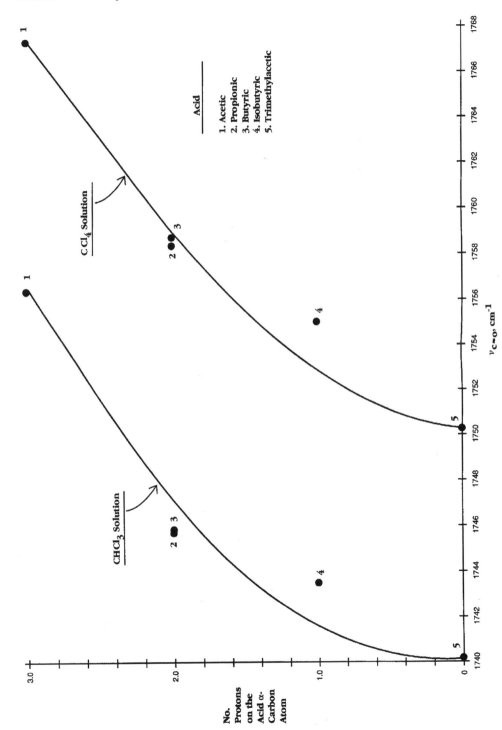

FIGURE 10.1 Plots of the number of protons on the acid α-carbon atom of the carboxylic acid vs νC=O for alkyl carboxylic acids in 2% wt./vol. in CCl_4 and in 2% wt./vol. $CHCl_3$ solutions.

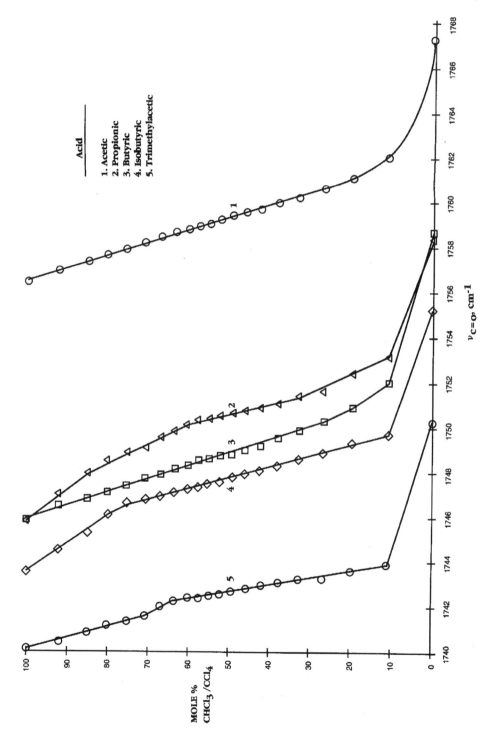

FIGURE 10.2 Plots of the mol % CHCl$_3$/CCl$_4$ vs νC=O for alkyl carboxylic acids in 2% wt./vol. CHCl$_3$ and/or CCl$_4$ solutions.

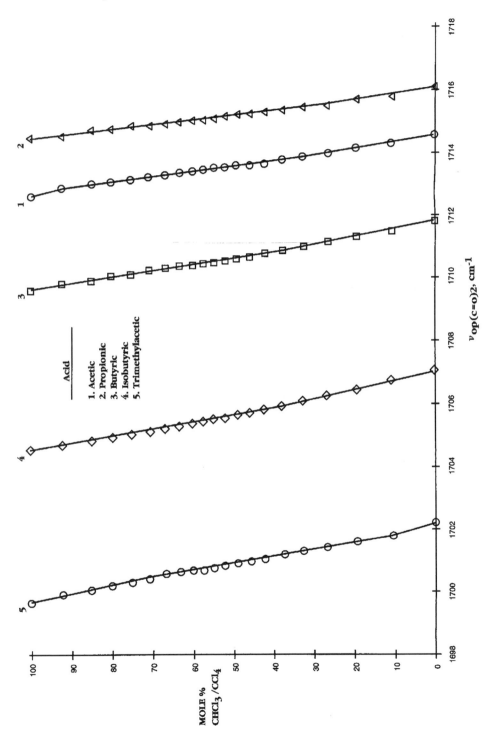

FIGURE 10.3   Plots of the mole % $CHCl_3/CCl_4$ vs $\nu_{op}(C=O)_2$ for alkyl carboxylic acids in 2% wt./vol. $CHCl_3$ and/or CCl_4 solutions.

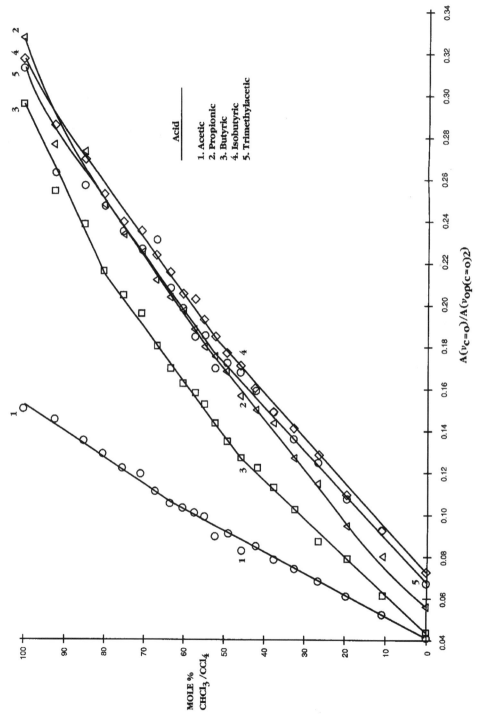

FIGURE 10.4 Plots of the mole % CHCl$_3$/CCl$_4$ vs absorbance [ν(C=O)]/absorbance [ν_{op}(C=O)$_2$] for the alkyl carboxylic acids in 2% wt./vol. CHCl$_3$ and/or CCl$_4$ solutions.

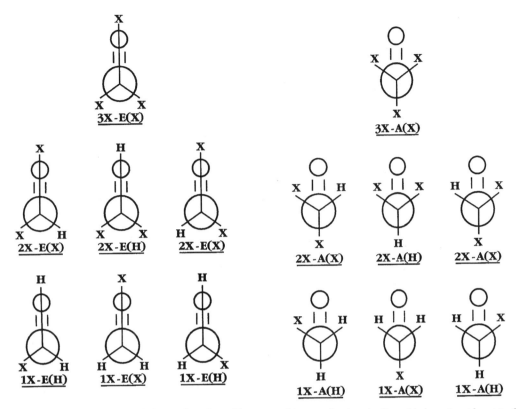

FIGURE 10.5   A schematic of the number of possible rotational isomers for the tri-, di-, and haloacetic acids not in the form of intermolecular hydrogen-bonded dimers. X is a halogen atom, 2X is two halogen atoms, 3X is three halogen atoms.

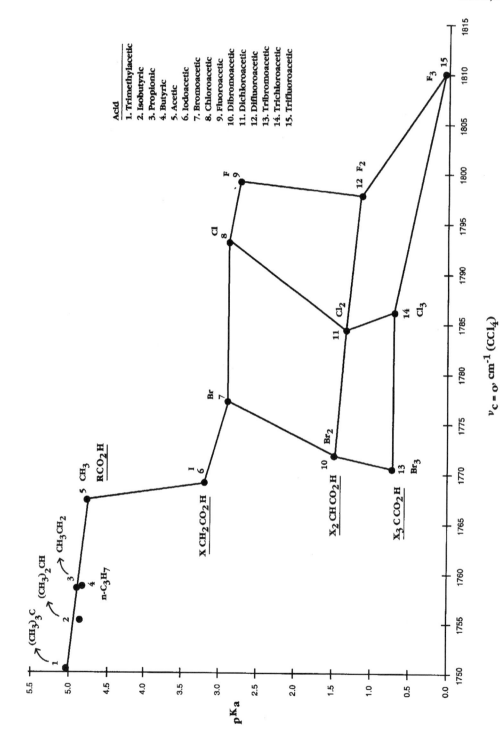

FIGURE 10.6 Plots of νC=O for each carboxylic acid in CCl$_4$ solution vs the pK$_a$ value for each carboxylic acid.

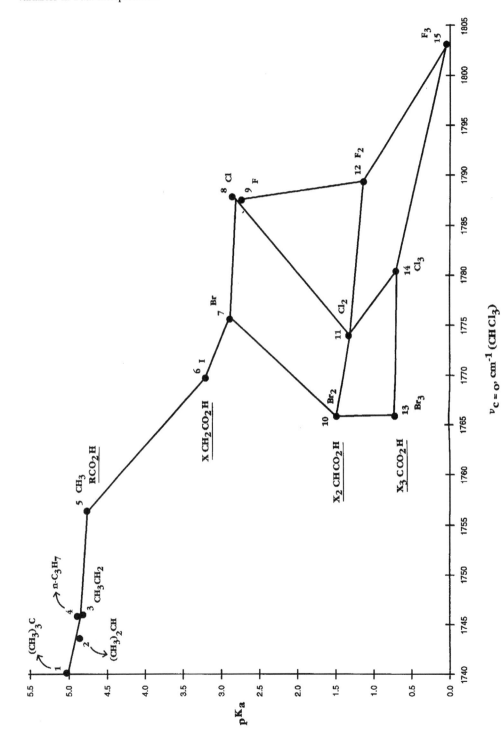

FIGURE 10.7   Plots of νC=O for each carboxylic acid in $CHCl_3$ solution vs the pK_a value for each carboxylic acid.

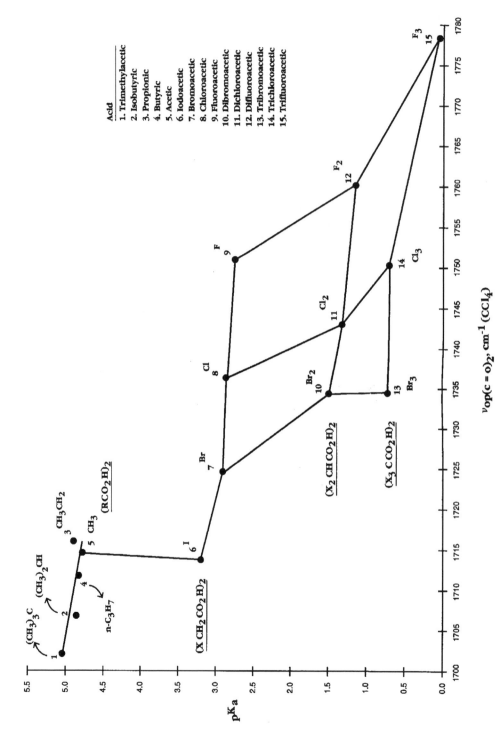

FIGURE 10.8   Plots of ν_{op}(C=O)$_2$ for each carboxylic acid in CCl$_4$ solution vs the pK$_a$ value for each carboxylic acid.

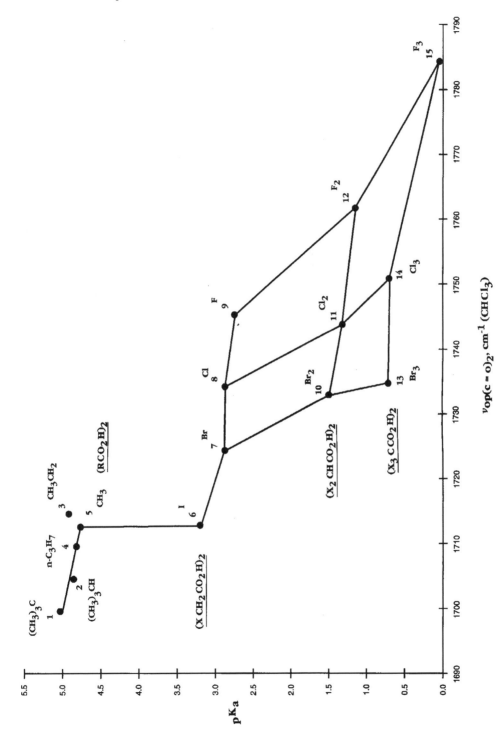

FIGURE 10.9 Plots of $v_{op}(C{=}O)_2$ for each carboxylic acid in CHCl$_3$ solution vs the pK$_a$ value for each carboxylic acid.

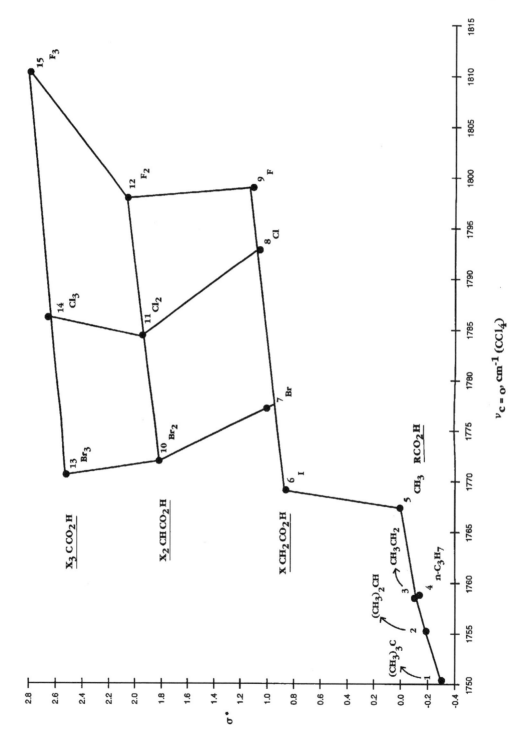

FIGURE 10.10   Plots of νC=O for each carboxylic acid in CCl$_4$ solution vs the σ^* value for each carboxylic acid.

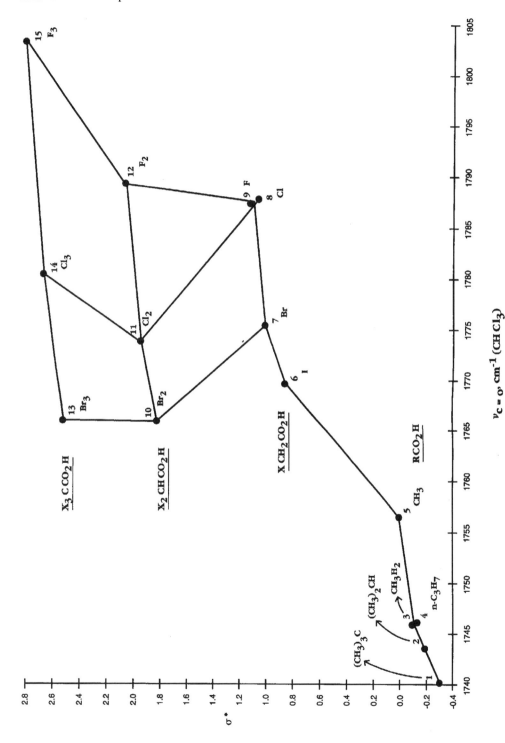

FIGURE 10.11   Plots of $\nu(C{=}O)$ for each carboxylic acid in $CHCl_3$ solution vs the σ^* value for each carboxylic acid.

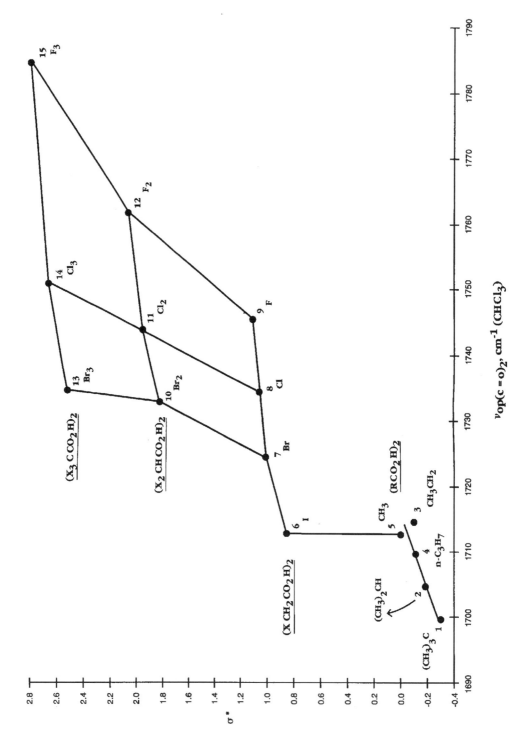

$\nu_{op}(c=o)_2$, cm^{-1} (CHCl$_3$)

FIGURE 10.12   Plots of ν_{op}(C=O)$_2$ for each carboxylic acid in CHCl$_3$ solutions *vs* the σ^* value for each carboxylic acid.

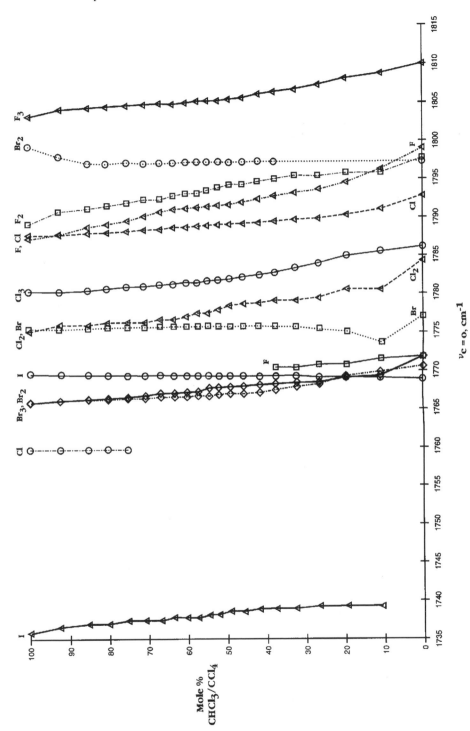

FIGURE 10.13 Plots of $\nu(C=O)_2$ for trihalo, dihalo, and haloacetic acids vs mole % $CHCl_3/CCl_4$.

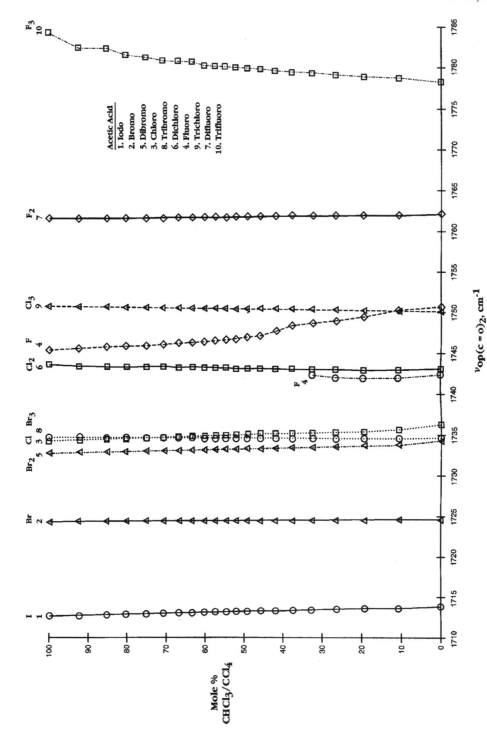

FIGURE 10.14 Plots of ν_{op}(C=O)$_2$ for trihalo, dihalo, and haloacetic acids vs mole % CHCl$_3$/CCl$_4$.

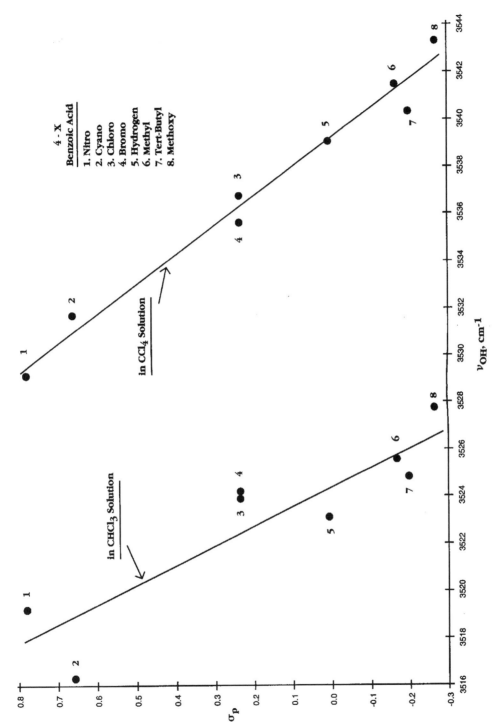

FIGURE 10.15   Plots of νOH mode for 4-X-benzoic acids in CCl_4 solution and in $CHCl_3$ solution vs σ_p values for the 4-X atom or group.

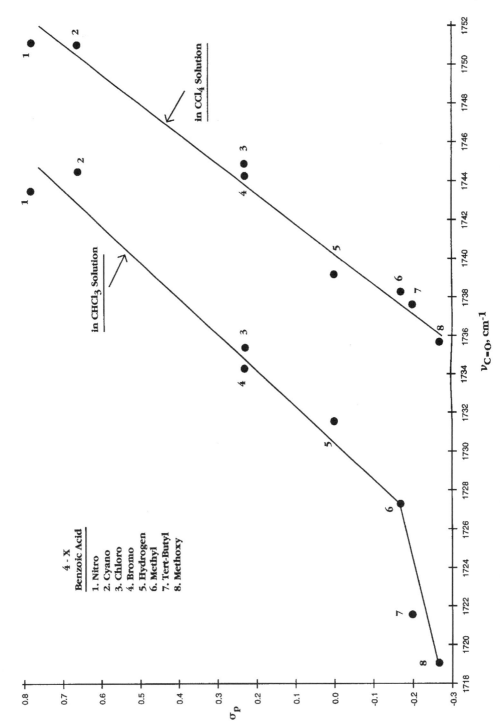

FIGURE 10.16 Plots of ν(C=O) mode for 4-X-benzoic acids in CCl_4 solution and in $CHCl_3$ solution vs σ_p values for the 4-X atom or group.

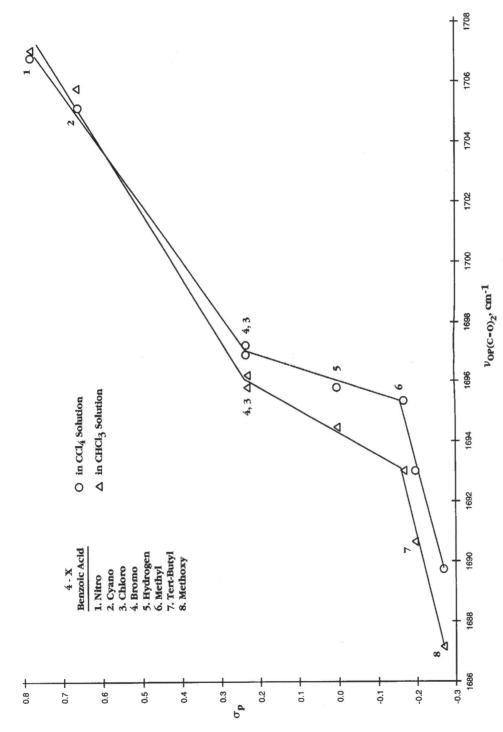

FIGURE 10.17 Plots of $\nu_{op}(C{=}O)_2$ frequencies for 4-X-benzoic acids in CCl_4 solution and $CHCl_3$ solution vs σ_p values for the 4-X atom or group.

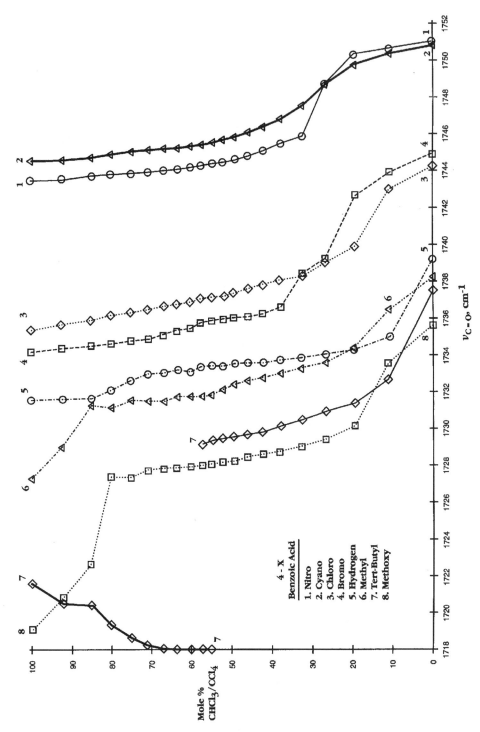

FIGURE 10.18   Plots of $v_{c=o}$ frequencies for 4-X-benzoic acids vs mole % $CHCl_3/CCl_4$.

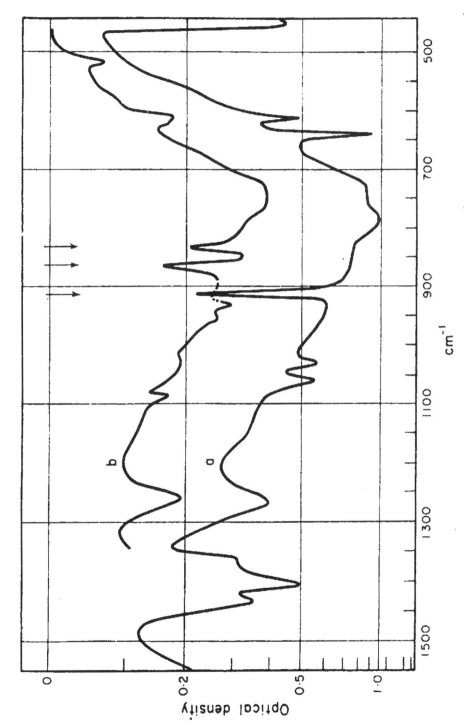

FIGURE 10.19 (a) Parts of the infrared spectrum of mulls containing sodium hydrogen diacetate; above $1330 \, \text{cm}^{-1}$ Fluorolube was used and below $1330 \, \text{cm}^{-1}$, Nujol. (b) A Nujol mull spectrum of $NaH(CD_4COO)_2$.

TABLE 10.1   IR vapor-phase data, CCl$_4$ and CHCl$_3$ solution-phase data, and assignments for aliphatic carboxylic acids

Acid	O−H str. vapor	O−H str. CCl$_4$ soln.	O−H str. CHCl$_3$ soln.	[vapor]-[CCl$_4$ soln.]	[vapor]-[CHCl$_3$ soln.]	[CCl$_4$ soln.]-[CHCl$_3$ soln.]
Acetic	3580(0.310)	3335.0(0.007)	3518.4(0.016)	45	61.6	16.6
Propionic	3579(0.460)	3535.7(0.015)	3520.7(0.032)	43.3	58.3	15
Butyric	3579(0.291)	3533.9(0.015)	3518.9(0.024)	45.1	60.1	10
Isobutyric	3579(0.411)	3534.1(0.012)	3518.6(0.026)	44.9	60.4	15.5
Trimethyl-acetic	3579(0.291)	3534.3(0.011)	3521.0(0.028)	44.7	58	13.3
	C=O str. vapor	C=O str. Cl$_4$ soln.	C=O str. CHCl$_3$ soln.			
Acetic	1789(1.230)	1767.3(0.056)	1756.3(0.110)	21.7	32.7	11
Propionic	1780(1.150)	1758.4(0.052)	1745.8(0.180)	21.6	34.2	12.6
Butyric	1780(1.250)	1758.7(0.037)	1745.9(0.116)	21.3	34.1	12.8
Isobutyric	1778(1.242)	1755.2(0.043)	1743.6(0.153)	22.8	34.4	11.6
Trimethyl-acetic	1769(1.142)	1750.3(0.047)	1740.1(0.174)	18.7	28.9	10.2
	o.p.(C=O)$_2$ str. vapor	o.p.(C=O)$_2$ str. CCl$_4$ soln.	o.p.(C=O)$_2$ str. CHCl$_3$ soln.			
Acetic		1714.6(1.343)	1712.5(0.735)			2.1
Propionic		1716.1(0.927)	1714.4(0.552)			1.7
Butyric		1711.8(0.839)	1709.5(0.393)			2.3
Isobutyric		1707.0(0.598)	1704.5(0.484)			2.5
Trimethyl-acetic		1702.2(0.696)	1699.6(0.556)			2.6
	C−C−O str. vapor	gamma C=O vapor		[vp-vp]		
Acetic	1178)9.744)	580(0.268)		598		
Propionic	1145(0.858)	588(0.250)		557		
Butyric	1150(0.667)	590(0.189)		560		
Isobutyric	1105(0.755)	585(0.353)		520		
Trimethyl-acetic	1115(1.245)	571(0.311)		544		

TABLE 10.2 Raman data and assignments for out-of-phase and in-phase (C=O)$_2$ stretching modes and C=C stretching modes for carboxylic acids

ACID	o.p.(C=O)$_2$ str.	i.p.(C=O)$_2$ str.	C=C str.	Acid	o.p.(C=O)$_2$ str.	i.p.(C=O)$_2$ str.
Acrylic	1727(2)	1661(2)	1638(9)	Acetic*	1718(sh)	1668(7) 1680(9)
Methacrylic		1660(3)	1640(4)	Dichloroacetic*		1664(6)
2-Chloroacrylic		1656(2)	1614(1)	Iodoacetic*		
Trichloroacrylic		1650(1)	1557(9)	4-Methoxy Benzoic*		1688(m)
Itaconic	1712(1)	1655(7)	1690(9)			
cis-Aconititic	1730(2)	1671(8)				
trans-Aconititic		1668(9)				
Tiglic		1653(9)	1625(1)			
2,6-Naphthalene-dicarboxylic		1639(5)				
Benzoic		1635(1)				
Terephthalic		1632(4)				
Isophthalic		1638(3)				
1,3,5-Tricarboxy benzene		1654(4)				
Succinic		1657(2)				
Thioglycolic		1687(20				
Thiopropionic		1646(2)				
3-(4-Hydroxy-3-methoxyphenyl)		1630(5)				
2-Propionic						
Polyacrylic		1678(3)				
Polymethacrylic		1694(1)				

* Reference 6.

TABLE 10.3 IR group frequency data for acetic acid and its derivatives in the vapor and solution phases

Acetic acid	sigma*	pKa 25°C	C=O str. vapor	C=O str. CCl₄ soln.	C=O str. CHCl₃ soln.		C=O str. CCl₄ soln. (9)	o.p.(C= number of O)₂ str. CCl₄ soln.	o.p.(C=O)₂ str. CHCl₃ soln.	o.p.(C=O)₂ str. CCl₄ soln. (9)	[vapor]-[CCl₄ soln.]	[vapor]-[CHCl₃ soln.]	[CCl₄ soln.]-[CHCl₃ soln.]
Trimethyl	[−0.300]	[5.03]	1769.1(1.142)	1750.30(0.047)	1740.1(0.174)			1702.2(0.696)	1699.6(0.556)		18.7	28.9	10.2
Dimethyl	[−0.190]	[4.85]	1788(1.242)	1755.2(0.043)	1743.6(0.153)			1707.0(0.595)	1704.5(0.484)		22.8	34.4	11.6
Methyl	[−0.100]	[4.88]	1780(1.150)	1758.4(0.053)	1745.8(0.180)			1716.1(0.927)	1714.4(0.552)		21.6	34.2	12.6
Hydrogen	[0.00]	[4.76]	1789(1.230)	1767.3(0.056)	1756.3(0.110)						21.7	32.7	11
Iodo	[0.85]	[3.19]	1781(1.240)	1769.1(0.020)	1769.6(0.045)	1739.2(0.029)*²	1771	1713.8(0.491)	1712.7(0.377)	1714	11.9	11.4	
Bromo	[1.00]	[2.87]		1777.1(0.020)	1775.5(0.061)		1772	1724.5(0.457)	1724.3(0.419)	1713	1.6		
Chloro	[1.05]	[2.85]	1815(0.585)	1792.8(0.030)	1787.7(0.071)*³	1759.7(0.100)*³	1772	1736.3(0.629)	1734.2(0.576)	1726	22.2	27.3	5.1
Chloro			1797(1.240)				1791			1737	4.2	9.3	
Fluoro	[1.10]	[2.72]		1798.9(0.034)	1787.3(0.488)*⁴	1771.8(0.023)*⁴	1797	1750.8(0.123)	1745.3(1.168)	1743			11.6
Dibromo	[1.82]	[1.48]	1804(0.540)										
Dibromo			1784(1.240)										
Dichloro	[1.94]	[1.30]		1784.3(0.027)	1775.0(0.098)		1784	1743.1(0.482)	1743.7(0.408)	1744			9.2
Difluoro	[2.05]	[1.13]		1797.7(0.073)	1789.2(0.197)		1794	1762.0(0.524)	1761.6(0.670)	1764			8.5
Tribromo	[2.52]	[0.72]		1770.5(0.039)	1765.9(0.094)		1772	1734.5(0.309)	1734.6(0.174)	1735			4.6
Trichloro	[2.65]	[0.70]		1786.1(0.066)	1780.3(0.149)		1789	1750.2(0.537)	1750.7(0.354)	1752			5.8
Trifluoro	[2.78]	[0.05]		1810.1(0.116)	1803.1(0.168)		1813	1778.3(0.462)	1784.3(0.288)	1780			7

Acetic acid	O—H str. CCl₄ soln.	O—H str. CHCl₃ soln.	[CCl₄ soln.]-[CHCl₃ soln.]	CCO str. vapor	gamma C=O vapor
Trimethyl	3534.3(0.011)	3521.0(0.028)	13.3	1115(1.245)	571(0.311)
Dimethyl	3534.1(0.012)	3518.6(0.026)	15.5	1105(0.755)	585(0.353)
Methyl	3535.5(0.015)	3520.7(0.032)	14.8	1145(0.858)	588(0.250)
Hydrogen	3535.0(0.007)	3518.4(0.016)	17	1178(0.744)	580(0.268)
Iodo	3527.0(0.006)	3507.9(0.014)	19.1	1189(0.487)	598(0.598)
Bromo	3524.4(0.008)	3504.2(0.017)	20.2		
Chloro	3524.7(0.008)	3503.7(0.022)	21		
Fluoro	3529.1(0.014)	3510.6(0.086)	18.5	1120(0.810)	616(0.290)
Dibromo	3517.1(0.007)	3497.7(0.0150)	19.4	1155(0.580)	599(0.415)
Dichloro	3508.0(0.019)	3492.4(0.023)	15.6		
Difluoro	3508.6(0.032)	3485.8(0.045)	22.8		
Tribromo	3506.1(0.008)	3481.4(0.020)	24.7		
Trichloro	3505.0(0.013)	3484.8(0.025)	20.2		
Trifluoro	3500.5(0.023)	3479.6(0.023)	20.9		

*² is for 19.4 mole % CHCl₃/CCl₄, and is assigned to the 1I–A(I) · HCCl₃ rotational conformer.
*³ is for 75.6 mole % CHCl₃/CCl₄ and is assigned to the 1Cl–A(Cl) · HCCl₃ rotational conformer.
*⁴ is also for CCl₄ solution and is assigned to the IF–A(F) rotational conformer.

TABLE 10.4 IR data and assignments for 4-X-benzoic acids in the vapor and solution phases

4-X-Benzoic Acid	O—H str. vapor	O—H str. CCl₄ soln.	O—H str. CHCl₃ soln.	[vapor]-[CCl₄ soln.]	[vapor]-[CHCl₃ soln.]	[CCl₄ soln.]-[CHCl₃ soln.]	sigma p
Nitro	3582	3529.2(0006)	3519.2(0.048)	52.8	62.8	10	0.78
Cyano	3584(0.4950)	3531.7(0.006)	3516.3(0.081)	52.3	67.7	15.4	0.66
Chloro	3585(0.290)	3536.7(0.016)	3523.9(0.044)	48.3	61.1	12.8	0.23
Bromo		3535.6(0.011)	3524.2(0.033)			11.4	0.23
Hydrogen	3585(0.350)	3539.1(0.014)	3523.1(0.035)	45.9	61.9	16	0
Methyl	3590(0.450)	3541.5(0.007)	3525.6(0.018)	48.5	64.4	15.9	-0.17
tert-Butyl	3582(0.300)	3540.3(0.012)	3524.8(0.027)	57.2	57.2	15.5	-0.2
Methoxy	3595(0.292)	3543.3(0.002)	3527.7(0.021)	51.7	67.3	15.6	-0.27
	C=O str. vapor	C=O str. CCl₄ soln.	C=O str. CHCl₃ soln.				
Nitro	1768	1751.0(0.029)	1743.4(0.406)	17	24.6	7.6	
Cyano	1768(1.230)	1750.9(0.026)	1744.4(0.485)	17.1	23.6	6.5	
Chloro	1762(1.243)	1744.1(0.074)	1735.3(0.333)	17.9	26.7	8.8	
Bromo		1744.8(0.056)	1734.2(0.250)			10.6	
Hydrogen	1762(1.250)	1739.2(0.073)	1731.5(0.188)	22.8	30.5	8	
Methyl	1760(1.231)	1738.2(0.020)	1727.3(0.092)	21.8	32.7	10.9	
tert-Butyl	1758(1.250)	1737.6(0.050)	1729.1(0.117)*¹	20.4	28.9	8.5	
			1718.0(0.120)*		40		
			1721.6(0.157)		36.4		
Methoxy [57.2 mol % CHCl₃/CCl₄]*¹	1760(1.243) [60.1 mol % CHCl₃/CCl₄]*²	1735.6(0.023)	1719.1(0.120)	24.4	40.9	16.5	
	o.p.(C=O)₂ str. vapor	o.p.(C=O)₂ str. CCl₄ soln.	o.p.(C=O) str. CHCl₃ soln.				
Nitro		1706.7(0.015)	1707.0(0.355)			-0.3	
Cyano		1705.0(0.040)	1705.7(1.832)			-0.3	
Chloro		1697.2(0.429)	1696.2(0.768)			1	
Bromo		1696.9(0.204)	1695.8(0.366)			1.1	
Hydrogen		1695.8(1.536)	1694.5(1.206)			1.3	
Methyl		1695.3(0.705)	1693.0(0.462)			2.3	
tert-Butyl		1693.0(1.442)	1690.7(0.901)			-0.7	
Methoxy		1689.7(0.260)	1687.2(0.559)			2.5	
	Ring-C=O str. vapor 1180	gamma C=O vapor 565					
Nitro	1172(0.530)	532(0.230)					
Cyano	1177(0.355)	570(0.060)					
Chloro	1180(0.590)	571(0.060)					
Hydrogen	1170(0.680)	582(0.239)					
Methyl	1178(0.450)	568(0.115)					
tert-Butyl	1166(0.742)	593(0.187)					

*¹ Two bands are present at 572 mol % CHCl₃/CCl₄, (14).

TABLE 10.4A   Factors affecting the CO_2H and $(CO_2)_2$ groups for 4-X-benzoic acids in $CHCl_3$ and/or CCl_4 solutions

Factor Hammett σp	O—H str. cm^{-1}	C=O str. cm^{-1}	op(C=O) str. cm^{-1}
NO_2 (0.78) to CH_3O (−0.27)	increase	decrease	decrease
CCl_4 solution	3529.2–3543.3	1751.0–1735.6	1706.7–1689.7
$CHCl_3$ solution	3519.2–3527.7	1743.4–1719.1	1707.0–1687.2
CH_3O (−0.27) to NO_2 (0.78)	decrease	increase	increase
CCl_4 solution	3543.3–3529.3	1735.6–1751.0	1689.7–1706.7
$CHCl_3$ solution	3527.7–3519.2	1719.1–1735.6	1687.2–1707.0
pKa values			
NO_2 (3.42) to CH_3O (4.47)	increase	decrease	decrease
CCl_4 solution			
$CHCl_3$ solution			
CH_3O (4.47) to NO_2 (3.42)	decrease	increase	increase
intermolecular hydrogen bond to 4-X group			
NO_2, CN, CH_3O	increase	increase	increase
OH : π system of phenyl group			
CH_3 is 0	increase	increase	increase
$(CH_3)3C$ is −1.54	decrease	decrease	decrease

TABLE 10.5   IR vapor-phase data and assignments for anthranilic acids

Compound	O–H str.	a.NH₂ str.	s.NH₂ str.	OH:N str.	C=O str.	NH₂ bending	O–C= str.	A[a.NH₂ str.]/ A[s.NH₂ str.]	a.NH₂ str.-s.NH₂ str.	gamma C=O
Anthranilic acid	3584(0.256)	3520(0.082)	3395(0.141)	3436(0.020)	1729(1.237)	1613(0.580)	1180(0.380)	0.58	125	561(0.152)
3-Methyl	3584(0.465)	3520(0.165)	3382(0.240)	3430(0.031)	1724(1.230)	1620(0.570)	1168(0.560)	0.69	138	564(0.260)
3-Ethyl	3584(0.505)	3521(0.180)	3382(0.260)	3435(0.030)	1725(1.230)	1620(0.620)	1155(0.560)	0.69	139	548(0.200)
5-Methyl	3584(0.320)	3519(0.098)	3395(0.170)	3435(0.020)	1726(1.240)	1625(0.290)	1162(0.922)	0.58	124	570(0.210)
5-Methoxy	3584(0.250)	3518(0.070)	3395(0.120)	3435(0.020)	1730(1.230)	1604(0.341)	1165(0.720)	0.58	123	580(0.120)
3,4-Dimethyl	3594(0.330)	3525(0.080)	3384(0.155)		1724(1.230)	1613(0.750)	1177(0.335)	0.52	141	570(0.150)
5-Chloro-3-methyl	3584(0.340)	3520(0.100)	3385(0.169)	3430(0.030)	1725(1.220)	1621(0.343)	1161(0.631)	0.59	135	571(0.170)
3,5-Dichloro	3584(0.360)	3515(0.200)	3395(0.200)	3430(0.050)	1732(1.230)	1618(0.380)	1178(1.110)	0.79	120	574(0.130)
		N–H str.		OH:N str.		NH bending				
N-Methyl	3584(0.357)	3392(0.212)		3480(0.0200)	1718(1.240)	1520(0.470)	1162(0.730)			559(0.210)
N-Phenyl	3584(0.235)	3348(0.150)		3445(0.030)	1720(1.209)	1518(0.650)	1162(0.641)			568(0.122)

TABLE 10.6   Infrared data for methacrylic acid and acrylic acid in $CHCl_3$ and/or CCl_4 solutions

Mole % $CHCl_3$ /CCl_4	Methacrylic acid s.CH_3 bending cm^{-1}	Methacrylic acid CH_2= wag cm^{-1}	Acrylic acid CH_2= bending cm^{-1}	Acrylic acid vinyl twist cm^{-1}
0	1375.5	948.7	1433.1	983.9
10.74	1375.6	948.8	1433	983.8
19.4	1375.7	948.9	1432.8	983.8
26.53	1375.8	949	1432.6	983.7
32.5	1375.9	949	1432.4	983.65
37.57	1376.1	949	1432.3	983.6
41.93	1376.2	949.1	1432.15	983.6
45.73	1376.3	949.1	1432	983.6
49.06	1376.4	949.2	1431.9	983.6
52	1376.5	949.2	1431.8	983.6
54.62	1376.6	949.2	1431.7	983.7
57.22	1376.7	949.2	1431.6	983.7
60.07	1376.8	949.2	1431.5	983.7
63.28	1376.9	949.2	1431.4	983.7
66.73	1377	949.2	1431.2	983.7
70.65	1377.1	949.2	1431	983.7
75.06	1377.2	949.2	1430.9	983.8
80.06	1377.3	949.2	1430.8	983.8
85.05	1377.4	949.2	1430.6	983.9
92.33	1377.8	949.2	1430.3	983.9
100		949.2	1429.9	984
delta cm^{-1}	2.3	0.5	−3.2	0.1

TABLE 10.7 Raman data and assignments for carboxylic acid salts

Compound	a.CO_2 str.	s.CO_2 str.	a.CH_2 str. / a.CH_2= str. / s.C–CO_2 str.	s.CH_3 str. / CH= str. / CCO_2 bend	s.CH_2 str. / s.CH_2= str.	CH_2 bend / CH= str.	i.p.CH_2 twist / C=C str.	cis CH s. rock / CH_2= bend	C–C str. / CH= bend	s.C–CO_2 str.
Strontium stearate		1468(3)	2923(3)	2884(9)	2852(5)	1447(4)	1296(3)		1062(1)	
Zinc stearate		1460(2)		2882(9)	2849(5)	1440(3)	1296(3)		1064(2)	
Stannous oleate		1435(5)		2897(9)	2853(9)		1302(3)	1267(2)		
K dichloroacetate*	1646(7)	1383(14)								
Lithium acrylate		1459(7)					1649(6)		1280(9)	904(3)
Sodium acrylate	1565(1)									
Zinc diacrylate	1603(1)?	1445(5)	3107(1)	3056(2)	3029(3)	2997(1)	1640(9)	138(1)	1280(5)	915(3)
Lead acrylate		1445(4)					1639(9)		1278(8)	920(2)
Barium methacrylate		1423(9)	853(6)	606(3)			1648(6)			
Lead methacrylate		1426(7)	866(9)	604(0)						
Zirconium methacrylate		1422(4)	857(4)	586(10)			1645(9)			
Zinc dimethacrylate	1600(0)?	1445(5)					1640(9)			

* Unpublished data from the Dow chemical company.

TABLE 10.8 The asymmetric and symmetric CO_2 stretching frequencies for carboxylic acid salts

Carboxylate ion	asym.CO_2 str. cm^{-1}	sym.CO_2 str. cm^{-1}
Formate	1538–1604	1342–1400
Acetate	1543–1585	1404–1457
Dichloroacetate	1646	1383
Propionate	1550	1415
Butyrate	1563–1578	1420–1430
Valerate	1561–1565	near 1437
Stearate	1539–1558	1438–1468
Cyanoacetate	1605	near 1390
Malonate	1570	1420
Succinate	1569	1437
Tartrate	1571–1600	1386–1455
Citrate	1575–1620	1390–1430
Acrylate	1565–1603	1445–1459
Methacrylate	1603	1422–1445
Benzoate	1515–1559	1392–1431
Salicylate	1540–1598	1365–1409
Phthalate	1565	1384
Cinnamate	1549	1424

TABLE 10.9   IR data for carboxylic acid salts

Compound	asym.CO_2 str. cm^{-1}	sym.CO_2 str. cm^{-1}	a.-s. delta cm^{-1}
Formate			
lithium	1604	1382	222
calcium	1596	1400	196
Acetate			
sodium	1585	1445	140
sodium iodo-	1586	1398	190
sodium fluoro-	1616	1449	167
sodium difluoro-	1640	1449	191
sodium mercapto-	1585	1400	185
sodium N,N-diethylamino-	1590	1400	190
Propionate			
cadmium	1550	1415	135
sodium 2-hydroxy	1590	1410	180
sodium 2,3-dichloro-2methyl-	1609	1395	214
Hexanoate			
sodium 2-ethyl	1555	1415	140
calcium 2-ethyl	1545	1424	121
Laurate			
sodium	1555	1421	134
Benzoate			
sodium	1545	1410	135
sodium 3-amino-	1559	1410	149
sodium 4-amino-	1545	1405	140
sodium 2-hydroxy	1582	1378	204
sodium 4-hydroxy-	1544	1415	129

Anhydrides

*Numbers in parentheses indicate in-text page reference.

Carboxylic acid anhydrides exhibit symmetrical and asymmetric $(C=O)_2$ stretching vibration [ν_{ip} $(C=O)_2$ and ν_{op} $(C=O)_2$], respectively. Open chain saturated aliphatic anhydrides exhibit ν_{ip} $(C=O)_2$ in the range 1815–1825 cm^{-1} and ν_{op} $(C=O)_2$ in the range 1745–1755 cm^{-1}. The ν_{ip} $(C=O)_2$ mode has stronger IR band intensity than ν_{op} $(C=O)_2$. Conjugation lowers both modes. The strong band in the range 1770–1780 cm^{-1} is assigned to ν_{ip} $(C=O)_2$, and the weaker IR band in the region 1715–1725 cm^{-1} is assigned to ν_{op} $(C=O)_2$. In the case of unconjugated 5-membered ring anhydrides the IR bands occurring in the range 1845–1870 cm^{-1} have relatively weak intensity and the IR bands in the range 1775–1800 cm^{-1} have strong intensity. These bands are assigned as ν_{ip} $(C=O)_2$ and ν_{op} $(C=O)_2$, respectively. Conjugated 5-membered anhydrides exhibit the weak IR band in the range 1850–1860 cm^{-1} and the strong IR band in the region 1760–1780 cm^{-1}, which are assigned to ν_{ip} $(C=O)_2$ and ν_{op} $(C=O)_2$, respectively (1,2).

It was found helpful in the discussions of the $\nu(C=O)_2$ modes of anhydrides to give letters and numbers to classify each type. For example, an open chain anhydride such as acetic is labeled (OC), succinic anhydride whose cyclic structure includes a 5-membered saturated ring (5SR), glutaric anhydride whose cyclic structure include a 6-membered saturated ring (6SR), maleic and phthalic anhydrides whose cyclic structures include a 5-membered unsaturated ring (5UR), naphthalic anhydride whose cyclic structure includes a 6-membered unsaturated ring (6UR), and 2,2-biphenyldicarboxylic anhydride whose cyclic structure includes a 7-membered unsaturated ring (7UR), (3).

Table 11.1 lists IR vapor phase and Raman data in the neat phase for anhydrides.

Ring strain and conjugation play a role in the $v(C=O)_2$ modes for anhydrides. The frequency separation between v_{ip} $(C=O)_2$ and v_{op} $(C=O)_2$ together with their band intensity ratio also can be used to classify anhydrides.

The inductive effect also affects the $v(C=O)_2$ modes for OC anhydrides. For example, v_{ip} $(C=O)_2$ decreases in the order 1830, 1825, 1824, and 1822 cm^{-1} and v_{op} $(C=O)_2$ decrease in the order 1775, 1769, 1761, and 1759 cm^{-1} for acetic, propionic, isobutyric and 2-ethylbutyric anhydrides, respectively. Moreover, when the inductive effect causes electrons to be withdrawn from the $(C=O)_2$ bonds such as in the case of trifluoroacetic anhydride, v_{ip} $(C=O)_2$ and v_{op} $(C=O)_2$ occur at the relatively high frequencies of 1881 and 1817, respectively (3).

Table 11.2 lists Raman data and assignments for carboxylic acid anhydrides. In the Raman, the v_{ip} $(C=O)_2$ mode is always more intense than v_{op} $(C=O)_2$. The OCU compounds exhibit v_{ip} $(C=O)_2$ and v_{op} $(C=O)_2$ in the range 1771–1788 cm^{-1} and 1715–1725 cm^{-1}, respectively, and occur at lower frequency than the OC anhydrides due to the effects of conjugation of the C=O and C=C groups.

PHTHALIC ANHYDRIDE

Figure 11.1 shows the IR spectra of phthalic anhydride in the region 2000–1600 cm^{-1}. The spectrum on the left is that for a saturated solution of phthalic anhydride in CCl$_4$ solution, the center spectrum is that for a saturated solution of phthalic anhydride in a 23.1% vol. CHCl$_3$/CCl$_4$ solution, and the spectrum on the right is that for a saturated solution of phthalic anhydride in CHCl$_3$ solution. All spectra were recorded using 0.2-mm KBr cells. The spectrum is more intense in going from left to right due to the increased solubility in CHCl$_3$. What is important in this case is to note that the 1789 cm^{-1} band is more intense than the 1776 cm^{-1} band in CCl$_4$ solution, in 23.1% vol. in CHCl$_3$/CCl$_4$ solution the two bands have equal intensity, and in CHCl$_3$ solution the 1775 cm^{-1} band has more intensity than the 1788 cm^{-1} band. Whereas phthalic anhydride is a planar molecule with C$_{2v}$ symmetry, changes in the band intensity ratio can not be attributed to the presence of rotational conformers because solvent techniques are also used to determine which bands in each set of band results from which rotational conformer. In the case of phthalic anhydride, the doublet near 1789 cm^{-1} and 1776 cm^{-1} is the result of v_{op} $(C=O)_2$ in Fermi resonance with a combination tone. It is common practice (but not correct) to assign as the fundamental the band in the doublet with the most intensity; in this case it is v_{op} $(C=O)_2$. The dilemma in this situation is apparent in the middle spectrum where both bands have equal intensity and only one v_{op} $(C=O)_2$ mode. The answer is that both bands are in Fermi resonance, and each band is a mixture of v_{op} $(C=O)_2$ and the combination tone. After correction for Fermi resonance, v_{op} $(C=O)_2$ shows a steady decrease in frequency in CCl$_4$ solution (1784.6 cm^{-1}) to (1779.4 cm^{-1}) in CHCl$_3$ solution. This is a decrease of 5.2 cm^{-1} for v_{op} $(C=O)_2$ going from CCl$_4$ to CHCl$_3$ solution, and this is reasonable from study of the other carbonyl-containing compounds included in this book. It is important to note that v_{ip} $(C=O)_2$ decreased only 3 cm^{-1} in going from CCl$_4$ solution (1856 cm^{-1}) to (1853 cm^{-1}) in CHCl$_3$ solution. The behavior of v_{op} $(C=O)_2$ and v_{op} $(C=O)_2$ in solvents such as CCl$_4$ and CHCl$_3$ will be discussed in Volume 2 of the book. The general decrease in frequency of v_{op} $(C=O)_2$ and v_{ip} C=O$_2$ is attributed to the field effect of the solvent. Hydrogen bonding between the Cl_3CH protons and the two carbonyl groups plays a role in lowering both $\nu(C=O)_2$ frequencies compared to where they occur in CCl_4 solution.

The lowest ν_{ip} $(C=O)_2$, ν_{op} $(C=O)_2$ and CT frequencies for phthalic anhydride are observed for solutions in dimethyl sulfoxide. These are $1850 \, cm^{-1}$, $1788 \, cm^{-1}$, and $1772 \, cm^{-1}$, respectively. After correction for F.R. the unperturbed ν_{op} $(C=O)_2$ is determined to be 1780.5 and CT at $1788.9 \, cm^{-1}$ (4).

MALEIC ANHYDRIDE

Vibrational assignments have been made for maleic anhydride (5). The combination tone $(560 \, cm^{-1}, B, +1235 \, cm^{-1}, A, = 1795 \, cm^{-1}, B,)$ was ruled out as the possibility of being in Fermi resonance with ν_{op} $(C=O)_2$, because it was noted that there is a strong dependence of the band intensity ratio on the nature of the solvent while the frequencies remain practically constant (6). Phthalic anhydride reported on in the preceding showed the same phenomena, but ν_{op} $(C=O)_2$ was shown to be in Fermi resonance.

Table 11.3 lists IR data for maleic anhydride in n-$C_6H_{14}/CHCl_3$, $CHCl_3/CCl_4$, and n-C_6H_{14}/CCl_4 mole % solutions. Maleic anhydride is of the type 5UR (3). The $\nu(C=O)_2$ frequencies for the 5% UR structure might be expected to be lower than those for 5SR structure due to conjugation of the C=C group with the two C=O groups. However, the ring strain is more in the case of maleic anhydrate than in the case of succinic anhydride, as the C=C bond distance is less than the C–C bond distance. Therefore, the two opposing effects essentially cancel each other in the case of maleic anhydride.

Maleic anhydride is a planar structure with C_{2v} symmetry. The ν_{ip} $(C=O)_2$ mode belongs to the A_1 symmetry species while ν_{op} $(C=O)_2$ belongs to the B_1 symmetry species. Therefore, ν_{op} $(C=O)_2$ can only be in Fermi resonance with a combination tone belonging to the B_1 symmetry species. It can not be in Fermi resonance with an overtone because any overtone is assigned to the A_1 symmetry species. A study of Table 11.3 shows that two bands occur in the range 1770 and $1793 \, cm^{-1}$ in each of the solvent systems, and in all cases the higher frequency band has more intensity than the lower frequency band. As in the case of phthalic anhydride, the ν_{op} $(C=O)_2$ mode is in Fermi resonance with the B_1 combination tone. The experimental data has been corrected for Fermi resonance, and unperturbed ν_{op} (C=O) is determined to be between 1789.8 and $1790 \, cm^{-1}$ in n-C_6H_{14} solution, $1787.1–1787.5 \, cm^{-1}$ in CCl_4 solution, $1785–1785.4 \, cm^{-1}$ in $CHCl_3$ solution, and $1778.9 \, cm^{-1}$ in $60.19 \, mol \, \%$ $(CH_3)_2SO/CCl_4$. These ν_{op} $(C=O)_2$ frequencies decrease in the order of increasing polarity of the solvent, and also decrease in frequency as the reaction field of the solvent is increased.

Table 11.4 lists the in-phase and out-of-phase $(C=O)_2$ stretching vibrations for hexahydrophthalic, tetrachlorophthalic, tetrabromophthalic, dichloromaleic (7), phthalic (4), and maleic anhydrides in different physical phases (6). In all cases the ν_{op} $(C=O)_2$ modes have been corrected for Fermi resonance. These data show that both ν_{ip} $(C=O)_2$ and ν_{op} $(C=O)_2$ decrease progressively in the order: vapor, and C_6H_{14}, CCl_4, $CHCl_3$ and $(CH_3)_2SO$ solution phases. In all cases the ν_{op} $(C=O)_2$ mode decreases more in frequency in each solvent pair than does the ν_{ip} $(C=O)_2$ mode.

In the case of styrene-maleic anhydride copolymer, the anhydride has a 5SR structure, and in CH_2Cl_2 solution ν_{ip} $(C=O)_2$ and ν_{op} (C=O) occur at 1856.6 and $1779.7 \, cm^{-1}$, respectively, and in $CHCl_3$ solution at 1856.5 and 1778 cm^{-1}, respectively. In this case, $v_{ip}(C{=}O)_2$ decreases 0.1 cm^{-1}, and v_{op} (C=O) decreases 1 cm^{-1}, a factor of 10 (7); this type of frequency difference was noted in the study of the other anhydride in different solvent systems.

REFERENCES

1. Dauben, W. S. and Epstein, W. W. (1959). *J. Org. Chem.*, **24**, 1595.

2. Bellamy, L. J., Connelly, B. J., Phillpots, A. R., and Williams, A. L. (1960). *Z. Elecktrochem.*, **64**, 563.

3. Nyquist, R. A. (1984). *The Interpretation of Vapor-phase Infrared Spectra: Group Frequency Data*, Vol. 1, Philadelphia: Sadtler Research Laboratories, Division of Bio-Rad.

4. Nyquist, R. A. (1989). *Appl. Spectrosc.*, **43**, 1374.

5. Mirone, P. and Chiorboli, P. (1962). *Spectrochim. Acta*, **18**, 1425.

6. Nyquist, R. A. (1990). *Appl. Spectrosc.*, **44**, 438.

7. Nyquist, R. A. (1990). *Appl. Spectrosc.*, **44**, 783.

8. Schrader, B. (1989). *Raman/Infrared Atlas of Organic Compounds*, 2nd edition, Germany, VCH.

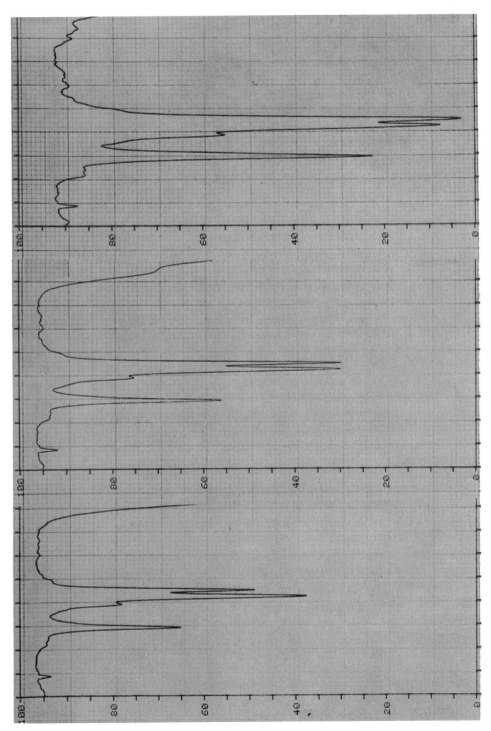

FIGURE 11.1 IR spectra of phthalic anhydride in the region 2000–1600 cm^{-1}. The spectrum on the left is for a saturated solution CCl$_4$, the center spectrum is for a saturated solution for in a 23.1% vol. CHCl$_3$/CCl$_4$, and the spectrum on the right is for a saturated solution in CHCl$_3$.

TABLE 11.1 IR vapor-phase data and Raman data in the neat phase for anhydrides

Anhydride type	sym.(C=O)$_2$ str. cm^{-1}	asym.(C=O)$_2$ str. cm^{-1}	(A)sym.(C=O)$_2$ str. (A)asym.(C=O)$_2$ str.	Frequency separation cm^{-1}
OC	1822–1830	1759–1775	0.96–1.31	55–63
OCU	[1771–1788]	[1715–1725]		[39–58]
5SR	1861–1880	1802–1812	0.11–0.23	59–72
6SR	1820–1830	1782–1790	0.35–0.51	36–40
5UR	1855–1880	1785–1813	0.07–0.35	51–85
6UR	1802	1768	0.72	34
7UR	1800	1772	0.66	28

[Raman data]

TABLE 11.2 Raman data and assignments for carboxylic anhydrides

Anhydride	ip(C=O)$_2$ cm^{-1}	op(C=O)$_2$ cm^{-1}	Type	Frequency separation cm^{-1}	C=C str. cm^{-1}	Ring breathing cm^{-1}
Allylsuccinic	1853(3)	1782(0)	5SR	71	1644(8)	
Acrylic	1788(3)	1730(2)	OCU	58	1630(9)	
Methacrylic	1782(3)	1725(2)	OCU	57	1639(9)	
Crotonic*	1771(27,p)	1732(vwk)	OCU	39	1648(80,p)	
Benzoic*	1771(32)	1715(13)	OCU	56		
Citraconic	1841(6)	1770(1)	5UR	71	1652(3)	
Propionic*	1812(7,p)	1743(3,p)	OC	69		
Butyric*	1813(7,p)	1751(3,p)	OC	62		
Trifluoroacetic*	1877(28,p)	1810(9,p)	OC	67		
4-Cyclohexene-1,2-dicarboxylic*	1834(2)	1729(1)	5SR	105	1629(3)	
Glutaric*	1780(18)	1755(vwk)	6SR	25		
Phthalic	1840(45)	1760(37)	5UR	80		1005(44)

* Reference 8.

TABLE 11.3 IR data for maleic anhydride in n-C_6H_{14}/$CHCl_3$, $CHCl_3$/CCl_4, and n-C_6H_{14}/CCl_4 solutions

Maleic anhydride mole % n-C_6H_{14}/$CHCl_3$	o.p.(C=O)$_2$ str. in FR cm^{-1}	B1 comb. tone in FR cm^{-1}	a[o.p.(C=O)$_2$ str. in FR]	A[B1 comb. tone in FR]	o.p.(C=O)$_2$ str. corrected for FR cm^{-1}	B1 comb. tone corrected for FR cm^{-1}	i.p.(C=O)$_2$ str. cm^{-1}	A[i.p.(C=O)$_2$ str.]
0	1781.4	1792.5	0.603	0.286	1785	1788.9	1851.7	0.069
23.57	1782.5	1793.2	0.493	0.324	1786.7	1789	1852.6	0.054
50.68	1782.7	1793.5	0.413	0.353	1787.7	1788.5	1853.3	0.044
75.51	1782.6	1783.8	0.381	0.332	1788.4	1787.9	1852.7	0.038
100	1793.7	1781.2	0.411	0.168	1789.8	1785.1	1852.6	0.025
delta cm^{-1}	12.3	−11.3			4.8	−3.8	0.9	
mole % $CHCl_3$/CCl_4								
0	1792.5	1781.4	0.521	0.489	1787.1	1786.8	1851.7	0.079
26.7	1781.7	1792.8	0.539	0.41	1786.5	1788	1852	0.08
52.2	1781.8	1792.9	0.553	0.354	1786.2	1788.6	1851.9	0.077
75.2	1781.9	1792.6	0.564	0.325	1785.8	1788.7	1851.9	0.079
100	1781.9	1792.6	0.573	0.274	1785.4	1789.1	1851.7	0.077
delta cm^{-1}	−10.6	11.2			−1.7	2.3	0	
Mole % n-C_6H_{14}/CCl_4								
0	1792.6	1781.8	0.546	0.497	1787.5	1787	1851.6	0.081
26.99	1793.1	1781.5	0.505	0.363	1788.3	1786.4	1852.4	0.055
51.37	1793.4	1781.5	0.479	0.302	1788.8	1786.1	1852.7	0.042
71.14	1793.7	1781.5	0.462	0.266	1789.2	1785.9	1852.9	0.037
100	1794	1781.2	0.3394	0.176	1790	1785.1	1852.6	0.027
delta cm^{-1}	1.4	−0.6			2.5	−1.9	1	
Mole % $(CH_3)_2SO$/CCl_4								
0	1792.5	1781.4	0.569	0.52	1787.2	1786.7	1851.7	0.084
28.99	1778.3	~1791.4	0.568	0.15	1781.1	1788.7	1847	0.099
48.78	1777.3	~1791.4	0.571	0.11	1779.6	1785.2	1847.6	0.102
60.19	1776.7	~1791.4	0.588	0.105	1778.9	1789.2	1847.3	0.105
100	1772.3	~1791.4	0.587		1772.3[*1]		1845.8	0.093
delta cm^{-1}	−20.2	10			[−14.9][*1]	−5.9	−5.9	
Mole % $(CH_3)_2SO$/$CHCl_3$								
0	1781.4	1792.5	0.603	0.286	1785	1788.9	1851.6	0.07
25.4	1780.5	~1791.4	0.545	0.183	1783.3	1788.7	1849.9	0.085
50.53	1779.1	~1791.4	0.5	0.121	1782.2	1789.2	1848.3	0.088
73.94	1776.6	~1791.4	0.466	0.07	1778.2	1789.5	1847.5	0.079
100	1772.3		0.449		1772.3[*1]		1845.8	0.076
delta cm^{-1}	−9.1	−1.1			[−12.7][*1]	−5.8	−5.8	

[*1] not corrected for Fermi resonance.

TABLE 11.4   The in-phase and out-of-phase (C=O)$_2$ stretching vibrations for carboxylic anhydrides in the vapor phase and in various solution phases

Anhydride	i.p.(C=O)$_2$ str. [vapor]	i.p.(C=O)$_2$ str. [n-C$_6$H$_{14}$]	i.p.(C=O)$_2$ str. [CCl$_4$]	i.p.(C=O)$_2$ str. [CHCl$_3$]	i.p.(C=O)$_2$ str. [(CH$_3$)$_2$SO]
Hexahydrophthalic	1865	1864.5	1861.4	1859.1	1852.3
Phthalic	1859	1858	1856	1853	1850
Tetrachlorophthalic			1845.9*	1845.5*	
Tetrabromophthalic			1867.2*	1863.4*	
Maleic		1852.6	1851.74	1851.68	1845.8
Dichloromaleic		1877.9*	1876.7*	1876.2*	

Anhydride	o.p.(C=O)$_2$ str. [vapor]	o.p.(C=O)$_2$ str. [n-C$_6$H$_{14}$]	o.p.(C=O)$_2$ str. [CCl$_4$]	o.p.(C=O)$_2$ str. [CHCl$_3$]	o.p.(C=O)$_2$ str. [(CH$_3$)$_2$SO]
Hexahydrophthalic	1802	1798.9	1792.7	1787.5	1781.4
Phthalic	1809	1789.7*	1784.6*	1779.4*	1771.1*
Tetrachlorophthalic			1790.7*	1786.8*	
Tetrabromophthalic			1795.6*	1790.3*	
Maleic		1790.0*	1787.1*	1785.4*	1772.3
Dichloromaleic		1801.7*	1799.0*	1796.4*	

Anhydride	i.p.(C=O)$_2$ str. [CCl$_4$]−[CHCl$_3$]	o.p.(C=O)$_2$ str. [CCl$_4$]−[CHCl$_3$]	i.p.(C=O)$_2$ str. [C$_6$H$_{14}$]−[CCl$_4$]	o.p.(C=O)$_2$ str. [C$_6$H$_{14}$]−[CCl$_4$]	i.p.(C=O)$_2$ str. [CCl$_4$]−[(CH$_3$)$_2$SO]	o.p.(C=)$_2$ str. [CCl$_4$]−[(CH$_3$)2SO]
Hexahydrophthalic	2.3	5.2	3.1	6.2	9.2	11.3
Phthalic	3	5.2	2	5.1	6	13.5
Tetrachlorophthalic	0.4	3.9				
Tetrabromophthalic	3.8	5.3				
Maleic	0.1	1.7	0.9	2.9	5.9	14.8
Dichloromaleic	0.5	2.6	1.2	2.7		

* corrected for Fermi resonance

Carboxamides, Ureas, Thioureas, Imidazolidinones, Caffeine, Isocaffeine, Uracils, Imides, Hydantoins, and s-Triazine(1H,3H,5H)-Triones

*Numbers in parentheses indicate in-text page reference.

There are several forms of carboxamides, and they are separated into three classes: primary, secondary, and tertiary. Primary carboxamides also exist in three forms: $H(C{=}O)NH_2$, $R(C{=}ONH_2$, and $\phi(C{=}O)NH_2$. Secondary carboxamides exist in the following forms: $H(C{=}O)NHR$, $R(C{=}O)NHR'$, $R(C{=}O)NH\phi$, $\phi(C{=}O)NHR$, $\phi(C{=}O)NH\phi$,

Tertiary carboxamides can have the basic structures as denoted for the secondary carboxamides with the replacement of the NH proton with R or ϕ. Each of these forms yields its own characteristic group frequencies, and each will be discussed in what follows.

Table 12.1 shows a comparison of primary, secondary, and tertiary amides in various physical phases.

In the case of the primary amides the νasym. NH_2, νsym. NH_2, and $\nu C{=}O$ modes occur at higher frequency in the vapor phase than in the neat phase. It is noteworthy that $\nu C{=}O$ for $\phi(C{=}O)NH_2$ (1719–1731 cm^{-1}) occurs at lower frequency than $\nu C{=}O$ for $R(C{=}O)NH_2$ (1732–1780 cm^{-1}) due to conjugation of the carbonyl group with the π system of the phenyl group (1). In the case of $R(C{=}O)NH_2$, the high frequency $\nu C{=}O$ mode at 1780 cm^{-1} is the result of the inductive effect of the CF_3 group for $CF_3(C{=}O)NH_2$, and the low frequency is the result of the inductive effect of the alkyl group ($C_{13}H_{26}$) for $R(C{=}O)NH_2$ (1). Evidence is also presented that in the vapor phase the $\alpha{-}X$ analogs are in an intramolecularly hydrogen-bonded trans configuration (1).

For example, in the vapor phase $\nu C=O$ occurs at 1740, 1749, and 1780 cm^{-1} for acetamide, fluoroacetamide, and trifluoroacetamide, respectively, while νasym. NH$_2$ occurs at 3565, 3552, and 3560 cm^{-1}, and νsym. NH$_2$ occurs at 3442, 3444, and 3439 cm^{-1}, respectively (1). Substitution of the first F atom on the α-carbon atom raises $\nu C=O$ by 9 cm^{-1} while substitution of three F atoms raises $\nu C=O$ by 40 cm^{-1}. Therefore, $\nu C=O$ is raised 31 cm^{-1} for the substitution of the second and third F atoms, or 15.5 cm^{-1} per F atom. These data indicate that the addition of the first α-fluorine atom on the α-carbon atom of acetamide forms an intramolecular hydrogen atom as depicted here, and lowers $\nu C=O$ by $\sim$6.5 cm^{-1}. This intramolecular N$-$H$\cdots$F bond is present in the F, F$_2$ and F$_3$ analogs, or otherwise each $\nu C=O$ mode would be $\sim$6.5 cm^{-1} higher in frequency (1).

Compounds such as 2-aminobenzamide and 2-methoxybenzamide in the vapor phase exhibit $\nu C=O$ at 1698 and 1709 cm^{-1}, respectively, while benzamide and 3-X and 4-X benzamides exhibit $\nu C=O$ in the range 1719–1731 cm^{-1}. In the case of 2-aminobenzamide and 2-methyoxybenzamide, the amide NH$_2$ proton is intramolecularly hydrogen bonded to the oxygen atom of the CH$_3$O group or the nitrogen atom of the amino NH$_2$ group.

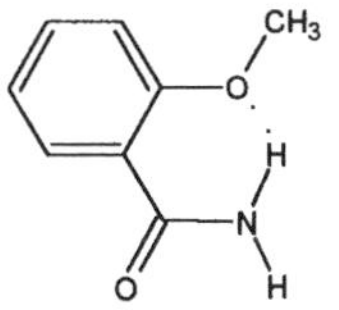

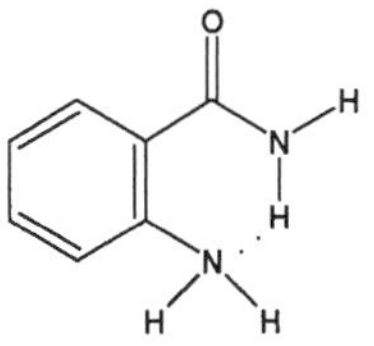

In the case of the 2-amino analog, the νasym. NH$_2$ and νsym. NH$_2$ for the amino NH$_2$ group occur at 3511 and ν 3370 cm^{-1}, respectively. The frequency separation of 141 cm^{-1} is evidence that an aniline type NH$_2$ group is intramolecularly hydrogen bonded (1). In the case of the 2-methoxy analog, νasym. NH$_2$ and νsym. NH$_2$ for the amide group are assigned at 3540 and 3420 cm^{-1}, which is lower than other members in the other benzamides studied in the vapor phase.

The 2-hydroxybenzamide in the vapor phase is a special case in that the OH proton is intramolecularly hydrogen bonded to the carbonyl group.

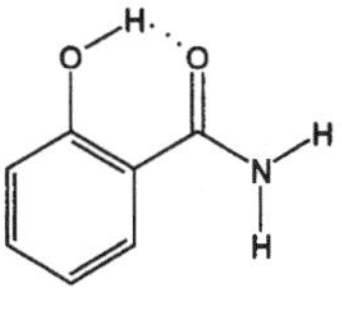

In this case $\nu C=O\cdots HO$ occurs at 1679 cm^{-1}, and $\nu OH\cdots N$ at 3270 cm^{-1} (1).

In the vapor phase, another unique case is 2,6-dichlorobenzamide, because $\nu C=O$ occurs at 1735 cm^{-1}. The high $\nu C=O$ frequency (1731 cm^{-1}) listed in Table 12.1 is for 3,5-dinitrobenzamide. In the case of the 2,6-Cl$_2$ analog, the phenyl and (C=O)NH$_2$ groups are not coplanar, and the reason that $\nu C=O$ occurs at such a relatively high frequency is that these two groups are no longer in resonance. The two groups are coplanar in the case of 3,5-dinitrobenzamide, but the high Hammett σ_m value for the two NO$_2$ groups contributes to the relatively high $\nu C=O$ frequency.

Methylmethacrylamide (νC=O, 1721 cm^{-1}, vapor) exhibits νC=O at lower frequency than acrylamide (νC=O, 1731 cm^{-1}, vapor) due to the inductive electron donation of the methyl group to the C=O group (1).

DILUTION STUDIES OF N-ALKYLACETAMIDE AND N-ALKYLCHLOROACETAMIDE

Table 12.1a lists the N−H stretching frequencies for N-methyl acetamide in mol/l concentrations ranging from 1.37×10^{-3} to 1.37. The νNH mode decreases from 3476 to 3471 cm^{-1}, and the absorbance increases from 0.2 to 8.9. At 1.37×10^{-3} mol/l no intermolecular association is noted between C=O$\cdots$HN. However, the % νN−H$\cdots$C=O increases steadily to ~95% at 1.37 mol/l. In the case of N-methyl chloroacetamide in mole/liter concentrations ranging from 9.21×10^{-3} to 9.21, the νN−H frequency range of from 3450–3448 cm^{-1} in a decreasing order with increasing concentration while absorbance increases from 0.2 to 5. In this case, the intermolecular hydrogen bonding between N−H$\cdots$O=C varies from 1–64% (5). It has been shown that the N-alkyl chloroacetamide and α-halo-p-x-acetanilides in dilute solution exist in an intramolecular hydrogen bonded-form (6,7).

Figure 12.1 shows plots of the concentration vs the absorption maxima at each of the νN−H$\cdots$O=C frequencies shown in Table 12.1a. At low concentrations (below 0.01 M for N−methyl acetamide and below 0.2 M for N-methyl chloroacetamide) the frequency of absorption maximum of the intermolecular νN−H$\cdots$O=C band is independent of concentration, indicating that in addition to a monomer only one N−H$\cdots$O=C intermolecular species is present, presumably a dimer (5). At higher concentrations the frequency of absorption decreases rapidly with increasing concentration. In the case of N-methyl acetamide at concentration above ~0.3 M the νN−H$\cdots$O=C frequency is independent of concentration and the average size of the intermolecular (N−H$\cdots$O=C)$_n$ complex is about seven or more. The N-methyl chloroacetamide is different, because in saturated CCl$_4$ solution only one-third of the molecules exist in an intermolecular hydrogen-bonded form (N−H$\cdots$O=C)$_n$. At concentrations of less than 0.2 M, N-methyl chloroacetamide molecules exist in the intramolecular hydrogen-bonded form (NH$\cdots$Cl) (5).

Table 12.1b lists amide I, II, and III frequencies for N-methyl acetamide in various mole/liter in Br$_3$CH solution. Bromoform was used in this case instead of chloroform in order to measure the amide II frequencies. The data show that as the concentration increases νN−H increases from 3456 to 3457 cm^{-1} while amide II (or N−H bending) increases from 1531 to 1537 cm^{-1} in 0.685 mol/l. At 0.685 mol/l concentration νN−H$\cdots$O=C bonding is observed at 3325 cm^{-1} and amide II is noted at 1560 cm^{-1}. With increasing concentration (2.74 mol/l) amide II is observed at 1563 cm^{-1}, νN−H(O=C at 3305 cm^{-1}. With increasing concentration (2.74 mol/l) νN−H is observed at 1563 cm^{-1} and νN−H$\cdots$O=C at 3305 cm^{-1}. As the νC=O and νC=O$\cdots$HN frequencies decrease from 1668–1649 cm^{-1} the amide II (or N−H bending) frequencies increase from 1531–1563 cm^{-1} with increasing concentration. The amide III mode is apparently independent of concentration, as it occurs near 1278–1279 cm^{-1}.

Table 12.1c lists IR data for the first overtone of amide II for N-methyl acetamide and N-methyl chloroacetamide in varying mole/liter concentrations in CCl$_4$ solution. These data show that the band intensity of the first overtone of amide II (or N−H bending) increases as the mole/liter concentration is increased (0.6 to 28×10^8 for N-methyl acetamide and 1.7 to 15×10^8 for N-methyl chloroacetamide). As the vN−H$\cdots$O=C frequencies decrease the amide II frequency increases with an increase in concentration. Thus, the first overtone of amide II should increase in frequency with an increase in the mole/liter concentration, and the vN−H$\cdots$O=C frequency shows decreases in frequency. Thus, the amount of Fermi resonance interaction between these two modes shows decrease, because the two modes are moving in opposite directions. As a consequence of this behavior, this absorption has no significant intensity in those amides, which are only slightly associated. Of the compounds studied, only Cl, Br, and CF$_3$ N-alkyl acetamide exhibited significant absorption in this region of the spectrum (5).

TRANS AND/OR CIS SECONDARY AMIDES

In the case of 4-, 5-, 6-, and 7-membered lactams the N−H group is cis to the carbonyl group. Table 12.1 shows that vN−H for the 5-, 6-, 7-membered lactams occurs at 3478, 3438, and 3442 cm^{-1}, respectively. It has been reported that a compound such as N-methyl acetamide exists 95% in the trans form and 5% in the cis form.

Moreover, N-tert-butyl phenylacetamide was reported to exist as 30% in the trans form and 70% in the cis form. The possibility of Fermi resonance between vN−H and the first overtone of vC=O was excluded on the grounds that the anharmonicity factor would be much too negative (8,9).

Figure 12.2 shows IR spectra of N-methyl acetamide, N-ethyl acetamide, N-isopropyl acetamide, N-tert-butyl acetamide, and acetanilide in CCl$_4$ solution in the region 3800–3300 cm^{-1}. In these cases, the vN−H stretch mode decreases in frequency in the order 3479, 3462, 3451, 3453, and 3449 cm^{-1}, respectively, and the vC=O frequency increases in the order: 1686, 1686, 1687, 1686, and 1708 cm^{-1}, respectively. The weak bands in the range 3350–3400 cm^{-1} are readily assigned as 2(vC=O) (6). The difference between the N-methyl acetamide vN−H frequency and those for the N-ethyl, N-isopropyl, N-tert-butyl, and N-phenyl analogs is 17, 28, 26, and 30 cm^{-1}, respectively. In addition, there is no other band present to indicate the presence of another rotational conformer, and the vN−H frequency decrease follows closely the inductive contribution of the N-alkyl group to the carbonyl group. Therefore, it was concluded that the N-alkyl acetamides exist in the trans configuration in dilute CCl$_4$ solution. There is no positive evidence for the existence of the cis conformer (6).

Table 12.2 lists IR data and assignments for N-alkyl acetamides and N-alkyl-α-substituted acetamides.

The data on the N-alkyl acetamides showed that v-H is sensitive to the nature of the N-alkyl group. Thus, when comparing data for N-alkyl α-haloacetamides vs those for N-alkyl acetamides it is necessary to compare data for the same N-alkyl analogs.

Comparison of the vN−H frequencies for N-butyl acetamide ($3460\,\mathrm{cm}^{-1}$) vs N-butyl α-chloroacetamide ($3433\,\mathrm{cm}^{-1}$), N-isopropyl acetamide ($3451\,\mathrm{cm}^{-1}$) vs N-isopropyl α-chloroacetamide ($3429\,\mathrm{cm}^{-1}$), and N-tert-butyl acetamide ($3453\,\mathrm{cm}^{-1}$) vs N-tert-butyl α-chloroacetamide ($3421\,\mathrm{cm}^{-1}$) shows that vN−H occurs at lower frequency by 22–$32\,\mathrm{cm}^{-1}$ in the case of the α-chloro analog. Comparison of the vC=O frequencies for N-butyl acetamide ($1688\,\mathrm{cm}^{-1}$) vs N-butyl α-chloroacetamide ($1684\,\mathrm{cm}^{-1}$), N-isopropyl acetamide ($1687\,\mathrm{cm}^{-1}$) vs N-isopropyl α-chloroacetamide ($1682\,\mathrm{cm}^{-1}$), and N-tert-butyl acetamide ($1688\,\mathrm{cm}^{-1}$) vs N-tert-butyl α-chloroacetamide ($1684\,\mathrm{cm}^{-1}$) shows that vC=O decreases in frequency by 4 to $5\,\mathrm{cm}^{-1}$.

Comparison of the vN−H frequencies for N-butyl α-chloroacetamide ($3433\,\mathrm{cm}^{-1}$) vs N-butyl α,-dichloroacetamide ($3439\,\mathrm{cm}^{-1}$), isopropyl α-chloroacetamide ($3430\,\mathrm{cm}^{-1}$) vs isopropyl α-α-dichloroacetamide ($3439\,\mathrm{cm}^{-1}$), and tert-butyl α-chloroacetamide ($3421\,\mathrm{cm}^{-1}$) vs N-tert-butyl α,α-dichloroacetamide ($3435\,\mathrm{cm}^{-1}$) shows that vN−H increases in frequency by 6 to $14\,\mathrm{cm}^{-1}$ with the addition of the second α-chloro atom.

Comparison of the vC=O frequencies for N-butyl chloroacetamide ($1684\,\mathrm{cm}^{-1}$) vs N-butyl α,α-dichloroacetamide ($1705\,\mathrm{cm}^{-1}$), N-isopropyl α-chloroacetamide ($1682\,\mathrm{cm}^{-1}$) vs N-isopropyl α,α-dichloroacetamide ($1703\,\mathrm{cm}^{-1}$), and N-tert-butyl α-chloroacetamide ($1684\,\mathrm{cm}^{-1}$) vs N-tert-butyl α,α-dichloroacetamide ($1702\,\mathrm{cm}^{-1}$) shows that vC=O increases in frequency by 18–$22\,\mathrm{cm}^{-1}$ with the addition of the second chlorine atom.

Comparison of the vN−H frequencies for N-butyl α,α-dichloroacetamide ($3439\,\mathrm{cm}^{-1}$) vs N-butyl α,α,α-trichloroacetamide ($3445\,\mathrm{cm}^{-1}$), N-isopropyl α,α-dichloroacetamide ($3430\,\mathrm{cm}^{-1}$) vs N-isopropyl α,α,α-trichloroacetamide ($3439\,\mathrm{cm}^{-1}$), and N-tert-butyl α,α-dichloroacetamide ($3427\,\mathrm{cm}^{-1}$) vs N-tert-butyl α,α,α-trichloroacetamide ($3435\,\mathrm{cm}^{-1}$) shows that vN−H increases in frequency by another 6 to $9\,\mathrm{cm}^{-1}$ with the addition of the third chlorine atom.

Comparison of the vC=O frequencies for N-butyl α,α-dichloroacetamide ($1705\,\mathrm{cm}^{-1}$) vs N-butyl α,α,α-trichloroacetamide ($1726\,\mathrm{cm}^{-1}$), N-isopropyl α,α,-dichloroacetamide ($1703\,\mathrm{cm}^{-1}$) vs N-isopropyl α,α,α-dichloroacetamide ($1725\,\mathrm{cm}^{-1}$), and N-tert-butyl α,α-dichloroacetamide ($1702\,\mathrm{cm}^{-1}$) vs N-tert-butyl α,α,α-trichloroacetamide ($1725\,\mathrm{cm}^{-1}$) shows that vC=O increases in frequency by another 21 to $23\,\mathrm{cm}^{-1}$ with the addition of the third chlorine atom.

These comparisons show that with the addition of the first α-chlorine atom on N-alkyl acetamide the vN−H frequencies decrease by 22–$32\,\mathrm{cm}^{-1}$ and the vC=O frequencies increase by 4 to $5\,\mathrm{cm}^{-1}$. With the addition of the second α-chlorine atom the vN−H frequencies increase by 6 to $14\,\mathrm{cm}^{-1}$ and the vC=O frequencies increase by 18–$22\,\mathrm{cm}^{-1}$. With the addition of the third α-chlorine atom vN−H increase by another 6–$9\,\mathrm{cm}^{-1}$ and vC=O increase by another 21–$23\,\mathrm{cm}^{-1}$.

There is only one explanation for the behavior of the vN−H and vC=O data for the N-alkyl chloroacetamide analogs. Upon addition of the first α-chlorine atom an intramolecular hydrogen bond is formed between the N−H proton and the α-chloroatom as depicted here:

The formation of the Cl$\cdots$HN bond causes vN$-$H to decrease in frequency while the vC$=$O mode is only increased by 4 to 5 cm^{-1}. With the addition of the second and third chlorine atoms the vN$-$H frequencies increase progressively by 6 to 14 cm^{-1} and 6 to 9 cm^{-1} while the vC$=$O frequencies increase progressively by 18 to 22 cm^{-1} and 21 to 23 cm^{-1}. The vN$-$H frequency increases with the addition of the second and third α-chlorine atoms and is the result of the inductive effect. In the case of the vC$=$O frequencies substitution of the second and third chlorine atoms is a combination of the inductive effect and the field effect between the carbonyl oxygen atom and the two gauche chlorine atoms (6).

Comparison of the vapor and CCl$_4$ solution data for N-isopropyl acetamide (3460 vs 3451 cm^{-1} and 1714 vs 1687 cm^{-1}) and N-tert-butyl α,α,α-trichloroacetamide (3444 vs 3435 cm^{-1} and 1743 vs 1725 cm^{-1}) shows that both vN$-$H and vC$=$O occur at higher frequency in the vapor. This most likely reflects the effect of N$-$H$\cdots$ClCCl$_3$ interaction in the solution phase.

The vN$-$H vibration occurs at the lowest frequency for the N-phenyl analogs of the forementioned four series of acetamides. The N-phenyl analogs of these series exhibit vN$-$H at the lowest frequency in the order: CH$_3$ (3449 cm^{-1}), ClCH$_2$ (3409 cm^{-1}), Cl$_2$CH (3419 cm^{-1}), and Cl$_3$C (3425 cm^{-1}), and exhibit vC$=$O at the highest frequency in the order: CH$_3$ (1708 cm^{-1}), ClCH$_2$ (1692 cm^{-1}), Cl$_2$CH (1713 cm^{-1}), and CCl$_3$ (1731 cm^{-1}). The inductive effect of the phenyl group tightens the C$=$O bond, thereby raising its vC$=$O frequency, and the N$-$H group becomes more acidic, causing it to form a stronger intramolecular N$-$H$\cdots$Cl bond. The vN$-$H and vC$=$O frequency behavior with α-chloro substitution is comparable to that exhibited by the N-alkyl α-chlorinated analogs.

N$-$H INTENSITY

A study of Table 12.2 shows that the apparent intrinsic integrated absorption (B $\times$ 10^8)* for the N$-$H stretching vibration is significantly raised by the intramolecular N$-$H$\cdots$X bond, while the intensity is influenced only to a relatively small extent by the inductive effect.

In saturations of 10% or $<$10%, CCl$_4$ solutions bonded vN$-$H$\cdots$O$=$C occur in the region 3280$-$3400 cm^{-1} (see Table 12.2a). For those compounds that exhibited significant "amide II" overtone absorption, both the observed frequencies and frequencies corrected for Fermi resonance with the "amide II" overtone are listed.

The frequency separation between the vN$-$H frequency and the intermolecular vN$-$H$\cdots$O$=$C frequency varies between 60 and 195 cm^{-1}. The intramolecularly hydrogen-bonded secondary amides exhibit less intermolecular H-bonding than the simple N-alkyl acetamides, and exhibit higher bonded vN$-$H frequencies.

In contrast, pyrrolidone is a cyclic secondary amide whose N$-$H group is cis with the C$=$O group. In this case, the vN$-$H frequency is nearly independent of concentration, ranging from

---

*B $\times$ 10^8 [Intensity B $=$ (1/CI) ln (I$_o$/I)dv in CM^2 molecule^{-1} s^{-1}, and represents the apparent intrinsic absorption for vN$-$H].

3207–3219 cm^{-1} at concentrations ranging from 0.01–1.0 M in CCl$_4$ solution. This indicates that only one bonded species is present: the cis dimer (10).

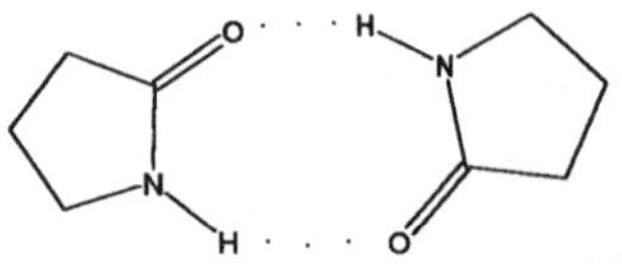

pyrrolidone cis intermolecular hydrogen bonded dimer

Table 12.3 lists IR data for N-alkyl α-substituted acetamides in 10% wt./vol. and 0.002 M CCl$_4$ solutions. In all cases, except for tert-butyl phenoxy-acetamide and tert-butyl trichloroacetamide, the νC=O or "Amide I" mode occurs at higher frequency in dilute solution. Within each series, the tert-butyl analogs show the least frequency shift in going from 10% wt./vol. to 0.002 M in CCl$_4$ solution; and this is most likely due to the steric effect of the tert-butyl group, which prevents the N–H proton of another amide molecule from forming as strong an intermolecular hydrogen bond as in the case of the other N-alkyl analogs, or prevents intermolecular hydrogen bonding as in the case of tert-butyl trichloroacetamide.

Table 12.4 shows a comparison of primary and secondary amides in the solid phase. In this case, the νasym. NH$_2$ frequencies occur in the range 3310–3380 cm^{-1} and νsym. NH$_2$ frequencies occur in the range 3150–3195 cm^{-1} while νN–H for the secondary amides occur in the range 3242–3340 cm^{-1}. The primary amides exhibit NH$_2$ bending in the region 1620–1652 cm^{-1} while the secondary amides exhibit N–H bending in the region 1525–1550 cm^{-1}. Inductive and resonance effects upon the νC=O frequency are also apparent in the solid phase.

Table 12.4a lists IR data for the νC=O frequencies for tertiary amides in the neat or solid phase. Formamides have the basic structure:

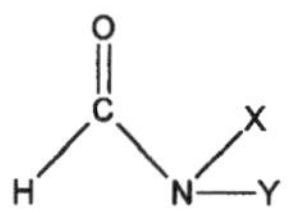

In the neat phase, when X and Y are C$_2$H$_5$ groups, νC=O occurs at 1665 cm^{-1}. When X is C$_2$H$_5$ and Y is C$_6$H$_5$, νC=O occurs at 1670 cm^{-1}. When X is C$_6$H$_5$ and Y is C$_6$H$_5$, νC=O occurs at 1690 cm^{-1}. The increase in the νC=O frequency when C$_6$H$_5$ is substituted for C$_2$H$_5$ is attributed to the larger inductive effect of C$_6$H$_5$ vs C$_2$H$_5$. Tertiary amides have the following basic structure:

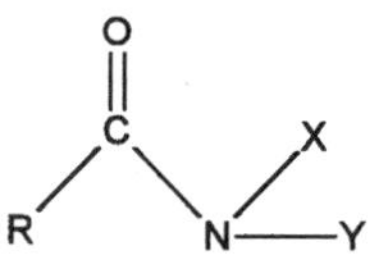

The tertiary acetamides exhibit νC=O in the region 1650–1670 cm^{-1}, and the inductive effect of the X or Y alkyl and phenyl groups is also apparent. In the vapor phase, νC=O for tertiary amides occur 40 ± 5 cm^{-1} higher in frequency than they occur in the neat or solid phase. The lower

vC=O frequencies in the condensed phases most likely result from dipolar interaction between molecules such as the one illustrated here; this interaction weakens the C=O bond.

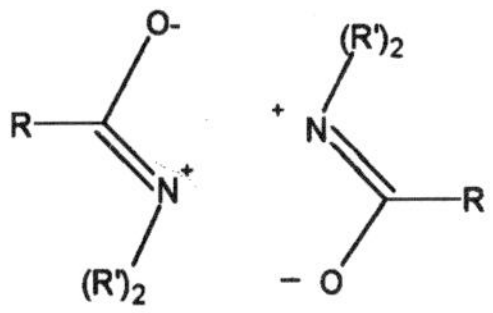

Table 12.5 shows a comparison of Raman data and assignments for acrylamide, methacrylamide, and their polymers. Study of Table 12.5 indicates that as the average molecular weight of polyacrylamide increases, the vC=O frequency increases from 1658 to 1668 cm^{-1}, while the Raman relative band intensity decreases. The N-alkyl acrylamides exhibit a weak Raman band in the region 1657–1659 cm^{-1} and a strong Raman band in the region 1627–1629 cm^{-1}, which are assigned to vC=O and vC=C, respectively. The N,N-dialkyl or N-alkyl, N-vinyl acrylamides exhibit a weak to strong Raman band in the region 1648–1666 cm^{-1} and a strong Raman band in the region 1609–1625 cm^{-1}, which are assigned to vC=O and vC=C, respectively. The inductive effect of the N-vinyl group apparently affects both the frequency and Raman band intensity of vC=O.

Methacrylamide exhibits a weak Raman band at 1673 cm^{-1}, which is 9 cm^{-1} lower in frequency than vC=O for acrylamide. This decrease in the vC=O frequency is attributed to the inductive effect of the α-methyl group. The medium-strong Raman band at 1647 cm^{-1} is assigned as vC=C for methacrylamide while the strong Raman band at 1639 cm^{-1} is assigned as vC=C for acrylamide. Apparently the inductive effect of the α-methyl group in polymethacrylamide causes vC=O to occur at lower frequency than vC=O for polyacrylamide.

Table 12.5a lists Raman data for N-alkyl or N-aryl acrylamides and methacrylamides in the neat phase. The most distinguishing features between the N-alkyl or N-aryl acrylamides and the N-alkyl or N-aryl methacrylamides is the relative Raman band intensities of vC=O/vC=C. In the case of the acrylamides it varies between 0.06 and 0.8 and in the case of the methacrylamides it varies between 0.11 and 3. Moreover, the frequency separation between vC=O and vC=C varies between 29–35 cm^{-1} for the acrylamides and varies between 35–39 cm^{-1} for the methacrylamides.

A strong Raman band in the range 875–881 cm^{-1} most likely results from a symmetric C−C−C skeletal stretching vibration of the group for the methacrylamide while a medium-strong Raman band in the range 1236–1256 cm^{-1} results from a skeletal vibration of the =C−C(=)−N group for the acrylamides.

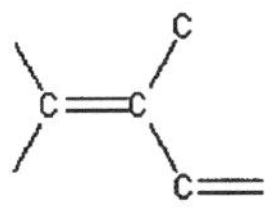

Table 12.6 lists IR data for p-x-acetanilides and p-x, α-haloacetanilides. Study of Table 12.6 shows that in general all vC=O modes occur at lower frequency in CHCl$_3$ solution than in CCl$_4$ solution. Part of the vC=O frequency decrease is the result of intermolecular hydrogen bonding (C=O$\cdots$HCCl$_3$). It should be noted that the vC=O frequencies increase in the order α-bromo-p-x-acetanilide, p-x-acetanilide, and α,α,α-trichloro-p-x-acetanilide. The α-bromo analogs and the

α,α,α-trichloro analogs in dilute solution exist in the intramolecular hydrogen bonded form as depicted here:

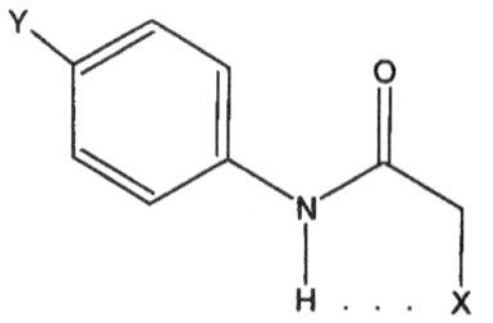

As discussed previously, substitution of the first α-halogen atom on the α-carbon atom decreases the νC=O frequency. Addition of the second and third halogen atoms on the α-carbon atom additively increases the νC=O frequency due to both inductive and field effects (7).

Figure 12.3 shows plots of the νC=O frequencies of α-bromo-p-x-acetanilide, p-x-acetanilide, and α,α,α-trichloro-p-x-acetanilide in 0.002 M solutions or less in CCl$_4$ and CHCl$_3$ vs the Hammett σ_p values. All six plots show that the νC=O mode increases linearly in frequency as the σ_p value increases. This is the result of the greater tendency of the para substituent to attract electrons (higher σ_p values), the smaller the nitrogen nonbonding electron pair. The carbonyl stretching frequency decreases as the electron density on the nitrogen atom increases as a result of an increased tendency toward shifting the mesomeric equilibrium:

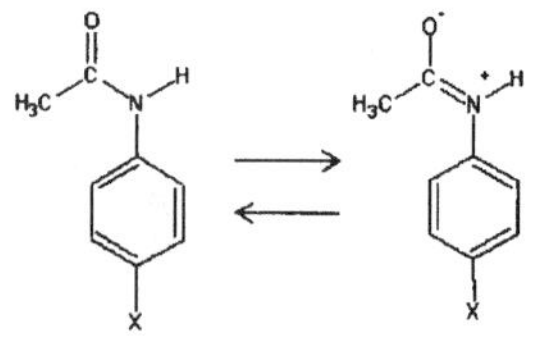

For example, the higher the nitrogen electron density, the greater the tendency to impart double-bond character to the C—N bond at the expense of double-bond character of the C=O bond, thus reducing the C=O force constant (7).

ACETANILIDES

It has been shown that the IR νC=O band intensities for p- and m-substituted acetanilides are a function of Hammett σ_p and σ_m values (11). The higher the σ_p or σ_m values the lower the νC=O IR band intensities. Examples given here are from Reference (7):

p-x	p-x-acetanilide	α-bromo-p-x-acetanilide	α,α,α-trichloro-p-x-acetanilide
CH$_3$O	14.4	15.10	13.42
Cl	12.56	14.41	12.93
NO$_2$	9.40	11.15	11.55

These absorbance values are $\times 10^7$ cm^2 molecule^{-1} s^{-1}

Table 12.7 lists IR data for the N—H stretching frequencies for p-x-acetanilides and p-x, α-haloacetanilides and their absorbance values ($\times 10^7$ cm^2 molecule^{-1} s^{-1}) obtained in CHCl$_3$ solution. The αN—H frequencies generally occur at lower frequency in CHCl$_3$ solution than in

CCl_4 solution. The νN–H frequencies in CCl_4 solution decrease in the order p-x-acetanilide (3436–3445 cm^{-1}), α,α,α-trichloro-p-x-acetanilide (3415–3422 cm^{-1}), and α-bromo-p-x-acetanilide (3391–3405). In $CHCl_3$ solution they occur in the same order, 3429–3440 cm^{-1}, 3406–3415 cm^{-1}, and 3387–3395 cm^{-1}.

Study of the absorbance values shows that in general they increase progressing in the order p-x-acetanilide, α-bromo-p-x-acetanilide, and α,α,α-p-x-acetanilide. There is a general but not systematic increase in the absorbance values as the σ_p values increase. These data support that an intramolecular hydrogen bond is formed between N–H$\cdots$X as already depicted here. In the case of the α-Br analog, the Br atom is larger than say a Cl atom, and the H$\cdots$Br distance would be shorter than the H$\cdots$Cl distance. A stronger intramolecular hydrogen bond would be formed as the distance is decreased, causing νN–H$\cdots$Br to decrease in frequency by approximately 40 cm^{-1}. These data show that as the N–H proton becomes more acidic, the νN–H frequency decreases (7).

N-ALKYL BENZAMIDES

Table 12.8 lists the NH and C=O stretching frequencies for N-alkyl p-methoxybenzamide, N-alkyl p-chlorobenzamide, and N-alkyl methyl carbamate in CCl_4 solutions (6). The absorbance values are $\times 10^8$ cm^2 molecule^{-1} s^{-1} for the NH group.

The N-alkyl p-x-benzamides are also affected by the nature of the N-alkyl group. The νN–H frequencies generally decrease with branching on the N-alkyl α-carbon atom.

The νC=O frequencies for the N-alkyl p-methoxybenzamides occur in the range 1670–1676 cm^{-1} and for N-alkyl p-chlorobenzamides in the range 1671–1681 cm^{-1}. Closer examination shows that in all cases νC=O for n-alkyl p-methoxybenzamide occurs at a lower frequency than that for the comparable N-alkyl p-chloro-benzamide (N-methyl vs N-methyl, etc.). This is the result of Hammett σ_p values. The σ_p values also appear to affect the band intensities, because a higher σ_p value increases the band intensity.

N-ALKYL METHYL CARBANATES

The N-alkyl methyl carbanates exhibit νN–H in the region 3449–3478 cm^{-1}; and the frequencies tend to decrease with branching on the α-carbon atom. The N-phenyl analog exhibits νNH at 3450 cm^{-1}. The νC=O frequencies for the N-alkyl analogs occur in the range 1730–1738 cm^{-1} while νC=O for the N-phenyl analog occurs at 1748 cm^{-1} due to the inductive effect of the phenyl ring (6).

N,N'-DIALKYL OXAMIDES AND N,N'-DIARYL OXAMIDE

Table 12.9 lists IR and Raman data for both N,N'-dialkyl oxamides and N,N'-diaryl oxamides (12). In the case of IR, the samples were prepared as either split mulls or KBr pellets. Raman spectra were recorded of the solid samples. The out-of-phase (N–H)$_2$ stretching vibration (a B$_{1u}$ mode) is IR active and in the case of the N,N'-dialkyl analogs occurs in the range 3279–

$3311\,\text{cm}^{-1}$ while for the N,N'-diaryl analogs it occurs in the range $3295\text{--}3358\,\text{cm}^{-1}$. The in-phase $(\text{N-H})_2$ stretching vibration (an Ag mode) is Raman active and in the case of the N,N'-dialkyl analogs it occurs in the range $3302\text{--}3325\,\text{cm}^{-1}$ while for the N,N'-diaryl analogs it occurs in the range $3320\text{--}3349\,\text{cm}^{-1}$. The IR band occurring in the range $1643\text{--}1660\,\text{cm}^{-1}$ for N,N'-dialkyl oxamides is assigned to the amide I mode (in this case, out-of-phase $(\text{C=O})_2$ stretching) while for the N,N'-diaryl analog it occurs in the range $1662\text{--}1719\,\text{cm}^{-1}$. In the solid state, a weak shoulder appears in the region $1628\text{--}1638\,\text{cm}^{-1}$ of the IR spectrum. It is the result of crystal splitting, as it is not present in solution phase or vapor-phase spectra (12). The in-phase $(\text{C=O})_2$ stretching A_g mode is assigned to the Raman band occurring in the range $1686\text{--}1695\,\text{cm}^{-1}$ in the case of the N,N'-dialkyl analogs while in the case of the N,N'-diaryl analogs it occurs in the range $1682\text{--}1741\,\text{cm}^{-1}$.

The Amide II mode—the out-of-phase $(\text{N-H})_2$ bending (B_{1u}) mode and the in-phase $(\text{N-H})_2$ bending (A_{1g}) mode—occur in the range $1508\text{--}1541\,\text{cm}^{-1}$ and $1547\text{--}1567\,\text{cm}^{-1}$ for the N,N'-dialkyl oxamides, respectively, while for the N,N'-diaryl analogs they occur in the range $1484\text{--}1520\,\text{cm}^{-1}$ and $1537\text{--}1550\,\text{cm}^{-1}$, respectively. In all cases discussed here the A_g modes occur at higher frequency than the corresponding B_{1u} mode. These IR and Raman data indicate that the N,N'-diaryl oxamide exist in an intermolecularly hydrogen-bonded trans configuration in the solid state where each oxamide group can be viewed as having C_{2h} symmetry (12). In the vapor phase N,N'-dimethyloxamide also exists in the trans configuration (12).

The N,N'-dialkyloxamide exhibits B_{1u} modes in the range $1223\text{--}1251\,\text{cm}^{-1}$ (Amide III), $725\text{--}782\,\text{cm}^{-1}$ (Amide IV), $532\text{--}619\,\text{cm}^{-1}$ (Amide VI), and A_g modes in the region $1298\text{--}1315\,\text{cm}^{-1}$ (Amide III), and $1130\text{--}1210\,\text{cm}^{-1}$ $((\text{C-N})_2$ stretching) (12).

DIMETHYLACETAMIDE AND TETRAALKYLUREA

Table 12.10 lists IR data and assignments for dimethylacetamide and tetraalkylurea in different physical phases (13,14). These data show that the C=O stretching frequency of dimethylacetamide or tetraalkylurea is sensitive to the physical phase, and that in $\text{n-C}_6\text{H}_{14}/\text{CHCl}_3$ solutions clusters of tetramethylurea exist that exhibit $v\text{C=O}$ at 1650.9 and $1635.4\,\text{cm}^{-1}$ in 14 mol % $\text{n-C}_6\text{H}_{14}/\text{CHCl}_3$ and at 1640.2 and $1629.5\,\text{cm}^{-1}$ in 80.6 mol % $\text{n-C}_6\text{H}_{14}/\text{CHCl}_3$ solutions.

In CCl_4 and/or CHCl_3 solutions, the behavior of the C=O stretching mode is explained on the basis of hydrogen-bonded complexes between solvent-solvent and solute-solvent and the bulk dielectric effects of the solvents. Figure 12.4 shows plots of $v\text{C=O}$ for 1 wt % solutions of acetone, dimethylacetamide, and tetramethylurea vs mole % $\text{CHCl}_3/\text{CCl}_4$. The plots become more complex proceeding in the series acetone, dimethylacetamide, and tetramethylurea. In the case of acetone, the nonlinear segment is attributed to the complex $(\text{CH}_3)_2\,\text{C=O}\cdots\text{HCCl}_3$, and the first nonlinear segments in the other two plots are attributed to $\text{C=O}\cdots\text{HCCl}_3$ complexes with dimethylacetamide and tetraalkylurea. The second break in the case of dimethylacetamide is attributed to intermolecular hydrogen bonding between both $\text{Cl}_3\text{CH}\cdots\text{O=C-N}(\cdots\text{HCCl}_3$. The third break may be due to a complex such as $(\text{Cl}_3\text{CH}\cdots)_2\text{O=C-N}\cdots\text{HCCl}_3$ and/or $\text{Cl}_3\text{CH}\cdots\text{O=C-N}(\cdots\text{HCCl}_3)_2$. The additional break in N,N,N',N'-tetramethylurea is attributed to the additional −N-group, which also is capable of forming a $\text{Cl}_3\text{CH}\cdots\text{N}$ bond.

It is interesting to compare the $v\text{C=O}$ frequencies of tetramethylurea (TMU), tetraethylurea (TEU), and tetrabutyl urea (TBU) in CCl_4 solution (1652.9, 1646.2, and $1643.2\,\text{cm}^{-1}$, respectively, and in CHCl$_3$ solution (1627.3, 1620.1, and 1616.1 cm^{-1}, respectively). The frequency difference between these two solvents increases in the order 25.6, 26.1, and 27.0 cm^{-1} for TMU, TEU, and TBU, respectively. In addition, it is noted that the C=O stretching frequency decreases in the order TMU, TEU, and TBU. These data show that as the alkyl groups contribute more of an inductive contribution to the C=O group the vC=O mode decreases in frequency and the strength of the C=O$\cdots$HCCl$_3$ bond increases (14).

Table 12.11 lists IR vapor-phase data for urea, thiourea, and guanidine derivatives. These data show that the 1,1,3,3-tetraalkylureas exhibit a weak IR band in the range 3338–3358 cm^{-1}, which results from the first overtone of C=O stretching. As $2v$C=O occurs at lower frequency than the calculated $2v$C=O frequency in each case, the $2v$C=O mode exhibits positive anharmonicity.

The 1,1,3,3-tetraethylthiourea exhibits a band at approximately 1085 cm^{-1} that is assigned to C=S stretching, vC=S. The 1619 cm^{-1} band for 1,1,3,3-tetramethylquanidine is assigned to C=N stretching, vC=N. All of the compounds listed in Table 12.11 exhibit a band in the range 1239–1334 cm^{-1}, which most likely results from an antisymmetric NCN stretching mode.

1,1,3,3-TETRAMETHYLUREA VS 1,3-DIMETHYL-2-IMIDAZOLIDINONE

Table 12.12 lists IR data for 1,1,3,3-tetramethylurea and 1,3-dimethyl-2-imidazolidinone in various solvents at 1% wt./vol. (14–16). These two compounds have the following empirical structures:

TMU   DMI

In all solvents, the vC=O mode for DMI occurs at higher frequency than vC=O for TMU. The only chemical difference between TMU and DMI is that a (CH$_2$)$_2$ group has been substituted for two CH$_3$ groups, and the chemical difference is minimal. The C=O group occurs at a significantly higher frequency in the case of DMI due to geometric restrictions of the 5-membered ring, which makes it more difficult for the carbonyl carbon atom to move in and out of the ring during a cycle of vC=O. (This is often referred to as ring strain. In any vC=O mode, the bond angle must change to some degree, and the smaller the X−C−Y angle the more difficult it is for the C=O bond to vibrate.)

In Table 12.12 under the solvent heading, the neat phase is listed as 1, hexane 2, sequentially to methyl alcohol as 20. These numbers are used to show for which solvent or next phase the particular data were recorded or the data difference was determined.

Figure 12.5 shows plots of $\nu C{=}O$ of DMI and TMU in the neat phase or in 1 of the 19 solvents vs $\nu C{=}O$ (hexane) minus $\nu C{=}O$ (solvent). These two plots are linear, and any set of numbers treated in the same mathematical way will yield a linear relationship. The important point to note in these two plots is that the number in each plot is not in the identical sequence. These differences suggest that the solute-solvent interaction is not comparable in all cases (16). This suggests that the steric factor of the $(CH_3)_2$ or $(CH_3)_4$ and $(CH_2)_2$ groups alters the spatial distance between solute and solvent.

Figure 12.6 shows plots of $\nu C{=}O$ for TMU and DMI vs the solvent acceptor number (AN). The numbers 17' through 20' are for the $\nu C{=}O{\cdots}HOR$ frequencies for these compounds in tertiary butyl alcohol, isopropyl alcohol, ethyl alcohol, and methyl alcohol, respectively. The numbers 17 through 20 are for $\nu C{=}O$ in these same alcohols, but where the $C{=}O$ groups are not intermolecularly hydrogen bonded. This indicates that intermolecularly bonded alkyl alcohols (R−OH) can cluster in surrounding TMU or DMI without forming intermolecular hydrogen bonds with the solute. Projection of these points by dashed lines onto the lower lines indicates that the AN values for the alcohols are much lower than the values determined by NMR (17). These projected AN values for the alcohols are comparable to the AN values for alkyl ethers.

Figure 12.7 shows plots of $\nu C{=}O$ for DMI and TMU vs mole % $CHCl_3/CCl_4$. Both plots show similar a, b, c, and d segments. It has been suggested that the concentration of DMI or TMU molecules are in equilibrium with the concentration of $CHCl_3$ and/or CCl_4 molecules in regions (a) through (d). In addition, as the mole % $CHCl_3/CCl_4$ is increased from 0 to 100, the $(CHCl_3)$ one, two, three or four complexes are replaced by $(CCl_3H{:}ClCl_2CH)_x$ complexes:

Region (a) DMI or TMU$\cdots(HCCl_3)_1$
Region (b) DMI or TMU$\cdots(HCCl_3)_2$
Region (c) DMI or TMU$\cdots(HCCl_3)_3$
Region (d) DMI or TMU$\cdots(HCCl_3)_4$

Complexes between CCl_4 and TMU or DMI would also decrease the $\nu C{=}O$ frequencies.

Different complexes are reported to be formed between DMI or TMU and mole % n-C_6H_{14}/CCl_4 or mole % n-$C_6H_{14}/CHCl_3$ solutions (15,16). Data in Table 12.13 show that in mole % $CHCl_3/CCl_4$ solutions only one $\nu C{=}O$ frequency is observed, in mole % $CCl_4/$n-C_6H_{14} solutions two $\nu C{=}O$ frequencies are observed, and in mole % $CHCl_3/$n-C_6H_{14} solutions three $\nu C{=}O$ frequencies are observed. These are due to the presence of different DMI clusters in these various complex solution mixtures.

CAFFEINE, ISOCAFFEINE, 1,3,5-TRIMETHYLURACIL, 1,3,6-TRIMETHYLURACIL, AND 1,3-DIMETHYL-2,4-(1H, 3H) QUINAZOLINEDIONE

Table 12.14 lists the in-phase and out-of-phase $(C{=}O)_2$ stretching frequencies for caffeine, isocaffeine, 1,3,5-trimethyluracil, 1,3,6-trimethyluracil, and 1,3-dimethyl-2,4-(1H,3H) quinazolinedione in CCl_4 and $CHCl_3$ solutions (18). All of these compounds contain the dimethyl cyclic 6-membered ring as shown here.

The two C=O stretching modes couple into in-phase $(C=O)_2$, ν_{ip} $(C=O)_2$, and out-of-phase $(C=O)_2$ stretching, ν_{op} $(C=O)_2$.

The ν_{ip} (C=O) mode for caffeine occurs at $1721\ cm^{-1}$ in the vapor phase, at $1710.6\ cm^{-1}$ in CCl_4 solution, and at $1708.9\ cm^{-1}$ in $CHCl_3$ solution while the ν_{op} $(C=O)_2$ mode occurs at $1685\ cm^{-1}$ in the vapor phase, $1667.5\ cm^{-1}$ in CCl_4 solution, and at $1658.4\ cm^{-1}$ in $CHCl_3$ solution. With change in phase, the ν_{op} $(C=O)_2$ mode shifts more in frequency than the ν_{ip} $(C=O)_2$ mode (17.5, 26.6, and $9.1\ cm^{-1}$) vs (10.4, 12.1, and $1.7\ cm^{-1}$).

In CCl_4 solution, caffeine, isocaffeine, 1,3,5-trimethyl uracil, 1,3,6-trimethyluracil, and 1,3-dimethyl-2,4-(1H,3H)-quinazolinedione exhibit ν_{ip} $(C=O)_2$ in the range 1706.6–$1716.3\ cm^{-1}$, and in $CHCl_3$ solution in the range 1700–$1711.3\ cm^{-1}$.

The ν_{op} $(C=O)_2$ mode for 1,3,5-trimethyluracil and 1,3-dimethyl-2,4-(1H,3H)-quinazoline-dione is in Fermi resonance with a combination or overtone, and in these two cases ν_{op} $(C=O)_2$ was corrected for F.R. Therefore, for these five compounds $\nu_{op}(C=O)_2$ occurs in the range 1663.4–$1669\ cm^{-1}$ in CCl_4 solution and in the range 1652.4–$1664\ cm^{-1}$ in $CHCl_3$ solution (18).

Table 12.15 lists IR data for uracils in the solid phase. In the solid phase, the uracils not substituted in the 1,3-position (contain two N−H groups) exhibit ν_{ip} $(C=O)_2$ in the range 1704–$1750\ cm^{-1}$ and ν_{op} $(C=O)_2$ in the range 1661–$1704\ cm^{-1}$. The 1,3,5-trimethyl and 1,3,6-trimethyl analogs exhibit both ν_{ip} $(C=O)_2$ and ν_{op} $(C=O)_2$ at lower frequency in the solid phase than in either CCl_4 or $CHCl_3$ solution. The frequency separation between ν_{ip} $(C=O)_2$ and ν_{op} $(C=O)_2$ varies between 23 and $80\ cm^{-1}$ in the solid phase.

IMIDES

Table 12.16 lists IR vapor-phase data and assignments for imides (1).

The imides exhibit an in-phase $[\nu_{ip}$ $(C=O)_2]$ and out-of-phase $[\nu_{op}$ $(C=O)_2]$ stretching vibrations. The ν_{ip} $(C=O)_2$ mode has weaker absorbance than the absorbance for ν_{op} $(C=O)_2$. On this basis ν_{ip} (C=O), and ν_{op} $(C=O)_2$ for diacetamide, $[CH_3C(=O)]_2NH$, are assigned at 1738 and $1749\ cm^{-1}$, respectively, in the vapor phase. All of the other imides included in Table 12.16 exhibit ν_{ip} (C=O), at higher frequency than ν_{op} $(C=O)_2$. For example, N-phenyldibenza-mide, $C_6H_5(C=O)_2NC_6H_5$, exhibit ν_{ip} $(C=O)_2$ and ν_{op} $(C=O)_2$ at 1801 and $1702\ cm^{-1}$, respectively.

Succinimides, maleimides, and phthalimides contain 5-membered rings:

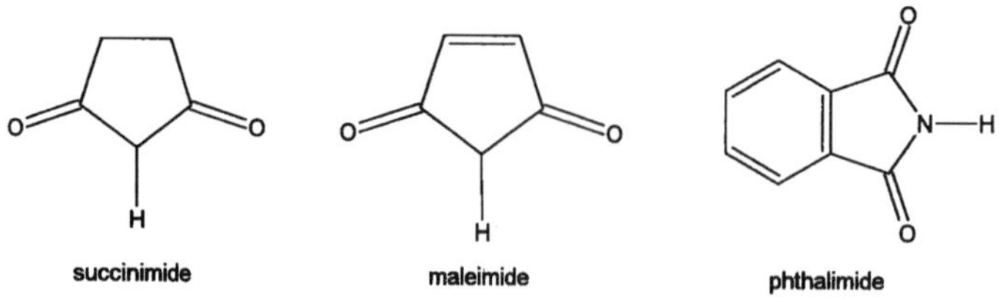

succinimide   maleimide   phthalimide

hydantoin also contains a similar type imide structure:

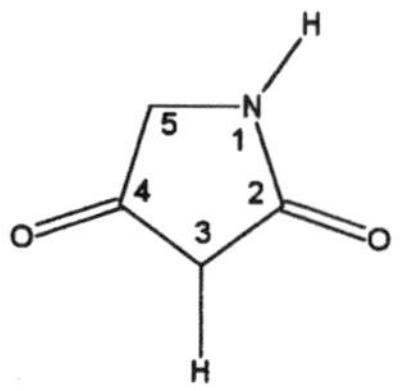

Succinimide exhibits ν_{ip} (C=O)$_2$ and ν_{op} (C=O)$_2$ at 1820 and 1772 cm^{-1}, respectively, while its N-(2,6-xylyl) analog exhibits ν_{ip} (C=O)$_2$ and ν_{op} (C=O)$_2$ at 1790 and 1741 cm^{-1}, respectively. In this case, the 2,6-xylyl group can not be coplanar with the N(−C=O)$_2$ group due to the steric factor of the 2,6-dimethyl groups. The N-alkylmaleimides and N-arylmaleimides exhibit ν_{ip} (C=O)$_2$ and ν_{op} (C=O)$_2$ in the range 1780–1820 cm^{-1} and 1730–1745 cm^{-1}, respectively.

Phthalimide exhibits ν_{ip} (C=O)$_2$ and ν_{op} (C=O)$_2$ at 1795 and 1766 cm^{-1} in the vapor phase, respectively. Substitution of a 3-or-4-nitro group raises both ν(C=O)$_2$ modes, which most likely is the result of the σ_p and σ_m effect. Substitution of a 3-amino group lowers both modes significantly, and this most likely is the combined effect of intramolecular hydrogen bonding HNH$\cdots$O=C and the σ_m effect.

N-alkylphthalimides exhibit ν_{ip} (C=O)$_2$ and ν_{op} (C=O)$_2$ in the range 1771–1789 cm^{-1} and 1734–1740 cm^{-1}, respectively, while the N-aryl analogs exhibit ν_{ip} (C=O)$_2$ and ν_{op} (C=O)$_2$ in the range 1795–1800 cm^{-1} and 1735–1746 cm^{-1}, respectively.

Hydantoins in the vapor phase exhibit ν_{ip} (C=O)$_2$ and ν_{op} (C=O)$_2$ at 1826 and 1785 cm^{-1}, respectively. The 5,5-dimethylhydantoin exhibits ν_{ip} (C=O)$_2$ at 1811 cm^{-1} and ν_{op} (C=O)$_2$ at 1775 cm^{-1} in the vapor phase; however, in the solid phase ν_{ip} (C=O)$_2$ occurs at 1779 cm^{-1}, and ν_{op} (C=O)$_2$ occurs as a doublet (1744 and 1716 cm^{-1}). Apparently in the solid phase the ν_{op} (C=O)$_2$ mode is split by crystalline effects.

The s-triazine (1H,3H,5H)-triones have the basic structure:

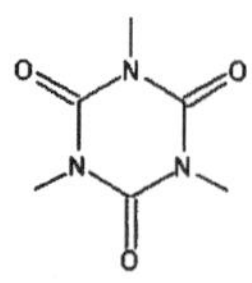

and the substituent groups are joined to the N atoms (hydrogen, aliphatic or aromatic groups). For the three compounds recorded in the vapor phase, ν_{op} (C=O)$_2$ occurs in the range 1710–1720 cm^{-1}. The ν_{ip} (C=O)$_2$ mode for the 1,3-diphenyl-5-octyl analog occurs at 1766 cm^{-1}.

All of these imide type compounds exhibit a weak band in the range 3470–3570 cm^{-1}, which is assigned to the combination tone v_{ip} (C=O)$_2$ + v_{op} (C=O)$_2$. In the vapor phase, imides that contain N—H exhibit a weak band in the range 3415–3455 cm^{-1} assigned to vN—H. In the case of hydantoins, vN—H occurs in the range 3482–3490 cm^{-1} in the vapor phase, and in the range 3205–3215 cm^{-1} in the solid phase. Of course in the solid phase hydantoins exist in an intermolecular hydrogen-bonded state.

4-BROMOBUTYLPHTHALIMIDE IN SOLUTION

Table 12.17 lists IR data for 4-bromobutylphthalimide in CCl$_4$ and/or CHCl$_3$ solutions (19). These data show that as the mole % CHCl$_3$/CCl$_4$ is increased from 0 to 100 both v_{ip} (C=O)$_2$ and v_{op} (C=O)$_2$ decrease in frequency. In addition, the v_{op} (C=O)$_2$ mode decreases more than v_{ip} (C=O)$_2$ by a factor of ~3.5.

HYDANTOINS

Table 12.18 lists IR data and assignments for hydantoins in the vapor and solid phases. The vN—H frequencies occur in the range 3485–3495 cm^{-1} in the vapor and in the range 3150–3330 cm^{-1} in the solid phase.

In the vapor phase, v_{ip} (C=O)$_2$ occurs in the range 1808–1825 cm^{-1} and in the range 1755–1783 cm^{-1} in the solid phase while v_{op} (C=O)$_2$ occurs in the range 1774-1785 cm^{-1} in the vapor phase and in the range 1702–1744 cm^{-1} in the solid phase. In the four cases where the same hydantoin was studied in both the vapor and solid phases, the frequency separation between v_{ip} (C=O)$_2$ and v_{op} (C=O)$_2$ is much less in the vapor phase (34–40 cm^{-1}) than it is in the solid phase (54–66 cm^{-1}).

TRIALLYL-1,3,5-TRIAZINE-2,4,6-(1H,3H,5H) TRIONE

Table 12.19 lists IR data and assignments for triallyl-1,3,5-triazine-2,4,6-(1H,3H,5H) trione in CHCl$_3$ and/or CCl$_4$ solutions (20). The empirical structure of this compound is

where R is allyl, and for simplicity we will name this compound T$_3$one. T$_3$one has been determined to have D$_{3h}$ symmetry, the v_{ip} (C=O)$_3$ mode belongs to the E' species, and the v_{ip} (C=O)$_3$ mode belongs to the A' species. The v_{op} (C=O)$_3$ mode is doubly degenerate, and is IR active while v_{ip} (C=O)$_3$ is Raman active. T$_3$one in 1% wt./vol. CCl$_4$ solution exhibits a strong Raman band at 1761.2 cm^{-1}, at 5% wt./vol. CCl$_4$ solution at 1762.3 cm^{-1}, and at 10% wt./vol.

CCl_4 solution at 1763.0 cm^{-1}. In CHCl$_3$ solutions at the same % wt./vol. concentrations, strong Raman bands occur at 1758.3, 1760.5, and 1761.4 cm^{-1}, respectively. Therefore, in both CCl$_4$ and CHCl$_3$ solution the Raman band increases in frequency with an increase in the % wt./vol. T$_3$one. A corresponding IR band is not observed, and this Raman band is assigned as v_{ip} (C=O)$_3$.

In the case of v_{op} (C=O)$_3$ for T$_3$one, 1 and 5% vol./wt. solutions in CCl$_4$ show that it decreases in frequency from 1698.6 cm^{-1} to 1698.5 cm^{-1} while at these same concentrations in CHCl$_3$ it decreases from 1695.7 cm^{-1} to 1695.5 cm^{-1}.

The v_{op} (C=O)$_3$ mode for T$_3$one decreases in a nonlinear manner as the mole % CHCl$_3$/CCl$_4$ is increased from 0 to 100 as shown in Figure 12.8. These data indicate that different T$_3$one $\cdots$(HCCl$_3$)$_n$ complexes are formed as the mole % CHCl$_3$/CCl$_4$ is increased, and that in general the v_{op} (C=O)$_3$ frequency decreases with increase in the mole % CHCl$_3$/CCl$_4$ are due to the dielectric effect of each particular solvent system.

A strong IR band assigned to v_{op} (C$_3$N$_3$) is assigned in the region 1455.2–1458.6 cm^{-1}. This mode increases in frequency in almost a linear manner as the mole % CHCl$_3$/CCl$_4$ is increased (see Figure 12.9).

The allyl groups in T$_3$one exhibit three characteristic frequencies. The vC=C mode is not affected by change in the solvent system, as it occurs at 1645.4 cm^{-1}. On the other hand, the C=CH$_2$ wag mode decreases in frequency in a nonlinear manner as the mole % CHCl$_3$/CCl$_4$ is increased (see Figure 12.10).

Table 12.20 lists IR data and assignments for T$_3$one in 1% wt./vol. solutions in various solvents (20). In these 18 solvents, v_{op} (C=O)$_3$ ranges from 1703.2 cm^{-1} in hexane to 1693.3 cm^{-1} in dimethyl sulfoxide. In the case of the four alcohols, v_{op} (C=O)$_3$ exhibits both v_{op} (C=O)$_3$ and v_{op} (C=O)$_3\cdots$HOR. The former occurs in the range 1701.5–1703.5 cm^{-1} and the latter in the range 1688–1690.6 cm^{-1}. Figure 12.11 shows a plot of the v_{op} (C=O)$_3$ frequency for T$_3$one vs the solvent acceptor number (AN). The solid triangles show the data points for the T$_3$one$\cdots$HOR molecules; the solid squares show the data points for T$_3$one molecules not intermolecularly hydrogen bonded in solution surrounded by intermolecularly hydrogen-bonded alcohol molecules. Projecting these points onto the broad linear line shows that the AN values for the alcohols when not hydrogen bonded to solvent molecules are very similar to the AN values for alkyl ethers. The AN values are not a precise measure of the solute-solvent interaction, as they do not accurately account for steric effects between solute and solvent, or do not distinguish between molecules that are or are not intermolecularly hydrogen bonded (20).

REFERENCES

1. Nyquist, R.A. (1984). *The Interpretation of Vapor-phase Infrared Spectra: Group Frequency Data*, Philadelphia: Sadtler Research Laboratories, Division of Bio-Rad Inc.

2. Richards, R.E. and Thompson, H.W. (1947). *J. Chem. Soc. London*, 1248.

3. Brown, T.L., Regan, J.E., Scheutz, D.R., and Sternberg, J. (1959). *J. Phys. Chem.*, 63, 1324.

4. Lin-Vien, D., Cotthrup, N.B., Fateley, W.G., and Grasselli, J.G. (1991). *The Handbook of Infrared and Raman Characteristic Frequencies of Organic Molecule*, San Diego: Academic Press, Inc.

5. McLachlan, R.D. and Nyquist, R.A. (1964). *Spectrochim. Acta*, 20, 1397.

6. Nyquist, R.A. (1963). *Spectrochim. Acta*, 19, 509.

7. Nyquist, R.A. (1963). *Spectrochim. Acta*, **19**, 1595.

8. Russell, R.A. and Thompson, H.W. (1956). *Spectrochim. Acta*, **8**, 138.

9. Moccia, R. and Thompson, H.W. (1957). *Spectrochim. Acta*, **10**, 240.

10. Klemperer, W., Cronyn, M., Maki, A., and Pimentel, G. (1954). *J. Am. Chem. Soc.*, **76**, 5846.

11. Thompson, H.W. and Jameson, D.A. (1958). *Spectrochim. Acta*, **13**, 236.

12. Nyquist, R.A., Chrisman, R.W., Putzig, C.L., Woodward, R.W., and Loy, B.R. (1979). *Spectrochim. Acta*, **35A**, 91.

13. Nyquist, R.A. and Luoma, D.A. (1991). *Appl. Spectrosc.*, **45**, 1501.

14. Nyquist, R.A. and Luoma, D.A. (1991). *Appl. Spectrosc.*, **45**, 1491.

15. Nyquist, R.A. and Luoma, D.A. (1991). *Appl. Spectrosc.*, **45**, 1497.

16. Nyquist, R.A., Putzig, C.L., and Clark, T.D. (1996). *Vib. Spectrosc.*, **12**, 81.

17. Gutmann, V. (1978). *The Donor-acceptor Approach to Molecular Interactions*, p. 29, New York: Plenum Press.

18. Nyquist, R.A. and Fiedler, S.L. (1995). *Vib. Spectrosc.*, **8**, 365.

19. Nyquist, R.A. (1990). *Appl. Spectrosc.*, **44**, 215.

20. Nyquist, R.A., Puehl, C.W., and Putzig, C.L. (1993). *Vib. Spectrosc.*, **4**, 193.

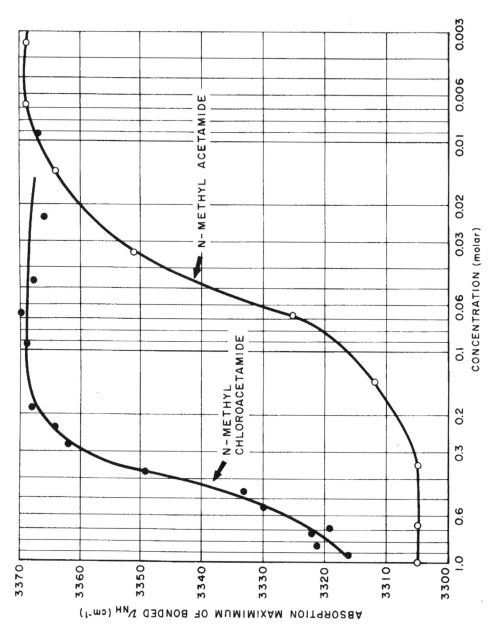

FIGURE 12.1   Plots of the νN–H$\cdots$O=C frequencies for N-methylacetamide and N-methyl chloroacetamide vs the absorption maximum at each of the $\nu\cdots$O=C frequencies.

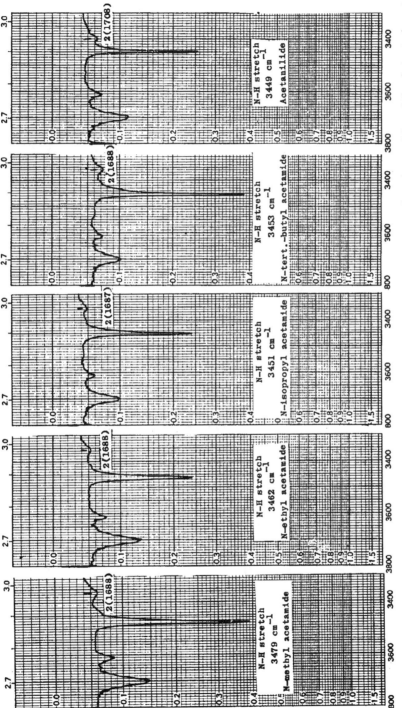

FIGURE 12.2 IR spectra of N-methyl acetamide, N-ethyl acetamide, N-isopropyl acetamide, N-tert-butyl acetamide, and acetanilide in CCl$_4$ solution in the region 3800–3300 cm^{-1}.

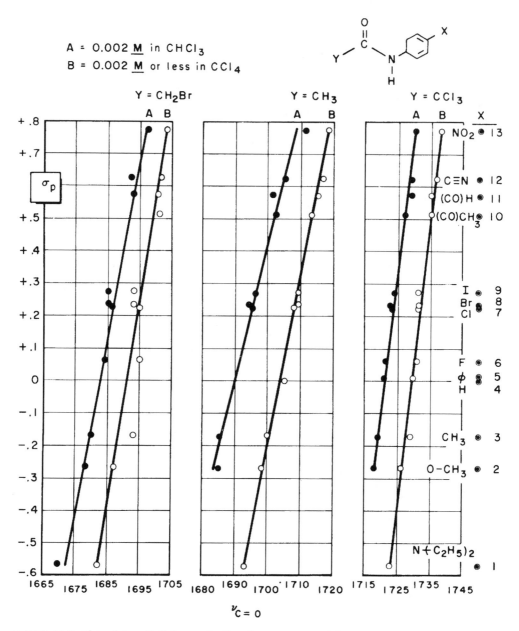

FIGURE 12.3 Plots of the νC=O frequencies for α-bromo-*p*-x-acetanilide, *p*-x-acetanilide, and α,α,α-trichloro-*p*-x-acetanilide in 0.002 M solutions or less in CCl₄ and CHCl₃ vs Hammett σ_p values.

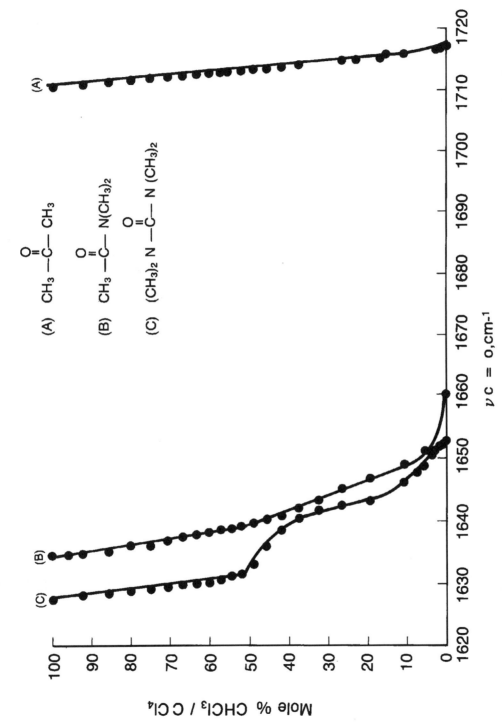

FIGURE 12.4 Plots of $\nu C{=}O$ for 1% wt./vol. solutions of acetone, dimethylacetamide, and tetramethylurea vs mole % $CHCl_3/CCl_4$.

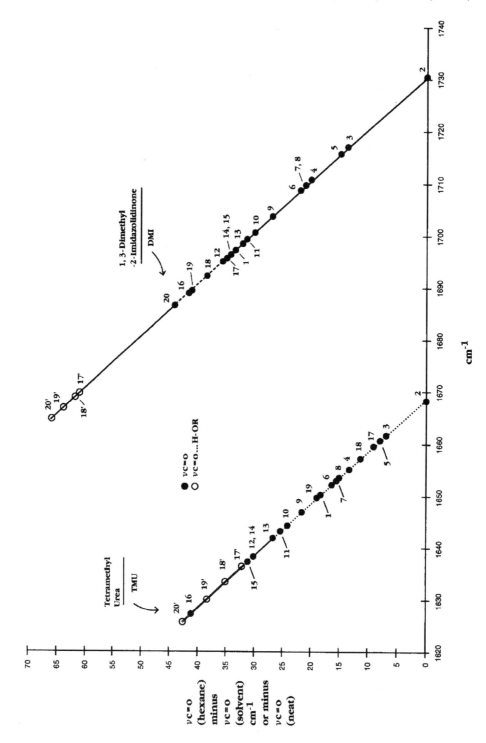

FIGURE 12.5　Plots of νC=O of 1,3-dimethyl-2-imidazolidinone (DMI) and 1,1,3-tetramethylurea (TMU) in the neat phase or in 1 of the 19 solvents vs νC=O (hexane) minus νC=O (solvent).

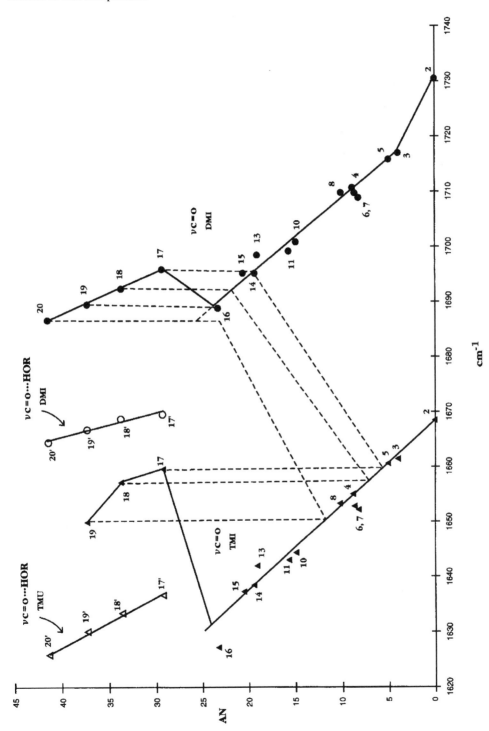

FIGURE 12.6 Plots of $\nu C{=}O$ for 1,1,3,3-tetramethylurea (TMU) and 1,3-dimethyl-2-imidazolidinone (DMI) vs the solvent acceptor number (AN).

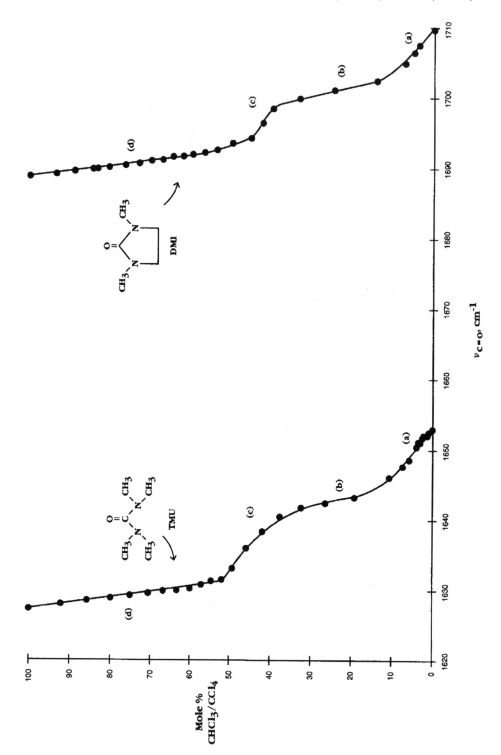

FIGURE 12.7   Plots of νC=O for 1,1,3,3-tetramethylurea (TMU) and 1,3-dimethyl-2-imidazolidinone (DMI) *vs* mole % CHCl$_3$/CCl$_4$.

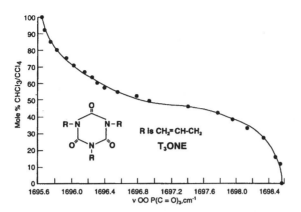

FIGURE 12.8   A plot of ν_{op} (C=O)$_3$ for triallyl-1,3,5-triazine-2,4,6-(1H,3H,5H) trione (T$_3$ one) vs mole % CHCl$_3$/CCl$_4$.

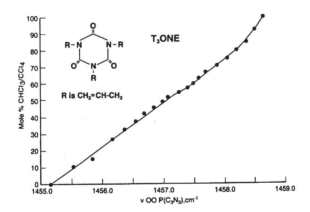

FIGURE 12.9   A plot of ν_{op} (C$_3$N$_3$) for triallyl-1,3,5-triazine-2,4,6-(1H,3H,5H) trione (T$_3$ one) vs mole % CHCl$_3$/CCl$_4$.

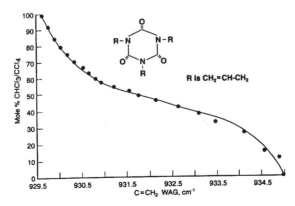

FIGURE 12.10   A plot of C=CH$_2$ wag for triallyl-1,3,5-triazine-2,4,6-(1H,3H,5H) trione (T$_3$ one) vs mole % CHCl$_3$/CCl$_4$.

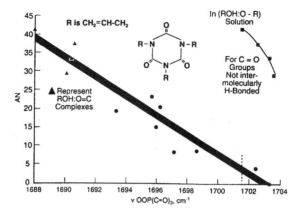

FIGURE 12.11 A plot of ν_{op} (C=O)$_3$ and ν_{op} (C=O)$_3 \cdots$ HOR for triallyl-1,3,5-triazine-2,4,6-(1H,3H,5H) trione vs the solvent acceptor number (AN).

TABLE 12.1   A comparison of primary, secondary, and tertiary amides in various physical phases

Primary amide assignments	Vapor phase [Ref. 1] R$-$(C=O)NH$_2$ cm^{-1}	Vapor phase [Ref. 1] C$_6$H$_5$$-$(C=O)NH$_2$ cm^{-1}	Neat phase nonbonded [Refs. 2,3,4] cm^{-1}	Intermolecular H-bonded [Refs. 2,3,4] cm^{-1}
asym.NH$_2$ str.	3548–3565	3556–3540	ca.3520	3350–3475
sym.NH$_2$ str.	3430–3444	3435–3448	ca.3400	3160–3385
C=O str.	1732–1780	1719–1731[*1]	1675–1715	1640–1680
NH$_2$ bending	1576–1600	1586–1600	1585–1620	1620–1640
C$-$N str.	1308–1400	1341–1355	1390–1430	1390–1430
NH$_2$ rock			1100–1150	1100–1150
NH$_2$ wag				600–750[*2]
N$-$C=O in-plane bend			550–600	550–600

Secondary amide assignments	Vapor phase[*3]		Neat phase nonbonded cm^{-1}	H-bonded cm^{-1}
NH str.,trans	3416–3419		3400–3490	3300
cis			3200	
[5 M.R.][*4]	3478			
[6 M.R.]	3438			
[7 M.R.]	3442			
C=O str.	1698–1720		1650–1700	1630–1680
C=O str.[4 M.R.]				1730–1780
C=O str.[5 M.R.]	1759			1700–1750
C=O str. [6 M.R.]	1715			
C= str. [7 M.R.]	1711			
CNH str.-bend,trans			1500–1550	1510–1570
,cis			ca. 1450	1400–1490
CNH str.-open,trans			1200–1250	1250–1310
,cis			ca.1350	1310–1350

Tertiary amide assignments				
C=O str.	1671–1731		1630–1680	

[*1] See text.

[*2] Broad.

[*3] See text.

*4 M.R. = membered ring or lactam.

TABLE 12.1A   The NH stretching frequencies and absorbance data for N-methyl acetamide in concentration ranging from 1.37×10^{-3} to 1.37 mol/liter in CCl_4 solution

Concentration mole/liter N-methylacetamide	N−H str.	A	Band halfwidth	% Bonded	Apparent intensity $B' \times 10(7)$
0.00137	3476	0.199	6.8	0	3.6
0.00342	3475	0.426	7	11	3.2
0.00685	3475	0.81	7.1	15	3
0.0137	3475	1.48	7.1	19	2.9
0.0342	3475	2.87	7.4	27	2.2
0.685	3474	3.99	7.3	56	1.6
0.137	3474	4.93	7.4	73	0.96
0.342	3472	6.63	7.4	86	0.52
0.685	3472	7.1	7.1	92	0.28
1.027	3471	8.1	7.4	94	0.21
1.37	3471	8.88	7.4	95	0.17

N-methyl chloroacetamide					
0.00921	3450	0.197	6.4	1	4.9
0.023	3450	0.479	6.4	4	4.8
0.046	3450	0.897	6.4	10	4.5
0.0681	3451	1.28	6.5	12	4.4
0.0921	3450	1.63	6.9	12	4.4
0.184	3450	2.6	7.2	26	3.7
0.23	3450	3.04	7.5	28	3.6
0.276	3449	3.41	7.5	33	3.3
0.368	3449	3.86	8	39	3.6
0.46	3449	3.97	8.2	50	2.5
0.553	3449	4.31	8.5	52	2.4
0.691	3448	4.37	9.1	58	2.1
0.727	3448	4.65	9.1	58	2.1
0.829	3447	4.89	9.2	61	2
0.921	3448	4.95	9.3	64	1.8

TABLE 12.1B   The NH stretching, C=O stretching, amide II, and amide III frequencies for N-methyl acetamide in $CHBr_3$ solutions

Concentration mole/liter [$CHBr_3$] N-methylacetamide	N−H str.	N−H str. bonded	C=O str.	Amide II	Amide III
0.00685	3456		1668	1531	1278
0.027	3456		1668	1531	1279
0.137	3457		1668	1532	1278
0.685	3457	3325	1660	1537,1560	1279
2.05	3454	3320	1656	1561	1279
2.74	3451	3305	1649	1563	1279

TABLE 12.1C The overtone for amide II of N-methyl acetamide and N-methyl chloroacetamide in varying concentrations in CCl$_4$ solutions

Concentration mole/liter [CCl$_4$] N-methyl acetamide	2(amide II)	A	Band halfwidth	Intensity B×10(8)
0.342		0.02	30	0.6
0.0685	3102	0.059	28	9
0.137	3103	0.206	30	16
0.342	3104	0.775	30	25
0.685	3105	1.65	31	26
1.027	3104	2.76	30	29
1.37	3105	3.55	30	28
N-methyl chloroacetamide				
0.276	3098	0.049	26	1.7
0.46	3098	0.33	26	6.7
0.553	3098	0.46	26	7.8
0.727	3098	0.82	27	10.5
0.829	3099	0.93	27	11
0.921	3099	1.38	28	15

TABLE 12.2   IR data and assignments for N-alkyl acetamides, and N-alkyl α- substituted acetamides in dilute solutes

Compound acetamide	N−H str.	A*[1]	C=O str.	2(C=O str.) or cis N−H str.	N−H bending	CCN str.
N-methyl	3478	21.5	1688	3365		
N-ethyl	3462	16.2	1688	3360		
N-propyl	3461	18.5	1690	3360		
N-n-butyl	3460	18.1	1688	3360		
N-isobutyl	3471	21.7	1689	3365		
N-isopropyl [vapor]	3460(0.051)		1714(1.240)	3420(0.020)	1490(0.803)	1244(0.372)
N-isopropyl	3451	18.9	1687	3360		
N-s-butyl	3450	23.2	1685	3360		
N-t-butyl	3453	25.4	1688	3360		
N-phenyl	3449	34.6	1708	3400		
α-chloro-acetamide						
N-n-butyl	3433	30.9	1684	3348		
N-isopropyl	3429		1682	3345		
N-s-butyl	3422		1680	3335		
N-t-butyl	3421	38.5	1684	3350		
N-phenyl	3409	56	1692	3350		
α,α-dichloroacetamide						
N-ethyl	3441	47.8	1707	3383		
N-propyl	3439	39.9	1706	3382		
N-n-butyl	3439	43.4	1705	3395		
N-isopropyl	3430	40.8	1703	3385		
N-t-butyl	3427	31.6	1702	3365		
N-phenyl	3419	51.3	1713	3400		
α,α,α-trichloroacetamide						
N-methyl	3462	58.9	1728			
N-ethyl	3449	45.4	1725			
N-propyl	3448	45.5	1724			
N-butyl	3445	41.6	1726			
n-isobutyl	3451	44.2	1725			
N-isopropyl	3439	39.3	1725			
N-s-butyl	3438	36	1724			
N-t-butyl[vapor]	3444(0.190)		1743(1.230)		1505(1.230)	1244(0.451)
N-t-butyl	3435	39.9	1725			
N-phenyl	3425	51.8	1731			
α-bromo-acetamide						
N-ethyl	3435	29.8	1680			
N-propyl	3432	30.9	1680			
N-n-butyl	3432	31.2	1681			
N-isobutyl	3433	32	1680			
N-isopropyl	3422	30.2	1679			
N-s-butyl	3420	30.6	1679			
N-t-butyl	3419	26.2	1680			

(continued)

TABLE 12.2 (*continued*)

Compound acetamide	N—H str.	A*[1]	C=O str.	2(C=O str.) or cis N—H str.	N—H bending	CCN str.
α,α-dibromo-acetamide						
N-methyl	3450	35.8	1701			
N-ethyl	3439	32.7	1700			
N-propyl	3438	30.5	1700			
N-n-butyl	3437	303.4	1700			
N-isobutyl	3440	30.2	1700			
N-isopropyl	3427	26.8	1699			
N-s-butyl	3422	26.8	1698			
N-t-butyl	3425	25.5	1701			
α,α,α-tribromoacetamide						
N-ethyl	3441	35.4	1712			
N-isopropyl	3430	28.2	1710			
N-t-butyl	3429	27.4	1717			
α,α,α-trifluoroacetamide						
N-methyl	3470	51.6	1741			
N-ethyl	3454	49.5	1736			
N-propyl	3456	46.6	1738			
N-n-butyl	3454	44.9	1736			
N-isobutyl	3456	56	1738			
N-isopropyl	3447	43.2	1735			
N-s-butyl	3443	38	1733			
N-t-butyl	3446	37.4	1736			
α-methoxy-acetamide						
N-ethyl	3438	30.7	1689			
N-propyl	3439	28.7	1690			
N-isobutyl	3440	38.3	1690			
N-isopropyl	3427	32	1687			
N-s-butyl	3422	31.1	1685			
N-t-butyl						
α-phenoxy acetamide						
N-ethyl	3446	35.5	1691			
N-propyl	3447	35.6	1692			
N-n-butyl	3446	32.9	1691			
N-isobutyl	3449	35.2	1692			
N-isopropyl	3430	32.7	1688			
N-s-butyl	3430	29.4	1688			
N-t-butyl	3428	28.4	1689			

TABLE 12.2A The NH stretching frequencies for N-alkyl acetamides, N-alkyl X-halo, X,X-dihalo-, X,X,X-trihaloacetamides, and N-alkyl methoxyacetamide in CCl_4 solutions

Compound [10%(wt./vol.) or saturated*] or [2%*1] [CCl4]	N−H str.	N−H str. H-bonded	N−H str [corrected for F.R.]	delta N−H str.	A (bonded)/ A (unbonded)
Acetamide N-alkyl					
ethyl*	3460	3300	3285	175	2.2
isobutyl*	3460	3288	3265	195	11
isopropyl*	3434	3280	3264	175	2.4
t-butyl*	3450	3303	3277	173	5.7
Bromoacetamide N-alkyl					
ethyl	3433	3299	3280	153	3.2
propyl	3430	3291	3268	162	4
butyl	3430	3294	3270	160	2.9
isopropyl	3420	3290	3269	151	2.7
t-butyl	3420	3272		148	148	0.7
Trifluoroacetamide N-alkyl					
methyl	3460	3325	3311	149	4.3
ethyl	3446	3310	3297	149	3.7
propyl	3451	3315	3305	146	3.5
isobutyl	3451	3320	3307	144	2.8
isopropyl	3440	3311	3295	148	2.3
s-butyl	3439	3310	3296	143	2.4
t-butyl	3438	3340	3313	125	0.6
Trichloroacetamide N-alkyl					
methyl*	3459	3390		69	0.15
ethyl	3443	3360		83	0.3
propyl	3442	3358		84	0.3
butyl	3447	3360		87	0.27
isobutyl	3448	3369		79	0.18
isopropyl	3437	3360		77	0.14
s-butyl	3430	3352		78	0.1
t-butyl	3433	3370		63	0.01
Dibromoacetamide N-alkyl					
ethyl*1	3439	3354		85	0.17
propyl*1	3437	3355		82	0.2
butyl*1	3433	3355		78	0.17
isobutyl*1	3440	3368		72	0.14
isopropyl*	3420	3360		60	0.04
s-butyl*1	3419	3343		76	0.17
Methoxyacetamide N-alkyl					
ethyl	3433	3345		88	0.3

(*continued*)

TABLE 12.2A   (*continued*)

Compound [10%(wt./vol.) or saturated*] or [2%*[1]] [CCl₄]	N−H str.	N−H str. H-bonded	N−H str [corrected for F.R.]	delta N−H str.	A (bonded)/ A (unbonded)
propyl	3431	3345		86	0.35
butyl	3438	3351		87	0.31
isobutyl	3439	3351		88	0.27
isopropyl	3421	3335		86	0.19
s-butyl	3422	3335		86	0.18
t-butyl	3418	3352		66	0.07
Phenoxyacetamide N-alkyl					
ethyl	3441	3349		92	0.3
propyl	3445	3354		91	0.3
isobutyl	3444	3356		88	0.24
isopropyl	3439	3352		87	0.21
s-butyl	3430	3340		90	0.2
t-butyl	3430	3355		75	0.05

*saturated at 2% in CCl₄.

TABLE 12.3   IR data for N-alkyl α-substituted acetamides in 10% wt./vol. CCl$_4$ and 0.002 M CCl$_4$ solutions

Compound	C=O str. 10% (wt./vol.) [CCl$_4$]	c=O str. (0.002 M) [CCl$_4$]	delta C=O str.	Amide II	2(amide II)
Bromoacetamide N-alkyl					
ethyl	1668	1680	12	1525,1555	3090
propyl	1666	1680	14	1545	3087
butyl	1666	1681	15	1540	3086
isopropyl	1664	1680	16	1521,1552	3081
t-butyl	1674	1680	6	1522,1550	3071
Trifluoroacetamide N-alkyl					
methyl	1724	1741	17	1564	3120
ethyl	1716	1736	20	1553	3100
propyl	1720	1738	18	1555	3109
isobutyl	1720	1738	18	1554	3108
isopropyl	1716	1735	19	1548	3094
s-butyl	1715	1733	18	1550	3099
t-butyl	1722	1736	14	1530	3076
Trichloroacetamide N-alkyl					
ethyl	1712	1725	13	1518	
propyl	1712	1724	12	1518	
butyl	1713	1726	13	1520	
isobutyl	1717	1725	8	1520	
isopropyl	1714	1721	7	1512	
s-butyl	1715	1724	9	1512	
t-butyl	1725	1725	0	1509	
Methoxyacetamide N-alkyl					
ethyl	1678	1689	11	1532	
propyl	1678	1690	12	1530	
isobutyl	1682	1690	8	1531	
isopropyl	1675	1687	12	1522	
s-butyl	1680	1685	5	1524	
t-butyl	1685	1690	5	1523	
Phenoxyacetamide N-alkyl					
ethyl	1683	1691	8	1531	
propyl	1684	1692	8	1531	
isobutyl	1688	1692	4	1533	
isopropyl	1686	1688	2	1527	
s-butyl	1684	1688	4	1524	
t-butyl	1690	1689	−1	1527	

TABLE 12.4   A comparison of primary and secondary amides in the solid phase

Compound solid [KBr pellet]	asym.NH_2 str. cm^{-1}	sym.NH_2 str. cm^{-1}	C=O str. cm^{-1}	NH_2 bending cm^{-1}	Gama NH_2 cm^{-1}
$H_2N-C=OCH_3$	3310	3160	1686	1652	720
$H_2N-C=OCH_2Cl$	3380	3180	1650	1620	645
$H_2N-C=OCHCl_2$	3320	3150	1670	1625	?
$H_2N-C=OCH_2\text{-}OCH_3$	3380	3195	1635	1645	725
$H_2N-C=OC_6H_5$	3362	3170	1659	1625	650
$H_2N-C=OC_6H_4(2\text{-}OH)$	3398	3185	1661	1629	615
	NH:O=C cm^{-1}	c=O:HN cm^{-1}	N—H bending cm^{-1}		
$4\text{-}CH_3C_6H_4-NH-C=OCH_3$	3295	1660	1530		
$C_6H_5-NH-C=OC_2H_5$	3242	1660	1546		
$C_6H_5-NH-C=OC_3H_7\text{-iso}$	3300	1664	1550		
$C_6H_5-NH-C=OC_6H_5$	3339	1648	1525		
$CH_3-NH-C=OC_6H_5$	3340	1635	1546		

TABLE 12.4A   A comparison of IR data for tertiary amides in the neat or solid phase

Compound	Phase	C=O str. cm^{-1}
$(C_2H_5)2N-C=OH$	neat	1665
$(C_2H_5)(C_6H_5)N-C=OH$	neat	1670
$(C_6H_5)2N-C=OH$	KBr pellet	1690
$(C_6H_5)(CH_3)N-C=OCH_3$	KBr pellet	1659
$(C_6H_5)(C_2H_5)-C=OCH_3$	KBr pellet	1650
$(C_6H_5)2N-C=OCH_3$	film	1670
$(C_6H_5)N-C=O(CH_2)_3$	melt	1685*
$(C_4H_9)2N-C=OCH_3$	neat	1640
$(C_2H_5)2N-C=OC_2H_5$	neat	1636
$(C_4H_9)2N-C=OC_2H_5$	neat	1635
$(CH_3)2N-C=OC_3H_7$	neat	1634
$(C_2H_5)2N-C=OC_3H_7$	neat	1634
$(CH_3)2N-C=OC_{11}H_{23}$	neat	1650
$1,2\text{-}C_6H_4(C=O-N(C_2H_5)_2)_2$	neat	1625

* 5-Membered ring.

TABLE 12.5   A comparison of Raman data and assignments for acrylamide, methacrylamide, and their polymers

Compound	A.M.W.	C=O str. monomer C=O str. polymer	C=O str.	C=C str.	C=O str. C=C str.	RI C=O str./ RI C=C str.	C=O str. Acrylamide C=O str. N-alkyl or N,N-dialkyl-acrylamide	Acrylamide C=O str. Methacrylamide C=O str. and for their polymers
Acrylamide			1682(2)	1639(9)	43	0.22	0	9
Polyacrylamide	18,000,000	14	1668(2)					15
		17	1665(3)					13
			1660(3)				7	
	6,000,000	23	1659(3)					6
	5,000,000	24	1658(4)					5
Acrylamide								
N-methyl			1659(1)	1628(9)	31	0.11	23	
N-ethyl			1657(2)	1628(9)	29	0.22	25	
N-tert-butyl			1659(1)	1629(9)	30	0.11		
N-(1,1-dimethyl-3-oxybutyl)			1658(3)	1627(9)	31	0.33		
Acrylamide								
N,N'-dimethyl			1648(1)	1612(9)	36	0.11	34	
N,N'-diethyl			1649(2)	1609(9)	40	0.22	33	
N-methyl			1666(8)	1625(7)	41	1.1	16	
N'-vinyl								
Methacrylamide			1673(1)	1647(5)				
Polymethacrylamide		20	1653(1)					

TABLE 12.5A Raman data and assignments for N-alkyl or N-aryl acrylamides and methacrylamides

N-alkyl or N-aryl acrylamide	C=O str.	C=C str.	[=C-C-N str.]	C=O str. C=C str.	RI C=O str./ RI C=C str.	N-alkyl or N-aryl methacrylamide	C=O str.	C=C str.	CH₂= bend	s.C-C-C str.	C=O str.- C=C str.	RI C=O str./ RI C=C str.
N'-methylene-bis	1663(1)	1632(9)	1252(5)	31	0.11	N'-methylene-bis	1661(9)	1622(3)	1421(6)	879(8)	39	3
methyl	1659(1)	1628(9)	1252(5)	31	0.11							
ethyl	1657(2)	1628(9)	1250(6)	29	0.22	ethyl	1656(9)	1621(5)		875(8)	35	1.8
trimethylene-bis	1651(4)	1617(5)	1245(9)	34	0.8							
butyl	1657(2)	1628(9)	1248(6)	29	0.22							
isopropyl	1656(2)	1622(9)	1248(6)	34	0.22							
cyclohexyl	1656(3)	1624(9)	1247(5)	32	0.33							
octadecyl		1625(2)				tert-butyl	1658(4)	1620(4)		881(9)	38	1
tert-octyl	1660(2)	1625(7)		35	0.29	benzyl	1653(4)					
benzyl	1652(2)				1							
N'-hexamethylene-bis	1654(5)	1620(9)	1240(9)	34	0.56	isobutoxy-methyl	1664(7)	1620(4)			36	1.8
isobutoxymethyl	1669(1)	1631(9)	1239(2)	38	0.11	2-hydroxy-propyl	1652(9)	1620(4)				
hydroxymethyl	1665(0.5)	1632(9)	1236(3)	33	0.06	butoxy	1671(1)	1632(9)			39	0.11
phenyl		1638(6)	1256(9)	29–35		phenyl	1657(3)	1620(3)			37	1
											35–39	

TABLE 12.6 IR C=O stretching frequencies for p-x-acetanilides and p-x, α-haloacetanilides in CCl₄ and CHCl₃ solutions

p-x group X	Sigma p	Acetanilide p-x C=O str. $[CCl_4]$	Acetanilide p-x C=O str. $[CHCl_3]$	α-Bromo-p-x acetanilide C=O str. $[CCl_4]$	α-Bromo-p-x acetanilide C=O str. $[CHCl_3]$	α,α,α-Trichloro-p-x-acetanilide C=O str. $[CCl_4]$	α,α,α-Trichloro-p-x-acetanilide C=O str. $[CHCl_3]$
$N(C_2H_5)_2$	−0.57	1693		1682	1670	1723	1718
OCH_3	−0.268	1698.5	1685	1687	1678	1726	1719
CH_3	−0.17	1700	1700	1693	1680	1729	
H	0	1705					
C_6H_5	0.01			1695	1684	1729.5	1721
F	0.062		1695	1695	1686	1731	1721
Cl	0.227	1708	1694	1693	1686	1731	1723
Br	0.323	1709	1696	1693	1685	1731	1723
I	0.276	1709	1702	1701	1685	1731	1724
$COCH_3$	0.516	1713	1701*	1701	1686	1736	1727
CHO	0.575	1715	1705	1701.5	1693*	1735	1729
CN	0.628	1716.5	1711	1703	1692	1737	1729
NO_2	0.778	1718			1697	1738	1730
$N(C_2H_5)_2:HCl$	?						1730
p-x group							
$COCH_3$		1685.5	1688	1687	1681	1689	1681
CHO		1701	1701	1700.5	1693	1706	1702

* not resolved from νC=O for the H−C=0 group.

TABLE 12.7 The NH stretching frequencies for p-x-acetanilides and p-x, α-haloacetanilides in CCl_4 and $CHCl_3$ solutions

p-x Group	Sigma p	p-x-acetanilide N–H str. [CCl_4]	p-x-acetanilide N–H str. [$CHCl_3$]	α-Bromo-p-x acetanilide N–H str. [CCl_4]	α-Bromo-p-x-acetanilide N–H str. [$CHCl_3$]	α,α,α-Trichloro-p-x-acetanilide N–H str. [CCl_4]	α,α,α-Trichloro-p-x-acetanilide N–H str. [$CHCl_3$]
N(C₂H₅)₂	−0.57	3443		3405	3395	3422	3415
OCH₃	−0.268	3445	3440	3403	3395	3421	3414
CH₃	−0.17	3443	3435	3402	3395	3422	
H	0	3441				3421	
C₆H₅	0.01			3401	3393	3422	3409
F	0.062	3441	3435	3402	3392	3420	3413
Cl	0.227	3441	3436	3400	3393	3420	3413
Br	0.232	3442	3437	3399	3392	3419	3411
I	0.276			3399	3388	3416	3413
COCH₃	0.515	3436	3431	3399	3390	3415	3409
CHO	0.575[*1]	3439	3442	3393	3388	3416	3407
CN	0.628	3437	3431	3391	3387	3409	3409
NO₂	0.778	3438	3430			3415	3406
N(C₂H₅)₂:HCl	?		3429				
N(C₂H₅)₂			A[*1]		A[*1]		A[*1]
OCH₃			2.89		4.12		4.55
CH₃			2.66		4.1		4.41
C₆H₅							4.81
F							5.12
Cl			2.68		4.68		5.06
Br			3.17		4.47		5.09
I			2.84		4.8		4.92
COCH₃			3.12		5.14		4.75
CHO			3.16		4.82		
CN			3.09		5.37		5.27
NO₂			3.66		5.3		5.36

[*1] Based on reading in Figure 12.3. See text.

TABLE 12.8   The NH and C=O stretching frequencies for N-alkyl p-methoxybenzamide, N-alkyl p-chlorobenzamide, and N-alkylmethyl carbamate in CCl$_4$ solutions

Compound [CCl$_4$ solutions] p-Methoxybenzamide	N−H str. cm^{-1}	A*[1]	C=O str. cm^{-1}
N-methyl	3481		1676
N-ethyl	3463	22.1	1669
N-propyl	3464	20.5	1670
N-n-butyl	3464		1671
N-isobutyl	3469	20	1671
N-isopropyl	3456	18.3	1665
N-s-butyl	3451	18.4	1668
N-t-butyl	3452	15.9	1672
p-Chlorobenzamide			
N-methyl	3481	24.7	1681
N-ethyl	3460	22.3	1675
N-propyl	3463	22.3	1675
N-n-butyl	3463	21.9	1677
N-isobutyl	3469	23.7	1679
N-isopropyl	3450	17.9	1671
N-s-butyl	3451	21.5	1674
N-t-butyl	3451	17.2	1676
Methyl Carbamate			
N-methyl	3478	41.6	1738
N-propyl	3464	32.3	1736
N-n-butyl	3461		1735
N-isobutyl	3467	27.9	1735
N-isopropyl	3452	28.2	1732
N-s-butyl	3449	28.5	1730
N-t-butyl	3455	30.5	1738
N-phenyl	3450	47.2	1748

*[1] Times 10^8 cm^2 molecule^{-1} sec^{-1}

TABLE 12.9 IR and Raman data for N,N'-dialkyl oxamide and N,N'-diaryl oxamide

Oxamide N,N'-dialkyl	o.p.(N–H)$_2$ str. IR	i.p.(N–H)$_2$ str. Raman	Amide I	Bu	Amide I Ag	Amide II Bu	Amide II Ag	Amide IV Bu
Methyl	3309	3325	1659	1639sh	1692	1533	1567	761
Butyl	3299	3315	1651	1632sh	1688	1532	1560	760
Octyl	3299	3314	1645	1629sh	1690	1508	1551	725
Dodecyl	3311	3325	1645	1628sh	1689	1512	1555	725
Hexadecyl	3304	3320	1645	1629sh	1690	1508	1554	725
Octadecyl	3304	3321	1643	1628sh	1690	1509	1554	752
2-Chloroethyl	3290	3310	1654	1636sh	1695	1536	1562	769
3-Chloropropyl	3306	3320	1654	1633sh	1689	1528	1559	782
2-Hydroxyethyl	3292	3308	1653		1691	1541	1557	764
3-Hydroxypropyl	3308	3324	1655	1635sh	1691	1538	1560	767
2-Methylallyl	3295	3305	1660	1634sh	1686	1524	1559	777
Benzyl	3279	3302	1653	1634sh	1689	1524	1557	781
o-Methylbenzyl	3302		1653	1631sh		1514		750
Range	(3279–3311)	(3302–3325)	(1643–1660)	(1628–1638)	(1686–1695)	(1508–1541)	(1547–1567)	(725–782)
N-(2-Chloroethyl) N'-(3-Hydroxypropyl)	3295	3314	1653		1691	1536	1562	764
N-2-(Chloroethyl) N'-(Allyl)	3299	3311	1657		1690	1535	1560	761
N,N'-diaryl								
Phenyl	3298	3320	1662	1682	1520	1550	730	
			1688sh	1697	1509sh			
3-Chlorophenyl	3308	3320	1669	1701	1518	1547	717	
				1690				
3,4-Dichlorophenyl	3358 3230sh	3335	1686	1704	1508	1545	693	
				1690	1514sh			
3,6-Dichloro-2-pyridyl	3325	3338	1711	1727	1488	1536	627	
3,5-(Trifluoromethyl)-2-pyridyl	3323	3349	1719	1741	1484	1537	638	
2,6-(Trifluoromethyl)-4-pyridyl	3290	3325	1692	1723	1510	1549	660	
Range	(3295–3358)	(3320–3349)	(16620–1719)	(1682–1741)	(1484–1520)	(1537–1550)	(628–750)	

TABLE 12.10 IR data and assignments for dimethylacetamide and tetraalkylurea in the vapor, neat and solution phases

Compound	C=O str. [vapor]	C=O str. [n-C$_6$H$_{14}$]	C=O str. [CCl$_4$]	C=O str. CHCl$_3$	C=O str. 14.0 mol % [n-C$_6$H$_{14}$/CHCl$_3$]	C=O str. 80.6 mol % [n-C$_6$H$_6$/CHCl$_3$]	C=O str. neat
Dimethylacetamide Urea	1690(1.250)		1660.5	1634.2			
Tetramethyl	1685(1.240)	1668.4(1.491)	1653.2 1652.9	1627.3	1650.9; 1635.4	1640.2; 1629.5	1649.6
Tetraethyl	1674(1.232)		1646.2	1620.1			
Tetrabutyl	1669(0.840)		1643.2	1616.1			1644
Dimethylacetamide Urea	[vapor]-[n-C$_6$H$_{14}$]	[vapor]-[CCl$_4$] 29.5	[vapor]-[CHCl$_3$] 55.8	[CCl$_4$]-[CHCl$_3$] 26.3	[vapor]-[neat]		
Tetramethyl	16.6	31.9	57.7	25.8	35.4		
Tetraethyl		27.8	53.9	26.1			
Tetrabutyl		25.8	52.9	27.1	25		

TABLE 12.11   IR vapor-phase data and assignments for urea, thiourea, and guanidine derivatives

Urea	N—H str.	2(C=O str.)	C=O str.	a.NCN?	N—H bending
1,1,3,3-Tetramethyl		3358(0.011)	1685(1.240)	1239(0.210)	
1,1,3,3-Tetraethyl		3338(0.006)	1674(1.240)	1261(1.240)	
1,1,3,3-Tetrabutyl			1669(0.790)	1255(0.159)	
1,3-Diethyl-1,3-diphenyl		3340(0.005)	1680(1.240)	1273(0.610)	
			1741(1.243)		
1,3-Dimethyl	3480(0.073)		1732(1.043)	1334(0.460)	1522(1.149)
			1726(1.143)		
Thiourea			C=S str.		
1,1,3,3-Tetraethyl			~1085(0.280)	1261(1.230)	
Guanidine			C=N str.		
1,1,3,3-Tetramethyl	3358(0.015)		1619(1.245)	1249(0.190)	

TABLE 12.12   IR data and assignments for tetramethylurea and 1,3-dimethyl-2-imidazolidinone in various solvents

Solvent	DMI C=O str. cm⁻¹	A	DMI C=O:HO str. cm⁻¹	[DMI C=O str.] [TMU C=O str.] cm⁻¹	TMU C=O str. cm⁻¹	TMU C=O str. [Hexane] C=O:HO str. cm⁻¹	TMU C=O str. [Hexane] [solvent] cm⁻¹	TMU C=O str. [Hexane] C=O:HO str. [solvent] cm⁻¹	DMI C=O str. [Hexane] C=O str. [solvent] cm⁻¹	DMI C=O str. [Hexane] C=O:HO str. [Hexane] cm⁻¹	AN
[Neat]	1697.1			47.4	1649.6		18.8		33.5		
Hexane	1730.6	0.922		62.2	1668.4		0		0		0
Diethyl ether	1716.8	0.974		55.3	1661.5		6.9		13.7		3.9
Tetrahydrofuran	1710.6	1.128		55.4	1660.7		13.2		20		8.8
Methyl t-butyl ether	1715.6	0.993		55	1660.7		7.7		14.9		5
Benzene	1708.7	1.091		56.6	1652.1		16.3		21.9		8.2
Carbon tetrachloride	1709.6	1.017		56.8	1652.9		15.5		21		8.6
Carbon disulfide	1709.6	1.321		56.2	1653.4		15		21		
1,2-Dichlorobenzene	1703.6	1.267		56.6	1646.9		21.5		27		10.1
Nitrobenzene	1700.6	0.935		56.2	1644.4		24		30		14.8
Benzonitrile	1699	1.017		55.9	1643.2		25.2		31.5		15.5
Nitromethane	1696.3	0.905		57.9	1638.4		30		34.3		
Acetonitrile	1698.3	1.04		56.4	1641.9		26.5		32.2		18.9
Dimethyl sulfoxide	1695.1	1.09		56.7	1638.4		30		35.5		19.3
Methylene chloride	1695	0.939		57.6	1637.4		31		35.6		20.4
Chloroform	1688.8	0.927		61.5	1627.3		41		41.7		23.1
t-Butyl alcohol	1695.6	0.901	1669.5	33	1659.5	1636.5	8.9	31.9	35	61.1	29.1
Isopropyl alcohol	1692.2	0.782	1668.7	35.3	1657.1	1633.4	11.3	35	38.4	61.9	33.5
Ethyl alcohol	1689.3	0.783	1666.8	39.2	1650.1	1630.1	18.3	38.3	41.3	63.8	37.1
Methyl alcohol	1686.5	0.599	1664.7	38.9		1625.8		43	44.1	65.9	41.3

TABLE 12.13 IR data and assignments for 1,3-dimethyl-2-imidazolidinone 1% wt./vol. in various solutions

Imidazolidinone [1% wt./vol.]		1,3-Dimethyl-2-imidazolidinone [1% wt./vol.]					1,3-Dimethyl-2-imidazolidinone [1% wt./vol.]						
Mole % CHCl$_3$/CCl$_4$	C=O str. cm^{-1}	Mole % CCl$_4$/n-C$_6$H$_{14}$	C=O str. cm^{-1}	A	C=O str. cm^{-1}	A	Mole % CHCl$_3$/n-C$_6$H$_{14}$	C=O str. cm^{-1}	A	C=O str. cm^{-1}	A	C=O str. cm^{-1}	A
0	1709.6	0	1730.6	0.826	1723	0.77	0	1730.6	0.855	1725.2	0.735	1698	0.397
3.51	1707.5	2.98	1729.8	0.774	1722.4	0.767	7.01	1727.9	0.309	1710.2	0.92	1697	0.57
4.67	1706.5	5.97	1729.4	0.748	1720.9	0.798	14.02	1728.3	0.193	1709.4	0.87	1696.3	0.699
7.01	1704.8	11.93	1729	0.675	1719.2	0.835	24.59	1728.7	0.154	1708.4	0.799	1695.3	0.79
14.02	1702.4	19.33	1728.7	0.621	1718.3	0.864	32.85	1729.5	0.122	1707.2	0.712	1694.6	0.84
24.59	1700.9	28.9	1728.6	0.602	1717.4	0.892	39.48	1729.1	0.092	1706.2	0.667	1694	0.865
32.85	1699.7	35.04	1728.3	0.595	1716.8	0.902	44.92	1729.5	0.085	1705.7	0.635	1693.7	0.886
39.48	1698.3	40.38	1727.9	0.592	1716.2	0.923	49.46	1729.5	0.085	1704.9	0.608	1693.2	0.912
42.2	1696.1	44.84	1727.5	0.475	1715.7	0.925	53.3	1729.1	0.08	1704.1	0.587	1692.9	0.928
44.92	1694.1	48.68	1727.5		1715.4		56.61	1729.1	0.074	1703.7	0.579	1692.6	0.883
49.46	1693.6	52.01	1727.5		1715		59.48	1729.1	0.06	1703.3	0.555	1692.4	0.925
53.3	1692.5	54.94	1727.1		1714.8		61.99					1692.3	0.929
56.61	1692.1	57.27			1714.4		64.4					1692	0.932
59.48	1691.9	60.09			1714		67.09					1691.7	0.94
61.99	1691.7	62.87			1713.8		69.97					1691.5	0.95
64.44	1691.5	65.93			1713.4		73.1					1691.2	0.952
67.09	1691.1	69.31			1712.9		78.53					1690.8	0.963
69.97	1690.0	73.04			1712.4		80.58					1690.4	0.967
73.1	1690.6	77.2			1711.7		84.69					1690	0.974
76.53	1690.4	81.87			1711.3		89.05					1689.4	0.955
80.58	1690.1	87.14			1710.5		93.52					1688.7	0.985
84.69	1689.8	93.13			1709.5		100						
89.05	1689.6	100											
93.52	1689.2												
100	1688.8												
delta cm^{-1}	20.8		-3.5		-13.5			-1.5		-21.9		-9.3	

TABLE 12.14 The in-phase and out-of-phase $(C=O)_2$ stretching frequencies for caffeine, isocaffeine, 1,3,5- trimethyluracil, 1,3,6-trimethyluracil, and 1,3-dimethyl-2,4-(1H,3H) 2, quinazolinedione in CCl_4 and $CHCl_3$ solutions

Compound [vapor]	i.p.$(C=O)_2$ str.+ o.p.$(C=O)_2$ str.	i.p.$(C=O)_2$ str.	o.p.$(C=O)_2$ str.	[vapor]-[CCl₄]	[vapor]-[CHCl₃]	[CCl₄]-[CHCl₃]
Caffeine	3395(0.005)	1721(0.731)	1685(1.250)			
[solution] Caffeine [CCl₄]		1710.6(0.193)	1667.5(0.298)	i.p.$(C=O)_2$ str. 10.4 o.p.$(C=O)_2$ str. 17.5	i.p.$(C=O)_2$ str. 12.1 o.p.$(C=O)_2$ str. 26.6	i.p.$(C=O)_2$ str. 1.7 o.p.$(C=O)_2$ str. 9.1
[CHCl₃] [solution]		1708.9(0.204)	1658.4(0.210)			
Isocaffeine [CCl₄]/[CHCl₃]						
10.74 mol %		1716.7	1673.2			i.p.$(C=O)_2$ str. 0.4* 5.4*
19.4 mol %		1716.3(0.016)	1669.0(0.019)			o.p.$(C=O)_2$ str. 4.2* 9.2*
[CHCl₃]		1711.3(0.487)	1664.0(0.487)			
[solution] Uracil 1,3,5-Trimethyl [CCl₄]		1706.6(0.128)	Corrected Fermi Res. 1663.4			i.p.$(C=O)_2$ str. 6.6 o.p.$(C=O)_2$ str. 11
[CHCl₃]		1700.0(0.124)	1652.4			
[solution] Uracil 1,3,6-Trimethyl [CCl₄]		1710.8(0.160)	1668.3			8.8
[CHCl₃] [solution]		1702.0(0.126)	1658.2			10.1
1,3-Dimethyl-2,4-(1H,3H)-quinazolinedione		Corrected				
[CCl₄]		1711.6(0.155)	Fermi Res. 1667			i.p.$(C=O)_2$ str. 7.9
[CHCl₃]		1703.7(0.105)	1657.4			o.p.$(C=O)_2$ str. 9.6

* differences from 10.74 and 19.4 mol% $CHCl_3$/CCl_4 from $CHCl_3$ solution.

TABLE 12.15   IR data for uracils in the solid phase

Uracil [solid phase]	i.p.(C=O)$_2$ str.	o.p.(C=O)$_2$ str.	[i.p.(C=O)$_2$ str.]-[o.p.(C=O)$_2$ str.]
5-Fluoro	1710	1661	49
5-Amino	1750	1670	80
5-Methyl	1745	1675	70
5-Bromo	1704	1680	24
5-Chloro	1718	1695	23
5-Nitro	1719	1695	24
5-Acetyl	1730	1704	26
6-Methyl	1720	1685	35
1,3-Dimethyl*	1733	1699	34
1,3-Dimethyl	1710	1658	52
1,3-Dimethyl-5- (morpholine carbonyl)	1709	1659	50
1,3-Dimethyl-5-nitro	1720	1668	52
1,6-Dimethyl-3-(p-chlorophenyl)-5-bromo	1700	1648	52
1,3-bis(2-Amino-ethyl-2HCl)	1701	1659	42
1,3,5-Trimethyl	1701	1667	34
1,3,6-Trimethyl	1689	1652	37
Range	1689–1750	1652–1704	23–80

* Vapor phase

TABLE 12.16   IR vapor-phase data and assignments for imides

Compound	i.p.(C=O)$_2$ str.+ +o.p.(C=O)$_2$ str.	A	N–H str.	A	i.p.(C=O)$_2$ str.	A	o.p.(C=O)$_2$ str.	A	A[i.p.(C=O)$_2$]/ A[o.p.(C=O)$_2$]/
Diacetamide					1738	0.941	1749	1.24	0.76
N-phenyldibenzamide	3485	0.011	3438	0.031	1801	0.101	1702	0.84	0.08
Succinimide	3522		3458	0.055	1820	0.089	1772	1.24	0.07
N-(2,6-xylyl)succinimide		0.011			1790	0.041	1741	1.24	0.03
4-Cyclohexene-1,2-di-carboximide	3518		3455	0.101	1799	0.201	1759	1.25	0.16
4-Methyl-N-(3',3'-trifluoromethyl-phenyl)-4-cyclohexene-1,2-dicarboximide		0.011			1795	0.051	1735	1.25	0.04
Maleimide	3499		3490	0.135			1755	1.24	
N-ethylmaleimide	3500	0.011			1820	0.019	1735	1.23	0.02
N-benzylmaleimide	3510	0.011			1810	0.021	1735	1.23	0.02
N-phenylmaleimide	3502	0.015					1739	1.23	
N-(4-iodophenyl)maleimide	3504	0.011			1795	0.091	1738	1.22	0.07
N-(3-chloro-4-methyl)phenylmaleimide	3504	0.011			1795	0.011	1738	1.23	0.01
N-(4-metoxyphenyl)maleimide	3504	0.011					1737	1.23	
N-(p-tolyl)maleimide	3510	0.011					1740	1.23	
N-(4-isopropylphenyl)maleimide	3504	0.011			1790	0.031	1735	1.24	0.03
N-(1-naphthyl)maleimide	3509	0.011			1790	0.031	1740	1.24	0.03
N-(2-chlorophenyl)maleimide	3518	0.011					1742	1.24	
N-(3,4-dimethoxyphenethyl)maleimide	3495	0.005			~1780	0.011	1730	1.23	0.01
N-[bis(3,5-trifluoromethyl)phenyl]maleimide	3510	0.011					1743	0.731	
N,N'-m-phenyldimaleimide	3504	0.011			1795	0.031	1739	1.24	0.03
N-[(dimethylamino)-o-tolyl]maleimide	3502	0.011			1880	0.026	1738	1.23	0.2
N-[(4-Chloro-2-trifluoromethyl)maleimide	3520	0.011			1785	0.101	1745	1.22	0.08
N-(dimethylamino)phenyl]maleimide	3508	0.011			1798	0.031	1739	1.23	0.03
N-(2,6-diisopropylphenyl)maleimide	3498	0.011					1731	1.24	
3,3-dimethylglutarimide							1745	1.24	
Phthalimide			3415 3479	0.093 0.111	1795	0.169	1766	1.23	0.14
4-Nitrophthalimide	3558	0.011	3470	0.281	1800	0.401	1769	1.23	0.33
3-Nitrophthalimide	3550	0.011	3470	0.205	1798	0.295	1768	1.23	0.24
3-Aminophthalimide	[a.NH$_2$ str.] [3520]	[A] [0.049]	3450 [s.NH$_2$ str.] [3402]	0.141 [A] [0.075]	1785	0.365	1745	1.24	0.29
N-methylphthalimide	3510	0.011			1785	0.111	1739	1.22	0.09
N-[(dimethylamino)methyl]phthalimide	3475	0.111			1771	0.201	1735	1.24	0.16
N-(2-bromoethyl)phthalimide	3509	0.021			1789	0.281	1739	1.23	0.23

Compound									
N-(2-chloroethyl)phthalimide	3510	0.011			1787	0.162	1740	1.222	0.13
N-(3-bromopropyl)phthalimide	3502	0.011			1788	0.171	1734	1.24	0.14
N-(3-chloro-2-hydroxypropyl)Phthalimide	3510	0.011			1785	0.151	1735	1.24	0.12
N-(2,5-dichlorophenyl)phthalimide	3524	0.011			1800	0.151	1746	1.22	0.12
N-(3-chloro-o-tolyl)phthalimide	3518	0.011			1795	0.121	1735	1.22	0.11
N-[(3,4-methylenedioxy)benzylidene)amino]-phthalimide	3522	0.011			1791	0.111	1740	1.21	0.09
N-(2,6-dimethylphenyl)phthalimide	3502	0.011			1761	0.129	1745	1.22	0.11
N-[(5-bromo-2-hydroxybenzylidene)amino]phthalimide	3539 [OH:N=C] [3475] [A][0.101]	0.011			1789	0.201	1759	1.22	0.16
2-Phthalimidoglutaric anhydride	3478	0.011			masked [Anhydride][1871]	[A][0.135]	1745 [Anhydride][1799]	0.245 [A][1.23]	
N-2-propynylphthalimide	3510 [CC–H str.] [3330] [3310]	0.011 [A][0.111] 0.031	[a',CCH bend][669]	[A][0.152]	1789 a'',CCH bend [629]	0.162 [A][0.138]	1741	1.24	0.13
Hydantoin	3564	0.011	3495(0.074)		1826(0.133)	0.221	1785(1.245)		
1-Methylhydantoin			3482	0.085	1815	0.231	1775	1.24	0.18
5,5-Dimethylhydantoin			3490	0.121	1811		1775	1.24	0.19
5,5-Dimethylhydantoin (solid phase)			3210(0.312)		1779(0.540)		1744(1.460) 1716(1.460)		
5-Methyl-5-propylhydantoin	3570	0.011	3490		1806	0.611	1775	1.21	0.55
5-Methyl-5-propylhydantoin (solid phase)			3215(0.271)	0.241	1761(0.415)		1707(1.170)		
5-Methyl-5-isopropylhydantoin	3570	0.011	3490		1805	0.682	1771	1.23	0.55
5-Methyl-5-isopropylhydantoin (solid phase)			3205(0.530)	0.241	1770(0.873)		1712(1.300)		
N-[(4,4-dimethyl-2,5-dioxoimidazolidinyl)-methyl]anthranilic acid, methyl ester	[N–H:O=O str.] [3395] [1719 ester]	[A] 0.075 0.501 0.011	3482	0.081	1810	0.201	1779	1.25	0.16
1,3-Diphenyl-5-octyl-s-triazine-(1H,3H,5H)-trione	3459	0.011			1766	0.161	1720	1.21	0.13
1-Benzyl-3,5-diallyl-s-triazine-2,4,6-(1H,3H,5H)-trione	3470	0.05					1711	1.24	
Triallyl-s-triazine-2,4,6-(1H,3H,5H)-trione	3470	0.05					1710	1.24	

TABLE 12.17 IR data for the in-phase and out-of-phase stretching frequencies of 4-bromobutyl phthalimide in CHCl$_3$/CCl$_4$ solutions

N-(4-bromobutyl)phthalimide [1% solutions] Mole % CHCl$_3$/CCl$_4$	i.p.(C=O)$_2$ str. cm^{-1}	o.p.(C=O)$_2$ str. cm^{-1}	[ip.(C=O)$_2$ str.]-[o.p.(C=O)$_2$ str.] cm^{-1}
0	1773.8	1718.7	55.2
26.53	1773.1	1716.9	56.2
52	1772.5	1715.6	56.9
70.65	1772.3	1714.2	58.1
100	1772.2	1713	59.2
delta cm^{-1}	−1.6	−5.7	4

TABLE 12.18   IR data and assignments for hydantoins in the vapor and solid phases

Hydantoin	NH str. cm⁻¹	A [NH str.]	i.p.(C=O)$_2$ str. cm⁻¹	o.p.(C=O)$_2$ str. cm⁻¹	A[i.p.(C=O)$_2$ str.]	A[o.p.(C=O)$_2$ str.]	A[i.p.(C=O)$_2$ str.]/A[o.p.(C=O)$_2$ str.]	[i.p.(C=O)$_2$ str.]-[o.p.(C=O)$_2$ str.] cm⁻¹
1,3,5,5-H,H,H,H	[3495]	[0.082]	[1825]	[1785]	[0.13]	[1.27]	[0.10]	[40]
H,H,H,H	3262	0.25	1783	1717	0.705	1.25	0.56	66
	3150	0.26						
H,H,CH₃,CH₃	3210	0.31	1779	1744	0.54	1.46	0.37	35
				1716		1.46		63
H,H,C₃H₇,C₃H₇	3200	0.312	1766	1717	0.622	1.3	0.48	49
H,H,C6H₁₁,C6H₁₁	3180	0.2	1755	1702	0.4	1.07	0.37	53
H,H,CH₃,C₃H₇	3215	0.271	1761	1707	0.415	1.17	0.35	54
H,H,CH₃,C₃H₇	[3485]	[0.234]	[1809]	[1774]	[0.613]	[1.24]	[0.49]	[35]
H,H,CH₃,iso-C₃H₇	3205	0.53	1770	1712	0.873	1.3	0.67	59
H,H,CH₃,iso-C₃H₇	[3485]	[0.224]	[1808]	[1774]	[0.690]	[1.26]	[0.58]	[34]
H,H,CH₃,iso-C₄H₉	3180	0.345	1780	1718	0.65	1.25	0.52	62
H,H,CH₃,iso-C₄H₉	[3485]	[0.210]	[1809]	[1774]	[0.55]	[1.25]	[0.44]	35
H,H,CH₃,C6H₅	3270	0.47	1758	1720	0.64	1.24	0.52	38
	3200	0.52		1709		0.96	0.67	49
H,H,CH₃,p-ClC₆H₄	3250	0.13	1777	1727	0.3	0.9	0.33	50
	3185	0.15		1720		0.8	0.38	57
H,H,C6H₅ * ,C₃H₅ *	3300	0.16	1781	1720	0.19	1.14	0.17	61
	3165	0.08	1761	0.15				41
C6H₅,C6H₅,H,H			1782	1720	0.23	0.86	0.27	62
C₂H₅,CH₃,C6H₅,C6H₅			1778	1721	0.2	0.95	0.21	57
			1771		0.21	0.22		50
iso-C₃H₇,CH₃,C6H₅,C6H₅			1770	1711	0.49	1	0.49	59
iso-C₄H₉,CH₃,C6H₅,C6H₅			1774	1720	0.39	1.05	0.37	54
iso-C₅H₁₁,CH₃,C6H₅,C6H₅			1779	1723	0.37	1.19	0.31	56
Range	3150–3300 [3485–3495]		1755–1783 [1808–1825]	1702–1744 [1774–1785]				35–66 [34–40]

[ ] Sadler vp at 280 °C

* = allyl.

TABLE 12.19   IR data and assignments for tri-allyl-1,3,5-triazine-2,4,6-(1H,3H,5H) trione in CHCl₃/CCl₄ solutions

Triallyl-1,3,5-triazine-2,4,6-(1H,3H,5H) trione Mole % CHCl₃/CCl₄	o.p.(C=O)₃ str. cm⁻¹	o.p.(CN)₃ str. cm⁻¹	C=C str. cm⁻¹	CH=CH₂ twist cm⁻¹	CH=CH₂ wag cm⁻¹
0	1698.6	1455.2	1645.4	992	935
10.74	1698.6	1455.5	1645.4	991.8	934.9
15.07	1698.5	1455.8	1645.4	991.7	934.6
26.53	1698.3	1456.2	1645.4	991.7	934.1
52	1696.8	1457.1	1645.4	991.6	931.6
70.65	1696	1457.9	1645.4	991.7	930.4
85.05	1695.8	1458.4	1645.4	991.8	929.9
100	1695.7	1458.6	1645.4	991.8	929.7
delta cm⁻¹	−2.9	3.4			−5.3

TABLE 12.20   IR data and assignments for tri-allyl-1,3,5- triazine-2,4,6-(1H,3H,5H) trione in various solvents

Triallyl-1,3,5-triazine-2,4,6-(1H,3H,5H) trione 1 wt./vol. % solutions	o.p.(C=O)₃ str. cm⁻¹	o.p.(C=O)₃ : HO str. cm⁻¹	AN
Hexane	1703.2		0
Diethyl ether	1702.3		3.9
Methyl t-butyl ether	1701.6		
Toluene	1698.7		
Carbon tetrachloride	1698.6		8.6
Carbon disulfide	1698.2		
1,2-Dichlorobenzene	1697.1		8.2
Acetonitrile	1695.5		
Nitrobenzene	1696		14.8
Benzonitrile	1694.5		
Methylene chloride	1696.2		20.4
Nitromethane	1695.9		
t-Butyl alcohol	1703.5	1690.1	29.1
Chloroform	1695.7		23.1
Dimethyl sulfoxide	1693.3		19.3
Isopropyl alcohol	1702.2	1690.4	33.5
Ethyl alcohol	1702.4	1690.6	37.1
Methyl alcohol	1701.5	1688	41.3

Aldehydes

*Numbers in parentheses indicate in-text page reference.

Aldehydes are easily oxidized to the corresponding carboxylic acid. Therefore, the reader is cautioned to look for the presence of carboxylic acid impurities when interpreting the IR or Raman spectra of aldehydes. Several published spectra of aldehydes contain acid impurity, and the impurity is not marked on the IR spectrum.

Aldehydes contain the O=C−H group; the empirical structure for the aliphatic forms is R−C(=O)H, for the conjugated form it is C=C−C(=O)H, and for the aromatic form it is $C_6H_5C(=O)H$. Characteristic vibrations of the aldehyde group are:

a) C=O stretching (IR, strong);
b) the first overtone of C=O stretching (IR very weak);
c) C−H bending (IR, medium); and
d) C−H stretching in Fermi resonance with C−H bending (IR, weak-medium bands).

Table 13.1 compares IR data and assignments for some nonconjugated aldehydes recorded in different physical phases. In all cases the vC=O frequency occurs at higher frequency in the vapor than in solution or neat phase. In the neat phase vC=O for the compounds of form R−C(=O)H exhibit vC=O in the range 1715–1731 cm^{-1}. Substitution of halogen atoms raises the vC=O frequencies. For example, in CCl_4 solution vC=O occurs at 1730, 1748, and 1768 cm^{-1} for the Cl, Cl_2, and Cl_3 analogs of acetaldehyde, respectively (1). In the vapor phase, vC=O occurs at 1760, 1778, and 1788 cm^{-1} for the Br_3, Cl_3 and CF_3 analogs of acetaldehyde, respectively (2,3). Therefore, these data show that vC=O for aldehydes increases in the order of Br, Cl, and F substitution. The inductive power of the halogen atom also increases in this same order.

The aldehydic in-plane (O=C−H plane) bending mode occurs in the region 1350–1400 cm^{-1}, and its first overtone (2δCH) in Fermi resonance with vC−H occurs in the regions 2800–2860 cm^{-1} and 2680–2722 cm^{-1}. Usually the higher frequency band has more intensity than the lower frequency band in the Fermi resonance doublet. However, sometimes the two bands have nearly equal or equal intensity. Where both bands of the Fermi doublet have equal intensity, both bands result from equal mixtures of vC−H and 2δC−H. It is not theoretically correct to assign the stronger of the two bands to only the vC−H fundamental.

Trichloroacetaldehyde does form a hydrate, and in its hydrate form it is no longer an aldehyde. It exists as a dihydroxide, CCl$_3$CH(OH)$_2$.

Table 13.1a list Raman data and assignments for several aldehydes. The Raman data were taken from Reference (4), and the assignments were made by Nyquist. Data in the parentheses indicate the relative Raman band intensities and p is polarized. These data show that the relative Raman band intensity for vC=O is less than that for the band intensities for vC−H and 2δCH in Fermi resonance.

Acrolein is a conjugated aldehyde, CH$_2$=CH−C(=O)H and the Raman bands at 1688 cm^{-1} and 1618 cm^{-1} are assigned to vC=O and vC=C, respectively. Benzaldehyde, C$_6$H$_5$C(=O)H, is also conjugated, and its vC=O mode occurs at 1701 cm^{-1} in the condensed phase. In the case of salicylaldehyde, or 2-hydroxybenzaldehyde, the OH group is intramolecularly hydrogen bonded to the C=O group, and its vC=O mode occurs at 1633 cm^{-1} in the condensed phase.

Table 13.2 lists IR data for 4-X-benzaldehydes in the vapor, CCl$_4$, and CHCl$_3$ solution phases (5). Many of the vC=O frequencies have been corrected for Fermi resonance (vC=O in Fermi resonance with 2γC−H). As shown in Fig. 13.1, the vC=O frequencies generally decrease as the Hammett σ_p values decrease. This figure shows plots of vC=O for 4-X-benzaldehydes vs Hammett's σ_p value for the 4-X atom or group. The points on each line correspond to vC=O frequencies assigned for the 0–100 mol % CHCl$_3$/CCl$_4$ solutions (5). Figure 13.2 shows plots of vC=O for 4-X-benzaldehydes corrected for Fermi resonance in CCl$_4$ solution vs the frequency difference between vC=O corrected for Fermi resonance in CCl$_4$ solution minus vC=O corrected for FR for each of the mole % CHCl$_3$/CCl$_4$ solutions for each of the 4-X-benzaldehydes (5). These plots again demonstrate the effect of the 4-X substituent upon vC=O as well as the effect of the solvent system. The mathematical treatment of the experimental data presented here always yields a linear relationship.

Figure 13.3 shows plots of vC=O and an overtone in Fermi resonance and their corrected frequencies vs mole % CHCl$_3$/CCl$_4$ for 4-(trifluoromethyl) benzaldehyde. After correction for FR, it can be readily seen that the corrected vC=O and OT frequencies are closer in frequency than their uncorrected observed frequencies. Figure 13.4 shows plots of vC=O and OT in Fermi resonance and their corrected frequencies vs mole % CHCl$_3$/CCl$_4$ for 4-bromobenzaldehyde. In this case it is noted that the corrected vC=O and OT frequencies converge at ~45 mol % CHCl$_3$/CCl$_4$. Below this point corrected vC=O occurs at higher frequency than corrected OT and above this point corrected vC=O occurs at lower frequency than corrected OT. Without correction for FR, the uncorrected vC=O mode occurs at higher frequency than uncorrected OT. Without correction for FR, both observed bands are some combination of vC=O and OT, which changes as the mole % CHCl$_3$/CCl$_4$ changes. Other examples of both types as just discussed are presented by Nyquist et al. (5). The general decrease in frequency in these plots is attributed to the bulk dielectric effect of the solvent system together with intermolecular hydrogen bonding of form C=O···HCCl$_3$.

Table 13.3 lists the unperturbed νC=O frequencies of 4-X-benzaldehydes in various solvent systems (6). The 4-X-benzaldehydes designated with an asterisk (*) list the νC=O frequency corrected for Fermi resonance. The overall νC=O frequency range for 4-X-benzaldehydes is 1664.4–1718.1 cm^{-1}. The highest νC=O frequency is exhibited by 4-nitrobenzaldehyde in hexane and the lowest νC=O frequency is exhibited by 4-(dimethylamino) benzaldehyde in solution in methyl alcohol.

Figure 13.4 shows plots of the νC=O frequency for each of the 4-X-benzaldehydes in a solvent vs the νC=O frequency difference between the νC=O frequency in hexane solution and each of the other solvents. These plots are similar to the plots shown in Fig. 13.2.

Figure 13.5 shows a plot of νC=O for each of the 4-X-benzaldehydes in dimethyl sulfoxide solution vs Hammett's σ_p values. Other plots of νC=O for each of the 4-X-benzaldehydes in each of the other solvents show similar plots with 4-OH, 4-CH$_3$-O, and 4-CH$_3$S analogs not correlating as well as the other 4-X analogs.

Figure 13.6 shows a plot of νC=O for 4-(dimethylamino benzaldehyde vs the solvent acceptor number (AN) for each of the solvents. Solvents are listed sequentially 1 through 17 starting with hexane as 1 and 17 for methyl alcohol. Two essentially linear relationships are noted in this figure. The plot 12, 15, 16, and 17 represents νC=O···HOR frequencies for tert-butyl alcohol, isopropyl alcohol, ethyl alcohol. and methyl alcohol, respectively. The 4-(dimethylamino) benzaldehyde would be expected to form the strongest C=O···HOR bond, because the (CH$_3$)N group is the most basic and the weakest is the case of the 4-NO$_2$ analog. The lower linear relationship is for νC=O frequencies. Other 4-X-benzaldehydes in these same solvents show similar plots with somewhat more scattered data points (see (6) for other plots). This led to the conclusion that the solute/solvent interactions are complex because the solvent also interacts with certain 4-X groups as well as with the π system of the phenyl group. The AN values give a rough prediction of νC=O and νasym NO$_2$, but do not take into account all of the solute/solvent interactions (e.g., steric factors, relative basic sites in the solute, intermolecular hydrogen bonding between solvent molecules vs between solute and solvent molecules).

In general, the Hammett σ_p values for the 4-X atom or group for 4-X-benzaldehydes appear to correlate with the νC=O frequencies. However, due to scattering of the data points in the plots in each of the other solvent systems (6), these values do not appear to take into account intermolecular hydrogen bonding, the relative basicity of the C=O group, and the interaction of the solvent with other sites in the 4-X-benzaldehydes.

The molecular geometry of both solvent and solute molecules, the basic and/or acidic sites in both solute and solvent molecules, the dipolar interactions between solute and solvent molecules, the steric factor of solute molecules, and the concentration of the solute most likely determine the overall solute/solvent interaction. Therefore, parameters such as AN values and Hammett σ_p values can not be expected to exhibit universal linear relationships between IR or Raman group frequencies. However, they do help in predicting the direction of frequency shifts within a class of compounds (6).

Table 13.4 lists the aldehydic C—H in-plane bending mode for 4-X-benzaldehydes in the CCl$_4$ and CHCl$_3$ solution and in the vapor phase (7). In CCl$_4$ solution, δC—H occurs in the range 1381.3–1392.1 cm^{-1} and in CHCl$_3$ solution in the range 1382.9–1394.3 cm^{-1} for the 4-X-benzaldehydes. The δC—H frequencies are higher by 1.5 to 4.1 cm^{-1} in CHCl$_3$ solution. It is well known that the aldehydic CH stretching mode, νCH, and the first overtone of the aldehydic CH in-plane bending mode, 2δCH, are in Fermi resonance (8–12). Because δCH frequencies are dependent upon the solvent system, one might expect that the amount of Fermi resonance interaction between νC—H and 2δCH would also be dependent upon the solvent system. Table 13.5 lists the δCH, calculated 2δCH, and 2δCH frequencies corrected for Fermi resonance in both CCl$_4$ and CHCl$_3$ solutions. The agreement between the calculated and corrected 2δCH frequencies varies between ~0.4 to 15 cm^{-1}.

Table 13.6 lists IR data and assignments for the perturbed and unperturbed νC—H frequencies for the 4-X-benzaldehydes (7). The unperturbed νC—H frequencies for 4-X-benzaldehydes in CCl$_4$ solution occur in the range 2768.7–2789.7 cm^{-1} and in CHCl$_3$ solution in the range 2776.6–2808.9 cm^{-1}.

Figure 13.7 shows plots of half of the Fermi doublet for each of the 4-X-benzaldehydes in the range 2726–2746 cm^{-1} vs mole % CHCl$_3$/CCl$_4$ and Figure 13.8 show plots of half of the Fermi doublet in the range 2805–2845 cm^{-1} vs mole % CHCl$_3$/CCl$_4$. Figure 13.9 shows plots of unperturbed νCH for each of the 4-X-benzaldehydes vs mole % CHCl$_3$/CCl$_4$. These plots show that the Fermi doublet and the unperturbed νCH frequencies increase in frequency as the mole % CHC$_3$/CCl$_4$ is increased. However, the νC—H or 2δCH frequencies apparently do not correlate with Hammett σ_p values because the plots show that the frequencies do not increase or decrease in the order 1 through 13.

Submaxima are also noted for 4-X-benzaldehydes in the range 2726–2816 cm^{-1}, and these bands are assigned to combination tones.

Benzaldehyde and 4-phenylbenzaldehyde are examples of how the unperturbed νCH and unperturbed 2δCH cross over with change in the mole % CHCl$_3$/CCl$_4$. In the case of benzaldehyde the crossover is at ~55 mol % CHCl$_3$/CCl$_4$ and for 4-phenylbenzaldehyde the crossover is at ~35 mol % CHCl$_3$/CCl$_4$ (see Figures 13.10 and 13.11).

REFERENCES

1. Bellamy, L. J. and Williams, R. L. (1958). *J. Chem. Soc. London*, 3645.

2. Nyquist, R. A. (1984). *The Interpretation of Vapor-Phase Infrared Spectra: Group Frequency Data*, Philadelphia: Sadtler Research Laboratories, Div. of Bio-Rad Inc.

3. Lin-Vien, D., Colthup, N. B., Fately, W. G., and Grasselli, J. G. (1991). *The Handbook of Infrared and Raman Characteristic Frequencies or Organic Molecules*, Boston: Academic Press, Inc.

4. Schrader, B. (1989). *Raman/Infrared Atlas of Organic Compounds*, 2nd ed., Weinheim, Germany: VCH.

5. Nyquist, R. A., Settineri, S. E., and Luoma, D. A. (1991). *Appl. Spectrosc.*, **45**, 1641.

6. Nyquist, R. A. (1992). *Appl. Spectrosc.*, **46**, 306.

7. Nyquist, R. A. (1992). *Appl. Spectrosc.*, **46**, 293.

8. Pozefsky, A. and Coggeshall, N. D. (1951). *Anal. Chem.*, **23**, 1611.

9. Pinchas, S. (1955). *Anal. Chem.*, **27**, 2.

10. Saier, E. L., Coussins, L. R., and Basillia, M. R. (1962). *J. Phys. Chem.*, **66**, 232.

11. Saier, E. L., Cousins, L. R., and Basillia, M. R. (1962). *Anal. Chem.*, **34**, 824.

12. Rock, S. L. and Hammaker, R. M. (1971). *Spectrochim. Acta*, **27**, 1899.

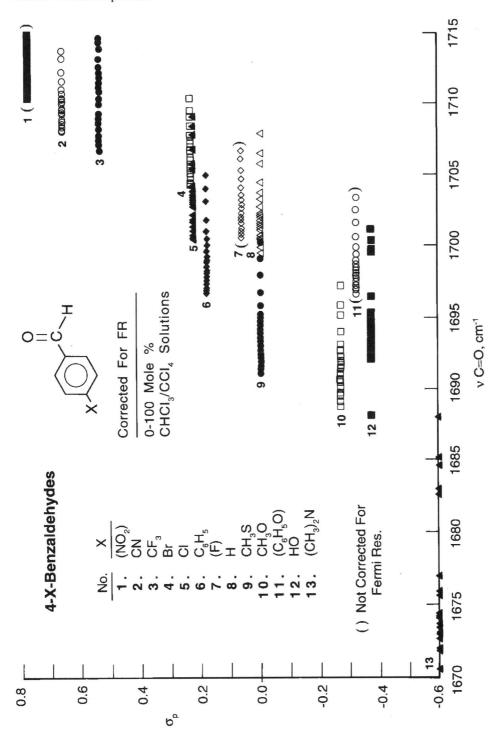

FIGURE 13.1 Plots of $\nu C{=}O$ corrected for Fermi resonance for each of the 4-X-benzaldehydes vs σ_p for the 4-X atom or group. The points on each line correspond to $\nu C{=}O$ frequencies for 0–100% $CHCl_3/CCl_4$ solutions.

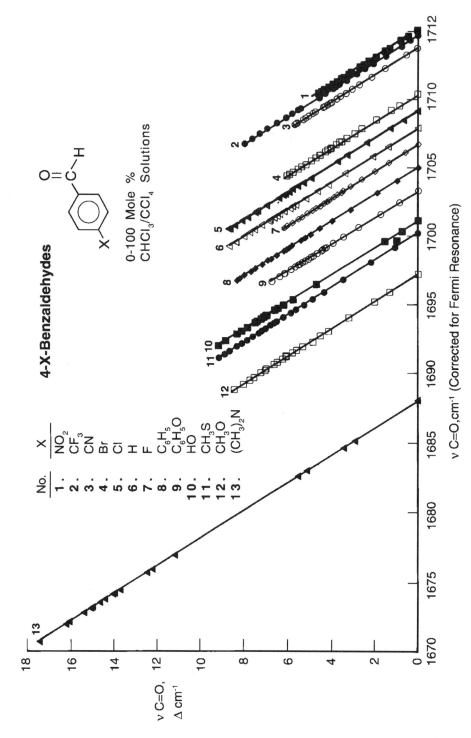

FIGURE 13.2 Plots of νC=O corrected for Fermi resonance in CCl₄ solution vs the frequency difference between (νC=O corrected for FR in CCl₄ solution minus νC=O corrected for FR for each of the mole % CHCl₃/CCl₄ solutions) for each of the 4-X-benzaldehydes.

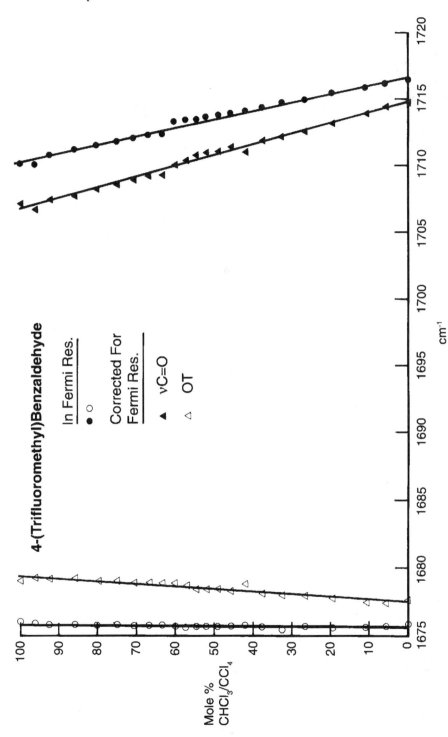

FIGURE 13.3 Plots of νC=O and an overtone in Fermi resonance and their corrected frequencies *vs* mole % CHCl₃/CCl₄ for 4-(trifluoromethyl) benzaldehyde.

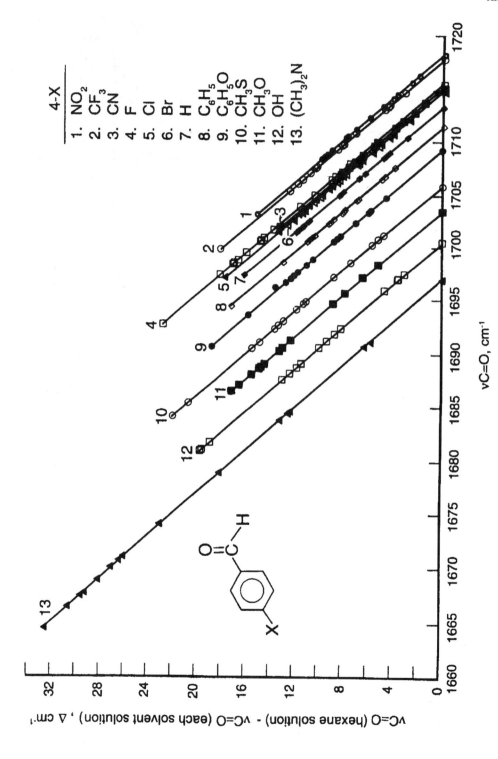

FIGURE 13.4   Plots of the $\nu C{=}O$ frequency for each of the 4-X-benzaldehydes in a solvent vs the $\nu C{=}O$ frequency difference between the $\nu C{=}O$ frequency in hexane solution and the $\nu C{=}O$ frequency in each of the other solvents used in the study.

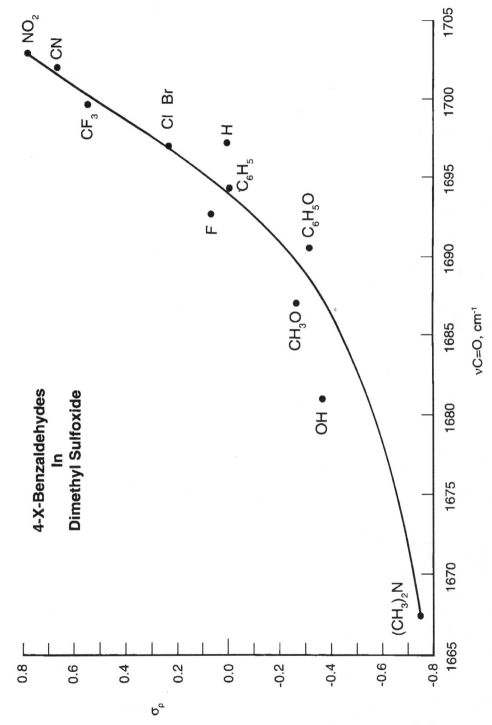

FIGURE 13.5   A plot of $\nu C{=}O$ vs Hammett σ_p for each of the 4-X-benzaldehydes in dimethyl sulfoxide solution.

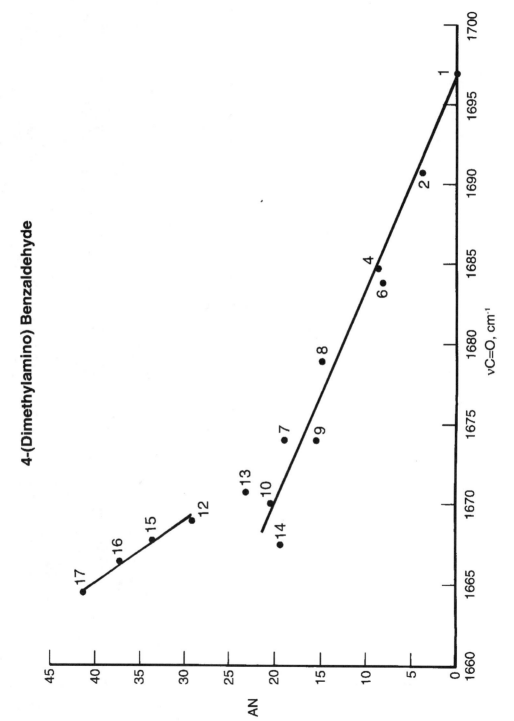

FIGURE 13.6   A plot of νC=O for 4-(dimethylamino) benzaldehyde *vs* the solvent acceptor number (AN) for each of the solvents.

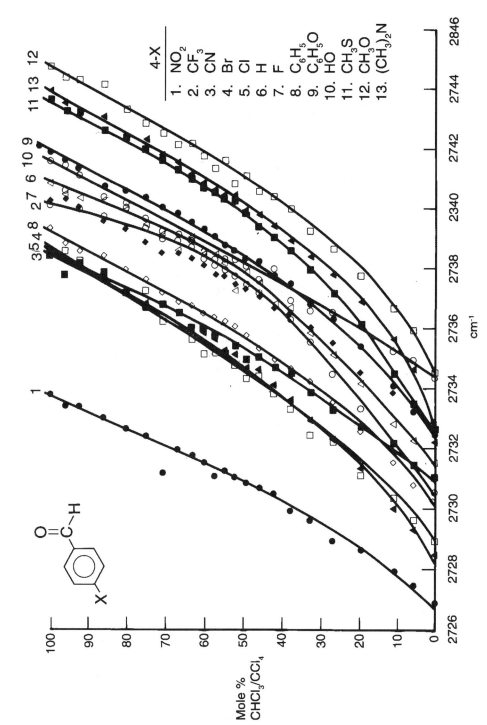

FIGURE 13.7 Plots of half of the Fermi doublet for each of the 4-X-benzaldehydes in the range 2726–2746 cm^{-1} vs mol % CHCl$_3$/CCl$_4$.

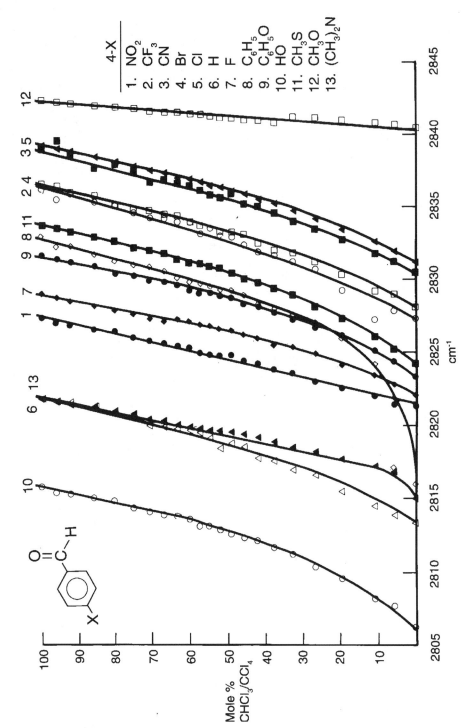

FIGURE 13.8   Plots of half of the Fermi doublet for each of the 4-X-benzaldehydes in the range 2805–2845 cm^{-1} vs mole % CHCl$_3$/CCl$_4$.

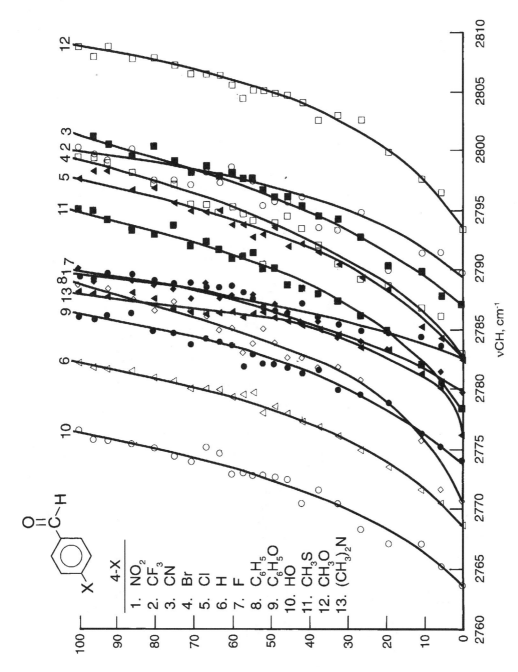

FIGURE 13.9   Plots of unperturbed νCH for each of the 4-X-benzaldehydes vs mole % $CHCl_3/CCl_4$.

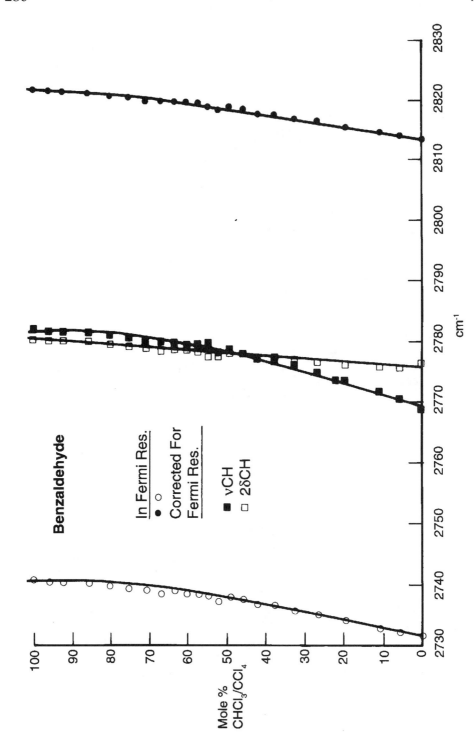

FIGURE 13.10   Plots of νCH and $2\delta CH$ in Fermi resonance and νCH and δCH corrected for FR for benzaldehyde vs mole % $CHCl_3/CCl_4$.

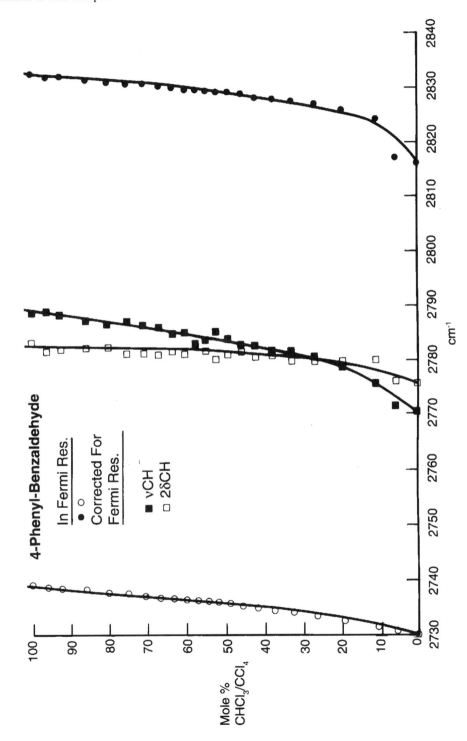

FIGURE 13.11   Plots of vCH and 2δCH in Fermi resonance and vCH and δCH corrected for FR for 4-phenylbenzaldehyde vs mole % CHCl$_3$/CCl$_4$.

TABLE 13.1 A comparison of IR data of nonconjugated aldehydes in different phases

Aldehyde	2(C=O str.) vapor cm^{-1}	C=O str. vapor cm^{-1}	C=O str. CCl$_4$ soln. cm^{-1}	2(C=O str.) neat cm^{-1}	C=O str. neat cm^{-1}	CH str. in F.R. neat cm^{-1}	2(CH bend) in F.R. neat cm^{-1}	CH bend neat cm^{-1}
Butyr				3430	1721	2820	2720	1390
Isobutyr				3430	1721	2810	2710	1397
3-Cyclohexenyl	3462	1741		3420	1731	2830	2700	1392
Phenethyl	3452	1743		3410	1715	2810	2722	1400
Chloroacet			1730*[1]					
Dichloroacet			1748*[1]			vapor	vapor	vapor
Trifluoroacet		1788						
Trichloroacet		1778	1768*[1]			2859	2700	1365
Tribromoacet	3501	1760				2841	2680	1352

*[1] See Reference 1.

TABLE 13.1A Raman data and assignments for aldehydes

Aldehyde	C=O str.	C—H str. in F.R.	2(delta-C—H) in F.R.	delta-C—H	gama-C—H	C=C str.
Propion*	1723(7,p)	2830(9,p)	2718(9,p)	1337(1,p)	849(11,p)	
Isobutyr*	1730(5,p)	2813(5,p)	2722(9,p)		798(14,p)	
Hexanal*	1723(10,p)		2723(15,p)	1303(7)	891(7,p)	
Heptanal*	1723(9,p)		2723(14,p)	1303(7)	895(5,p)	
Bromal	1744(10,p)	2850(7,p)		1353(2,p)	785(9,p)	
Acrolein*	1688(42,p)	2812(3,p)		1359(32,p)		1618(22,p)
Benz*	1701	2815	2732			
Salicyl*	1633(22,P)					

*Reference 4.

TABLE 13.2 IR data and assignments for 4-X-benzaldehydes in the vapor and CHCl$_3$ and CCl$_4$ solution

4-X-Benzaldehyde x-Group	C=O str. [vapor] cm^{-1}(A)	C=O str. [CCl$_4$] cm^{-1}(A)	C=O str. [CHCl$_3$] cm^{-1}(A)	[vapor]- [CCl$_4$] cm^{-1}	[vapor]- [CHCl$_3$] cm^{-1}	[CCl$_4$]- [CHCl$_3$] cm^{-1}
NO$_2$	1728(1.289)	1715.0(0.261)	1710.4(0.417)	13	17.6	5.4
CN		1713.7*	1708.1*			
CF$_3$		1714.6*	1707.1*			
Br	1720(1.250)	1710.4	1704.3*	19.6	15.7	6.1
Cl	1722(1.250)	1709.1*	1700.5*	12.9	21.5	8.6
C$_6$H$_5$		1705.0*	1696.7*			
F	1719(1.230)	1706.7(0.862)	1700.6(0.808)	12.3	18.4	6.1
H		1707.9*	1699.2*			
CH$_3$S		1700.2*	1691.1*			
CH$_3$O	1717(1.240)	1697.2*	1688.8*	19.8	28.2	8.4
HO	1715(1.250)	1701.1*	1688.2*	13.9	26.8	12.9
(CH$_3$)$_2$N	1711(1.042)	1688.1*	1671.8*	22.9	39.2	16.3

* Corrected for Fermi Res.

TABLE 13.3 The C=O stretching frequency for 4-X-benzaldehydes in various solvents

Solvent	4-NO$_2$ cm^{-1}	4-CF$_3$*† cm^{-1}	4-CN*† cm^{-1}	4-F cm^{-1}	4-Cl cm^{-1}	4-Br*† cm^{-1}	4-H cm^{-1}
Hexane	1718.1	1717.8	1715.2	1715.7*	1714.8	1714.8	1713.3
Diethyl ether	1714.1	1713.3	1711.9	1709.7	1710.4	1709.6	1709.1
Methyl t-butyl ether	1713.3	1713.4	1711.3	1706.4	1709.7	1708.1	1708.8
Carbon tetrachloride	1714.4	1714.6	1713.7	1706.7	1710.7	1710.4	1708.5
Carbon disulfide	1712.2	1713.1	1712.8	1705.1	1708.2	1709.5	1706.6
Benzene	1711.1	1710.7	1709.6	1703.9	1706.7	1707.8	1705.2
Acetonitrile	1709.2	1707.6	1707.6	1701.6	1703.9	1703.6	1702.5
Nitrobenzene	1708.6	1705.4	1707.1	1700.9	1703.1	1704.8	1702.1
Benzonitrile	1708.4	1707.6	1706.9	1700.7	1701.4	1703.3	1701.8
Methylene chloride	1710.3	1707.5	1708.7	1698.7*	1703.6	1704.3	1702.4
Nitromethane	1708.5	1706.4	1707.1	1698.5*	1702.6	1702.1	1701.5
t-Butyl alcohol	1716.2	1709.6	1707.9	1697.3*	1712.1	1706.2	1710.4
Chloroform	1710.4	1717.1	1708.1	1700.6	1703.4	1704.3	1701.9
Dimethyl sulfoxide	1703.1	1699.7	1702.1	1697.7	1697.1	1697.1	1697.2
Isopropyl alcohol	1716.2	1709.7	1709.9	1698.4*	1709.1	1706.9	1708.6
Ethyl alcohol	1715.6	1709.3	1708.7	1699.4*	1705.6	1706.4	1707.1
Methyl alcohol	1708.8	1706.1	1707.2	1697.5*	1704.6	1705.1	1705.1
Tetrahydrofuran	1709.9	1710.4		1704.8	1705.8	1707.4	1705.1
1,2-Dichlorobenzene	1703.7	1708.8		1702.7	1705.8	1705.4	1703.9
Range	(1718.2–1803.1)	(1717.8–1699.7)	(1715.2–1702.1)	(1715.4–1692.7)	(1714.8–1697.1)	(1714.8–1697.1)	(1713.3–1697.2)

	4-C$_6$H$_5$	4-C$_6$H$_5$O	4-CH$_3$S	4-CH$_3$O*†	4-OH	4-N(CH$_3$)$_2$*†	
Hexane	1711.4	1709.3	1705.9	1703.5		1696.9	
Diethyl ether	1707.6	1704.7	1700.8	1698.3	1696.9	1690.7	
Methyl t-butyl ether	1706.7	1703.5	1701.2	1697.1	1697.4	1691.1	
Carbon tetrachloride	1706.5	1703.4	1700.2	1695.6	1700.5	1684.6	
Carbon disulfide	1704.6	1701.1	1698.5	1694.7	1697.1	1684.4	
Benzene	1703.5	1700.7		1691.2	1695.9	1683.8	
Acetonitrile	1701.1	1697.5	1694.1	1690.6	1689.1	1674.1	
Nitrobenzene	1700.8	1697.3	1692.6	1690.6	1689.1	1678.9	
Benzonitrile	1700.8	1697.3	1692.3	1690.4	1688.2	1674.1	
Methylene chloride	1701.1	1697.1	1693.1	1690.3	1689.1	1670.1	
Nitromethane	1700.8	1696.2	1694.7	1689.1	1688.8	1671.1	
t-Butyl alcohol	1698.6	1693.5	1685.3	1686.4	1681.7	1668.9	
Chloroform	1700.5	1696.6	1691.1	1688.8	1687.6	1670.6	
Dimethyl sulfoxide	1694.3	1690.6		1687.1	1681.1	1667.4	
Isopropyl alcohol	1702.4	1702.3	1685.3	1688.1	1680.9	1667.7	
Ethyl alcohol	1704.5	1701.1	1684.1	1688.6	1692.3	1666.4	
Methyl alcohol	1702.6	1698.3	1690.5	1687.1	1690.6	1664.4	
Tetrahydrofuran	1703.4	1700.6	1697.2	1697.8	1691.9	1685.4	
1,2-Dichlorobenzene	1702.3	1696.8	1694.8	1694.4	1691.2	1681.3	
Range	(1711.4–1694.3)	(1709.3–1690.6)	(1705.9–1684.1)	(1703.5–1686.4)	(1696.9–1680.9)	(1696.9–1666.4)	
Overall Range in all solvents	(1718.1–1664.4)						
Delta C=O	53.7						

† See the explanatory text on page 269 for discussion of material designated by an asterisk.

TABLE 13.4   The CH bending vibration for 4-X-benzaldehydes in CCl$_4$ and CHCl$_3$ solutions and in the vapor phase

4-X-Benzaldehyde X	Vapor cm^{-1}	1%(wt./vol.) PCCl$_4$] cm^{-1}	1%(wt./vol.) [CHCl$_3$] cm^{-1}	[CHCl$_3$]–[CCl$_4$] cm^{-1}	[vapor]–[CHCl$_3$] cm^{-1}
NO$_2$		1382.92	1384.79	1.87	
CF$_3$		1386.36	1388.61	2.25	
CN		1381.28	1382.89	1.61	
Br	1387	1383.12			
Cl	1381	1383.6	1385.8	2.23	−4.8
O		1387.35	1390.11	2.76	
F	1386	1385.91	1388.39	2.48	−2.4
C$_6$H$_5$		1383.96	1385.45	1.49	
C$_6$H$_5$O		1386.52	1388.98	2.46	
OH		1393.46			
CH$_3$S		1387.34	1390.03	2.69	
CH$_3$O		1390.2	1394.31	4.11	
(CH$_3$)$_2$N	1385	1392.07	1394.28	2.21	−9.3
Range		(1381.28–1392.07)	(1382.89–1394.28)	(1.49–4.11)	

TABLE 13.5 The overtone of CH bending in Fermi resonance with =CH stretching for 4-X-benzaldehydes corrected for Fermi resonance

4-X-benzaldehydes 1%(wt./vol.)	4-NO₂ CH bend cm⁻¹	4-CF₃ CH bend cm⁻¹	R–CN CH bend cm⁻¹	4-Br CH bend cm⁻¹	4-Cl CH bend cm⁻¹	4-H CH bend cm⁻¹	4-F CH bend cm⁻¹	C₆H₅ CH bend cm⁻¹	C₆H₅O CH bend cm⁻¹	OH CH bend cm⁻¹	CH₃S CH bend cm⁻¹	CH₃O CH bend cm⁻¹	4-N(CH₃)₂ CH bend cm⁻¹
CCl₄	1382.92	1386.36	1381.28		1383.61	1387.35	1385.91		1386.52		1387.34	1390.2	1392.07
CHCl₃	1384.79	1388.61	1382.89		1385.84	1390.11	1388.39		1388.98		1390.03	1394.31	1394.28
delta-CH bend	1.87	2.25	1.61		2.23	2.76	2.48		2.46		2.69	4.11	2.21
[calculated 2(CH bend)]													
CCl₄	2765.84	2772.72	2762.56	2766.24	2767.22	2774.7	2771.82		2773.04		2774.68	2780.4	2784.14
CHCl₃	2769.58	2777.22	2765.78	2775.53	2771.68	2780.22	2776.78		2777.96		2780.06	2788.62	2788.56
delta-2(CH bend)	3.74	4.5	3.22		4.46	5.52	4.96		4.92		5.38	8.22	4.42
2(CH bend) corrected for Fermi Res.													
CCl₄	2765.47	2771.99	2774.31	2774.08	2782.93	2779.82	2775.08	2775.57	2781.68	2773.1	2778.39	2781.58	2770.98
CHCl₃	2771.7	2775.96	2777.44	2775.53	2788.78	2788.78	2779.13	2783.4	2787.33	2780.74	2782.08	2778.21	2777.53
delta-CH bend CCl₄ CHCl₃	6.23	3.97	3.13	7.84	5.85	8.96	4.05	7.83	5.65	7.64	3.69	-3.37	6.55

TABLE 13.6 IR data and assignments for the observed and corrected for Fermi resonance CH stretching frequencies for 4-X-benzaldehydes in CCl$_4$ and CHCl$_3$ solutions

4-X-Benzaldehydes 1% solutions X	obs. =CH str. CCl$_4$ cm^{-1}	obs. =CH str. CHCl$_3$ cm^{-1}	[=CH str. corrected for F.R.] CCl$_4$ cm^{-1}	[=CH str. corrected for F.R.] CHCl$_3$ cm^{-1}	obs. =CH str. [CHCl$_3$]-[CCl$_4$] cm^{-1}	cor. F.R. [CHCl$_3$]-[CCl$_4$] cm^{-1}	[obs.] [cor. FR.] CCl$_4$ cm^{-1}	[obs.] [cor. FR.] CHCl$_3$ cm^{-1}
NO$_2$	2821.3	2827.3	2782.7	2789.4	6	6.7	38.6	37.9
CF$_3$	2827.3	2836.2	2789.7	2800.3	8.9	10.6	37.6	35.9
CN	2830.4	2838.9	2787.1	2799.9	8.5	12.8	43.3	39
F	2822.1	2829	2779.8	2790.1	6.9	10.3	42.3	38.9
Cl	2828.1	2836.6	2782.9	2799.5	8.5	16.6	45.2	37.1
H	2813.3	2821.8	2768.7	2782.2	8.5	13.5	44.6	39.6
C$_6$h$_5$	2815.9	2832.8	2770.6	2788.8	16.9	18.2	45.3	44
OH	2806.2	2815.7	2763.6	2776.6	9.5	13	42.6	39.1
C$_6$H$_5$O	2823.3	2831.6	2774	2786.1	8.3	12.1	49.3	45.5
CH$_3$S	2824.2	2833.6	2778.4	2795.2	9.4	16.8	45.8	38.4
CH$_3$O	2840.5	2842.3	2793.4	2808.9	1.8	15.5	47.1	33.4
(CH$_3$)$_2$N	2814.9	2821.8	2776.1	2788.2	6.9	12.1	38.8	33.6

Ketones

*Numbers in parentheses indicate in-text page reference.

Acetone is the simplest member of the ketone series. Its empirical structure is $CH_3-C(=O)-CH_3$. Table 14.1 lists IR and Raman data for acetone and acetone-d_6 (1,2). The CD_3 frequencies and assignments are listed directly under those for the CH_3 frequencies and assignments. The frequency ratios for CH_3/CD_3 vary between 1.11 and 1.38. The B_1 and B_2 CH_3 rocking mode to CD_3 rocking mode frequency ratios are 1.132 and 1.108, respectively, and this indicates that these two modes are coupled with B_1 and B_2 modes, respectively. These date illustrate that both IR and Raman data and deuterated analogs are required to make detailed assignments of molecular compounds.

SOLVENT-INDUCED FREQUENCY SHIFTS

Hallam has reviewed the literature concerning attempts to develop an accurate quantitative and physical meaningful explanation of solvent-induced stretching frequencies (3). Kirkwood et al. (4) and Bauer and Magot (5) related the observed frequency shifts and the dielectric constant ε of the solvent. The (K)irkwood (B)auer (M)agot work resulted in the KBM equation:

$$(v_{vapor} - v_{solution}/v\text{vapor} = \Delta v/v = C[(\varepsilon - 1)/(2\varepsilon + 1)]$$

Josien and Fuson refined the equation to include a term based on the index of refraction of the solvent (6). Bellamy et al. found that $\Delta v/v$ for any solute plotted vs $(\Delta v/v)$ for any other solvent within a class of compounds produced a linear curve (7). They therefore predicted that group frequency shifts were local association effects between solute and solvent and not dielectric effects. Bellamy et al. proposed that v hexane should be substituted for v vapor in the KBM equation in order to negate the effects of phase change (7). Table 14.2 illustrates the application of the KBM equation using IR data for acetone $vC=O$ frequencies in various solvents. Table 14.2 shows that the KBM equation predicts the $vC=O$ mode within -2.1 to $+14.3\,cm^{-1}$. The best fit is for acetone in solution with acetonitrile and the worst fit is for acetone in water. In the case of $CHCl_3$ and the four alcohols, the $vC=O$ frequency differences between the calculated and observed range are between 4.3 and 7.7 cm^{-1}. The larger differences here compared to the other solvents are most likely the result of intermolecular hydrogen bonding ($C=O\cdots HCCl_3$ or $C=O\cdots HOR$).

Table 14.2a contains the calculated values for $A - 1/2A + 1$ and $X-Y/X$ where A and Y are equal to 0 to 85 and X equals 85 (see the KBM equation). These two sets of data are plotted in Fig. 14.1. These data show that any set of numbers using the equivalent of the KBM equation or Bellamy's proposal yields a mathematical curve. In particular, the linear plot $X-Y$ vs Y is thus meaningless in predicting $vC=O$ frequencies (8). Intermolecular hydrogen bonding and dipolar interaction between solute and solvent as well as dielectric effects must play a role in the $vC=O$ frequencies of carbonyl containing compounds. Steric factors, which also play a role between solute-solvent interaction, must also be considered in predicting $vC=O$ frequencies in any particular physical phase (see in what follows).

Table 14.3 lists the $C=O$ stretching frequencies for aliphatic ketones in the vapor phase and in 1% wt./vol. in various solvents (9).

In this series of ketones (dimethyl ketone through di-tert-butyl ketone) the steric factor of the alkyl group(s) and the basicity of the carbonyl group both increase. As the steric factor of the alkyl group increases the intermolecular distance between the carbonyl group and a solvent molecule increases. In the case of intermolecular hydrogen bonding between a solvent and a proton and a carbonyl group, the strength of the hydrogen bond depends upon at least four factors. These are, the basicity of the carbonyl group, the acidity of the solvent proton, the steric factor of the dialkyl groups of the dialkyl ketone, and the steric factor of the atoms or groups of the solvent molecules not involved directly with the intermolecular hydrogen bond.

All of the aliphatic ketones exhibit their νC=O mode at its highest frequency in the vapor phase (1699–1742 cm^{-1}). In solution, the highest νC=O frequencies are exhibited in hexane solution (1690.3–1727.2 cm^{-1}). With the exception of dimethyl ketone (acetone) in hexane, the νC=O frequency for the other dialkyl ketones decreases in frequency with increasing negative values for both σ^* (increasing electron release to the carbonyl group) and E_s (an increasing steric factor of the alkyl group) and the summation of σ^* times the summation of $E_s \cdot 10^{-2}$.

In the case of dimethyl ketone, its νC=O mode occurs at a higher frequency than νC=O for methyl ethyl ketone only in the following solvents, tert-butyl alcohol, chloroform, isopropyl alcohol. ethyl alcohol, and methyl alcohol. In these solvents there is intermolecular hydrogen bonding between solute and solvent (C=O$\cdots$HOR and C=O$\cdots$HCCl$_3$). Moreover, in these protic solvents the νC=O frequency order for methyl ethyl ketone and diethyl ketone is reversed from the sequence that they exhibit when these dialkyl ketones are in the other solvents.

SOLUTE-SOLVENT INTERACTION AFFECTED BY STERIC FACTORS

Table 14.3a shows a comparison of the carbonyl stretching frequency difference for ketones in hexane solution and in each of the other solvents (9). The strength of an intermolecular hydrogen bond (C=O$\cdots$HOR or C=O$\cdots$HCCl$_3$) is proportional to this frequency difference. The larger this frequency difference, the stronger the intermolecular hydrogen bond. Or, in other words, the stronger the intermolecular hydrogen bond between the ketone solute and the protic solvent, the more νC=O shifts to lower frequency when compared to νC=O for the same ketone in n-hexane solution. The most acid proton for the alcohol series is that for methyl alcohol, and the least acidic proton is that for tert-butyl alcohol. In addition, the steric factor is the largest for tert-butyl alcohol, and the least for methyl alcohol. The most basic ketone carbonyl group is in the case of di-tert-butyl ketone and the least basic carbonyl group is in the case of dimethyl ketone. Neglecting steric factors, these facts would predict that the strongest intermolecular hydrogen bonds would be formed between methyl alcohol and di-tert-butyl ketone, and the weakest between tert-butyl alcohol, and dimethyl ketone. Study of Table 14.3a shows that the strongest intermolecular hydrogen is actually formed between methyl alcohol and diisopropyl ketone, and the strength of the intermolecular hydrogen bond with diisopropyl ketone decreases in the order methyl alcohol, ethyl alcohol, isopropyl alcohol, and tert-butyl alcohol. The strength of the intermolecular hydrogen bond formed between diisopropyl ketone and chloroform falls between that for tert-butyl alcohol and isopropyl alcohol. In the alcohol series, the strength of the intermolecular hydrogen is also stronger between diisopropyl ketone than between di-tert-butyl ketone. The strength of the intermolecular hydrogen bond is less in the case of ethyl isopropyl ketone compared to diisopropyl ketone, but it also is stronger than in the case of di-tert-butyl ketone. These data show that steric factors increase the C=O$\cdots$H intermolecular hydrogen bond distance, thus weakening the possible strength of this intermolecular hydrogen bond.

The ketone νC=O frequency shifts in nonprotic solvents are also less in the case of di-tert-butyl ketone vs the other dialkyl ketones. Therefore, steric factors of the alkyl groups also play a role in the dielectric effects of the solvent upon the carbonyl group.

Table 14.3b shows a comparison of the carbonyl stretching frequency difference for dialkyl ketones in methyl alcohol and other protic solvents (9). In this case, the strength of the intermolecular hydrogen bond (C=O$\cdots$H) decreases as the number increases. This comparison shows that steric factors also affect the strength of the intermolecular hydrogen bond.

Table 14.3c shows a comparison of the differences in the carbonyl stretching frequencies of dialkyl ketones in hexane solution and in alcohol solution for solute molecules not intermolecularly hydrogen bonded (9). These data show that the frequency difference decreases for these ketones in alcohol solution, progressing in the series methyl alcohol through tert-butyl alcohol. With increased branching on the α-carbon atom of the C—OH group, the intermolecular polar effect due to the alcohol oxygen atom is decreased; thus, there is a lesser polar effect upon dialkyl ketone carbonyl groups surrounded by intermolecularly hydrogen-bonded alcohol molecules progressing in the series methyl alcohol through tert-butyl alcohol. The νC=O mode decreases in frequency as the polarity of the solvent increases.

Table 14.4 lists data for the C=O stretching frequencies for n-butyrophenone and tert-butyrophenone in 0–100 mol% $CHCl_3/CCl_4$ solutions (2% wt./vol.). The νC=O mode for both n-butyrophenone (1691–1682.6 cm^{-1}) and tert-butyrophenone (1678.4–1674.1 cm^{-1}) decreases in frequency as the mole % $CHCl_3/CCl_4$ is increased from 0–100 cm^{-1}. However, it is noted that the νC=O frequency for n-tert-butyrophenone decreases in frequency by only one-half as much as that for n-butyrophenone (4.3 cm^{-1}/9 cm^{-1} = 0.48) in going from solution in CCl_4 to solution in $CHCl_3$. The C=O group for the tert-butyro analog is more basic than the n-butyro analog, and on this basis one would expect a stronger intermolecular hydrogen bond to be formed between C=O$\cdots$HCCl$_3$ for the tert-butyro analog than for the n-butyro analog, and it is noted that this is not the case. The reason for this is that the steric effect of the tert-butyro group prevents the Cl$_3$CH proton from coming as close in space to the C=O oxygen atom in the case of tert-butyrophenone, which prevents it from forming as strong a C=O$\cdots$HCCl$_3$ bond as in the case of the n-butyro analog where the n-butyro group has a lesser steric factor.

Figure 14.2 show a plot of the νC=O frequency for tert-butyrophenone vs the mole % $CHCl_3/CCl_4$. The resulting curve is nonlinear due to the formation of C=O$\cdots$HCCl$_3$ hydrogen bonds. The general decrease in frequency is due to the dielectric effects of the solvent system. The carbonyl stretching frequencies for n-tert-butyrophenone occur at lower frequencies (1678.4–1674.1 cm^{-1}) compared to those for di-tert-butyl ketone (1685.9–1680.7 cm^{-1}) in 0–100 mol% $CHCl_3/CCl_4$ due to conjugation of the phenyl group with the carbonyl group, which weakens the C=O bond (10).

INDUCTIVE, RESONANCE, AND TEMPERATURE EFFECTS

Table 14.5 list IR data for acetone, α-chloroacetone, acetophenone, and benzophenone in CS_2 solution between ~29 and -100°C (11). Figure 14.3 shows plots of the carbonyl stretching frequencies for these four compounds vs the temperature of the CS_2 solution in °C. These plots show that all of the vC=O modes decrease in frequency as the temperature is lowered from room temperature. Two vC=O frequencies are noted in the case of α-chloroacetone, and both occur at a higher frequency than vC=O for acetone. The inductive effect of an α-Cl atom increases the vC=O frequency, and the inductive effect is independent of spatial orientation. There is a field effect of a Cl atom near in space to the carbonyl oxygen atom, and it also causes the vC=O mode to increase in frequency. Thus, rotational conformer I is assigned to the higher frequency vC=O band, while rotational conformer II is assigned to the lower frequency vC=O band in the case of α-chloroacetone. The concentration of rotational conformer I increases while the concentration of rotational conformer II decreases with decrease in temperature (11).

Substitution of one or two phenyl groups for one or two methyl groups of acetone yields acetophenone (vC=O, 1689.4 cm^{-1}) and benzophenone (vC=O, 1663.3 cm^{-1}), respectively. Thus, the first phenyl group causes vC=O to decrease in frequency by 28.1 cm^{-1} while the second phenyl group causes vC=O to decrease in frequency by an additional 26.1 cm^{-1}. The phenyl group(s) is (are) conjugated with the carbonyl group, and it weakens the C=O bond, which causes its vC=O mode to vibrate at lower frequency.

OTHER CHEMICAL AND PHYSICAL EFFECTS

Table 14.6 shows a comparison of the vC=O frequencies for 2% wt./vol. ketone in 0–100 mol% $(CH_3)_2$ SO/CCl$_4$ solutions. Figure 14.4 shows plots of vC=O vs mole % $(CH_3)_2$ SO/CCl$_4$ for the same six ketones as shown in Table 14.6. All six curves decrease in frequency in a linear manner as mole % $(CH_3)_2$ SO/CCl$_4$ is increased from ~30 to 100 mol% $(CH_3)_2$SO/CCl$_4$. This is in the order of the increasing polarity of the solvent system. All six ketones appear to be affected in the same manner because the linear portion of the curves is parallel. The vC=O frequencies decrease in the order acetone, 2,4,6-trimethylacetophenone, 4-nitroacetophenone, acetophenone, 4-methoxyacetophenone, and benzophenone. The effect of conjugation was discussed previously. In the case of 2,4,6-trimethylacetophenone, the carbonyl group and the phenyl group are not coplanar; therefore, the C=O group is not conjugated with the phenyl group. Thus, the C=O group is higher in frequency than that exhibited by acetophenone by 13 cm^{-1}, but lower in frequency than acetone by 14 cm^{-1}. Hammett σ_p values for 4-nitro and 4-methoxy benzophenone cause the vC=O mode to be higher and lower in frequency, respectively, than vC=O for acetophenone.

OTHER CONJUGATED CARBONYL CONTAINING COMPOUNDS

It is interesting to consider the possible molecular configurations of conjugated carbonyl containing compounds. Lin-Vien et al. (13) have reviewed the published studies of these compounds (14), and they report that a planar compound such as 3-buten-2-one exists in s-trans and s-cis configurations in CCl_4 solution. These two conformers for 3-buten-2-one are illustrated here:

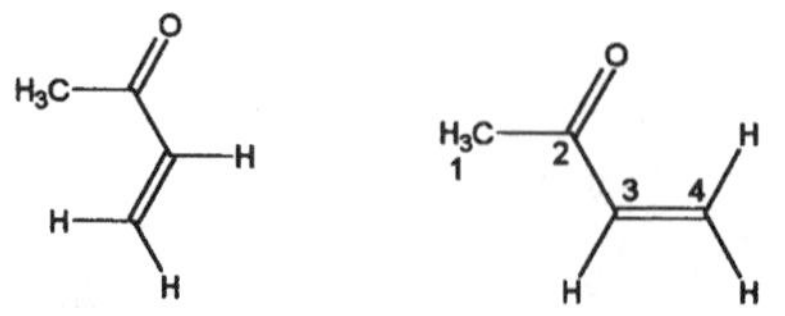

s-trans	s-cis
CCl₄ soln, (cm⁻¹)	CCl₄ soln. (cm⁻¹)
C=O stretching 1687 vs	C=O stretching 1707 vs
C=C stretching 1648 w	C=C stretching 1618 m, sh
(C=O str.)-(C=C str.) 39	(C=O str.)-(C=C str.) 89

These two conformers result from $180°$ rotation of the C=C group about the C^2-C^3 single bond. This notation adequately describes the molecular configurations in the forementioned case (13–15).

In the vapor phase the C=O str. and C=C str. frequencies are assigned at 1715 and $1627\,cm^{-1}$, respectively (16). The frequency separation between these two modes is $88\,cm^{-1}$. Corresponding modes for the s-trans isomer are not detected in the vapor phase at elevated temperature. Therefore, 3-butene-2-one exists only as the s-cis conformer at elevated temperature in the vapor phase. The IR bands at 986 and $951\,cm^{-1}$ confirm the presence of the $CH=CH_2$ group.

Similarly, a compound such as 3-methyl-3-buten-2-one can also be adequately defined as s-trans and s-cis conformers as illustrated here:

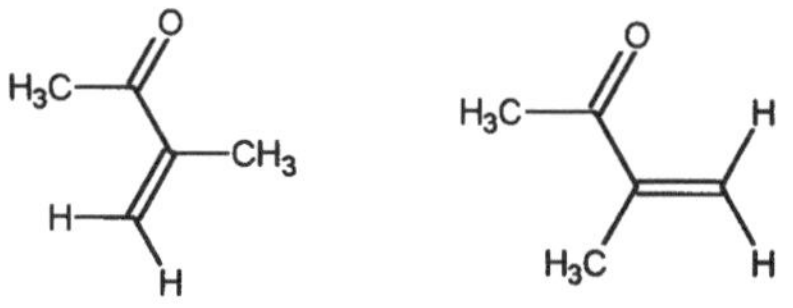

In the vapor phase at elevated temperature the IR bands at $1700\,cm^{-1}$ and $1639\,cm^{-1}$ are assigned to C=O stretching and C=C stretching, respectively (16). The frequency separation between these two modes is $61\,cm^{-1}$. Therefore, 3-methyl-3-buten-2-one exists only as the s-cis conformer in the vapor-phase at elevated temperature. The IR band at $929\,cm^{-1}$ confirms the presence of the $C=CH_2$ group.

Let us now consider the number of possible conformers for 4-methyl-3-buten-2-one:

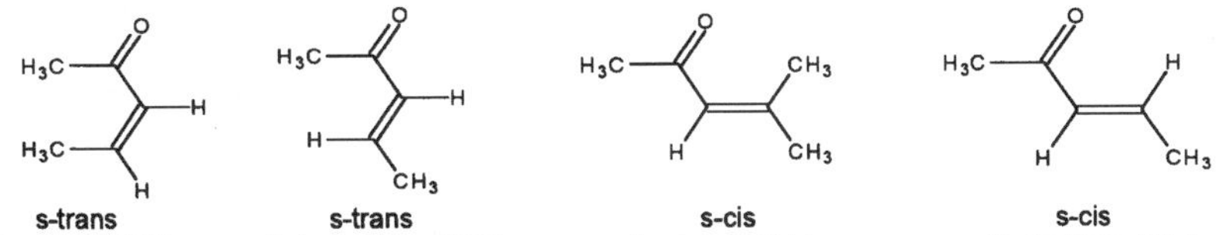

| s-trans | s-trans | s-cis | s-cis |
| [s-trans, cis CH₃] | [s-trans, trans CH₃] | [s-cis, cis CH₃] | [s-cis, trans CH₃] |

Here we note that s-trans and s-cis do not define the spatial position of the CH_3 group. Therefore, the additional term trans CH_3 and cis CH_3 must be used to adequately specify each of the four possible conformers for 4-methyl-3-buten-2-one as shown in brackets in the conformers shown here.

The C=O and C=C stretching frequencies for 4-methyl-3-buten-2-one in CCl_4 solution are given here:

s-trans conformer (cm⁻¹)		(C=O str.)-(C=C str.) (cm⁻¹)
C=O str. (vs)	C=C str. (m,sh)	
1674	1654	29

s-cis conformer (cm⁻¹)		
C=O str. (s)	C=C str. (s)	
1692	1632	60

These data support only the presence of the s-cis or s-trans part of the conformer. NMR data are needed to help establish the presence of a cis or trans CH_3 group, and these data were not available. Similar compounds containing the trans CH=CH group exhibit a weak-medium band in the region 974–980 cm⁻¹.

As already shown, the C=C and C=O stretching frequencies for s-cis and s-trans conformers are very different. The question to answer is why they are different. It is possible that in one case the C=C and C=O stretching vibrations couple into in-phase and out-of-phase stretching modes in one conformer and not in the case of the other conformer.

In the case of 3-methyl-1,3-pentadiene the two C=OC–C=C groups are coupled into an in-phase $(C=C)_2$ vibration and an out-of-phase $(C=C)_2$ vibration as depicted here:

in-phase $(C=C)_2$ str.   out-of-phase $(C=C)_2$ str.

In the case of 3-methyl-1,3-pentadiene, the in-phase str. mode occurs at 1650 cm⁻¹ and the out-of-phase str. mode occurs at 1610 cm⁻¹ in the vapor phase. In cases such as 2-methyl-2-pentene and 2,4,4-trimethyl-2-pentene the C=C bond is not conjugated and the C=C stretching mode occurs at 1665 and 1658 cm⁻¹, respectively. Therefore, it appears that the in-phase $(C=C)_2$ stretching vibration occurs near that expected for isolated C=C stretching vibrations while the out-of-phase $(C=C)_2$ stretching vibrations occur considerably lower than isolated C=C stretching vibrations.

The same behavior for the C=O and C=C stretching modes was already noted here for the s-cis conformers. The C=O str. mode occurred at a frequency expected for a conjugated carbonyl containing compound, while the C=C str. mode occurred at a lower frequency than expected for an isolated C=C double bond. On the other hand, the s-trans conformers exhibited frequencies for C=O and C=C stretching expected for conjugated carbonyl containing compounds while the C=C stretching frequency occurred at frequencies comparable to those exhibited by compounds containing isolated trans CH=CH groups. On this basis, we believe that these modes are best described as in-phase and out-of-phase C=C−C=O stretching vibrations in the case of the s-cis conformers, and as C=O and C=C stretching modes in the case of the s-trans conformers. Compounds such as 3-methyl-4-phenyl-3-buten-2-one and α-hexylcinnamaldehyde contain the C=CH group, and we are only able to establish that they are in the s-cis configuration. The cis [H, CH$_3$] and cis [H, C$_6$H$_{11}$] are one of the two possibilities for these two compounds. The bands in the region 867–870 cm^{-1} support the presence of the C=CH group.

Table 14.7 also lists IR vapor-phase data for chalcone and its derivatives. The IR data is recorded at elevated temperature, and all of the data indicate that these compounds exist only as the s-cis, trans CH=CH conformer.

CHALCONES

Chalcones have the following empirical planar structure:

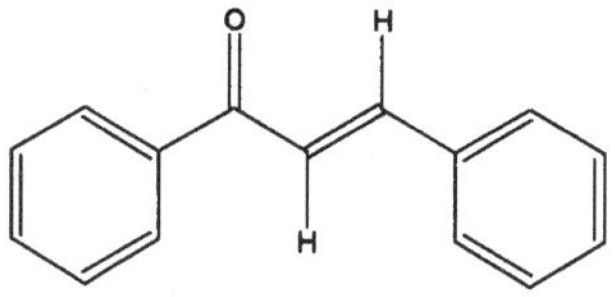

The phenyl group of the styryl group is numbered 2 through 6, and the phenyl group of the benzoyl group is numbered 2′ through 6′. Substitution in the 2,6-positions with Cl$_2$ or (CH$_3$)$_2$ would sterically prevent the styryl phenyl group from being coplanar with the rest of the molecule. Moreover, substitution of Cl$_2$ or (CH$_3$)$_2$ in the α,2-positions on the styryl group would also sterically prevent the styryl phenyl group from being coplanar with the rest of the molecule. Substitution of Cl$_2$ or (CH$_3$)$_2$ in the 2′,6′ -positions would sterically prevent the phenyl group of the benzoyl group from being coplanar with the rest of the molecule. The six chalcones studied, (see Table 14.6) exhibit vC=O in the region 1670–1684 cm^{-1}, and exhibit vC=C in the region 1605–1620 cm^{-1}. The frequency separation between vC=O and vC=C varies between 59 and 73 cm^{-1} (16). These data indicate that these chalcones exist in planar s-cis configurations. Noncoplanar chalcones were not available for study.

Table 14.7a lists some fundamental vibrations for the conjugated ketones studied. These group frequencies aid in identifying these compounds by additional spectra-structure identification.

INTRAMOLECULAR HYDROGEN BONDING

Table 14.8 lists IR data for 2-hydroxy-5-X-acetophenone in CCl$_4$ solution (3800–1333 cm^{-1}) and CS$_2$ solution (1333–400 cm^{-1}). The intramolecular vOH$\cdots$O=C and γOH$\cdots$O=C vibrations for 2-hydroxy-5-X-acetophenone were presented in Chapter 7.

The $\nu C=O\cdots HO$ frequencies for 2-hydroxy-5-X-acetophenones occur in the range 1641–1658 cm^{-1} (17). These compounds exhibit $\nu C=O\cdots HO$ at lower frequency by 40 ± 10 cm^{-1} compared to nonhydrogen bonded acetophenones due to the strength of the $C=O\cdots HO$ bond. In the solid phase (Nujol mull) the $\nu C=O\cdots HO$ mode occurs 13 to 17 cm^{-1} lower in frequency than in CCl$_4$ solution.

The 2-hydroxy-5-X-acetophenones exhibit characteristic vibrations in the range 954–973 cm^{-1} [C−C(=)−C stretching], 1283–1380 cm^{-1} [phenyl-0 stretching], and 1359–1380 cm^{-1} [symmetric CH$_3$ bending].

CYCLOALKANONES

In the vapor phase cycloalkanones exhibit $\nu C=O$ frequencies in the range 1719–1816 cm^{-1} (2). The frequencies decrease as the number of carbon atoms in the cycloalkanone ring increase from 4 to 8 and 10 (1816, 1765, 1732, 1721, 1720, and 1719 cm^{-1}, respectively). The behavior of the $\nu C=O$ frequency is attributed to changes in the C−C(=)−C bond angle. During a cycle of $C=O$ stretching, more or less energy is required to move the carbonyl carbon atom as the C−C(=)−C bond angle becomes smaller or larger than the normal C−C(=)−C bond angle for an open chain ketone such as dimethyl ketone (acetone). This is because during a cycle of $C=O$ stretching, the C−C(=)−C angle must increase, and as the size of the cycloalkanone C−C(=)−C angle decreases from normal bond angles (cyclohexanone for example), the more difficult it is for a normal $C=O$ vibration to occur. Conversely, in cases where the bond angle is larger than normal, the easier it is for the $\nu C=O$ vibration to occur.

Table 14.9 lists IR vapor-phase data and assignments for cyclobutanone and cyclopentanone. The $\nu C=O$ frequencies were already discussed here. It should be noted that νasym. CH$_2$, νsym. CH$_2$, CH$_2$ twisting, the ring deformation, and the first overtone of $\nu C=O$ also decrease in frequency while the CH$_2$ bending mode increases in frequency as the ring size is increased from four to five carbon atoms.

Table 14.10 lists the $C=O$ stretching frequencies for cyclopentanone and cyclohexanone in the vapor, neat, and solution phases (18).

Cyclopentanone exhibits $\nu C=O$ at 1765 (vapor) and 1739.2 cm^{-1} in neat phase after correction for Fermi resonance (18). In all solutions, $\nu C=O$ has been corrected for Fermi resonance. Cyclopentanone exhibits $\nu C=O$ at 1750.6 cm^{-1} in n-hexane solution and at 1728.8 cm^{-1} in water solution. Cyclohexanone exhibits $\nu C=O$ at 1723 cm^{-1} in n-hexane solution and at 1701 cm^{-1} in ethyl alcohol solution. After correction for Fermi resonance, $\nu C=O$ for cyclopentanone decreases in frequency by approximately 17.1 cm^{-1} progressing in the series of solvents hexane through methyl alcohol (18). Progressing in the same series of solvents, $\nu C=O$ for cyclohexanone decreases in frequency by approximately 22 cm^{-1}. The $C=O$ group for cyclohexanone is more basic than the $C=O$ group for cyclopentanone, and this is given as the reason that there is more of a solute-solvent interaction in the case of cyclohexanone than in the case of cyclopentanone (18). The $\nu C=O$ frequencies for these two cycloalkanones do not correlate well with the solvent acceptor numbers (AN), and this is attributed to steric factors of the solvents that hinder solute-solvent interaction.

Figure 14.5 shows a plot of $\nu C=O$ for cyclohexanone vs mole % CHCl$_3$/n-C$_6$H$_{14}$. Definite breaks in the plot are noted at $\sim$2.5 to 1, $\sim$6 to 1, $\sim$50 to 1, and $\sim$62.1 to 1 mol of CHCl$_3$ to 1 mol cyclohexanone. The cause of these $\nu C=O$ frequency shifts is most likely a result of different hydrogen bonding complexes between C=O and CHCl₃ which changes with increasing CHCl₃ concentration, that is,

$$(CH_2)_5C=O \cdots HCCl_3$$

$$(CH_2)_5C=O \cdots HCCl_3(HCCl_3)_n$$

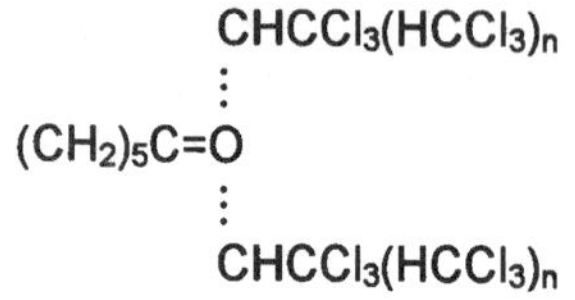

$$CHCCl_3(HCCl_3)_n$$
$$\vdots$$
$$(CH_2)_5C=O$$
$$\vdots$$
$$CHCCl_3(HCCl_3)_n$$

The general decrease in the νC=O frequency most likely is the result of continual change in solvent dielectric effect. Figure 14.6 shows a plot of νC=O for cyclohexanone vs mole % CCl₄/n-C₆H₁₄. This linear plot decreases in frequency as the mole % CCl₄/n-C₆H₁₄ increases. The dielectric effect of this solvent mixture increases as the mole % CCl₄/n-C₆H₁₄ increases, causing νC=O to occur at lower frequency in a linear manner. Unlike CHCl₃, there are no different CCl₄ solute complexes as noted.

Figure 14.7 show a plot of νC=O for 0.345 mol% acetone in CHCl₃/CCl₄ solution vs the mole % CHCl₃/CCl₄ (19). This plot shows that it is linear over the mole % CHCl₃/CCl₄ range of ~17 –100%. Extrapolation of the linear plot to zero mol% CHCl₃ indicates that the νC=O frequency for acetone in the range 0–17% ratio CHCl₃/CCl₄ varies from linearity by ~1 cm⁻¹. The mole fraction of CHCl₃ is in excess of the 0.345 mol% acetone present, even at the 1.49 mol% ratio CHCl₃/CCl₄ where the CHCl₃ protons forms weak hydrogen bonds between Cl atoms of other CHCl₃ molecules and CCl₄ molecules as well as with the carbonyl oxygen atom (19).

Figure 14.8 shows a plot of the νC=O frequency for acetone vs the reaction field for each of the mole % CHCl₃/CCl₄ solutions. Comparison of Fig. 14.7 with Fig. 14.8 shows that the curves are identical. The reaction field $|R| = \dfrac{(e-1)}{2e+n^2}$, with e the dielectric constant of each solvent, and n the refractive index of each solvent. A plot of mole % CHCl₃/CCl₄ vs the reaction field yields a linear curve (19). Therefore, it appears as though the refractive index of the solvent as well as the dielectric value of the solvent system together with intermolecular hydrogen bonding with C=O of the solute affects the induced frequency shift of νC=O in solution with CHCl₃/CCl₄. In summary, the frequency behavior of the solvent-induced ketone carbonyl stretching vibration, νC=O, is affected by the reaction field, inductive effects, and solute-solvent intermolecular hydrogen bonding (29).

Table 14.11 lists IR data for 14H-dibenzo[a, j]xanthen-14-one in CHCl₃/CCl₄ and in various solvents (20). This ketone has the following empirical structure:

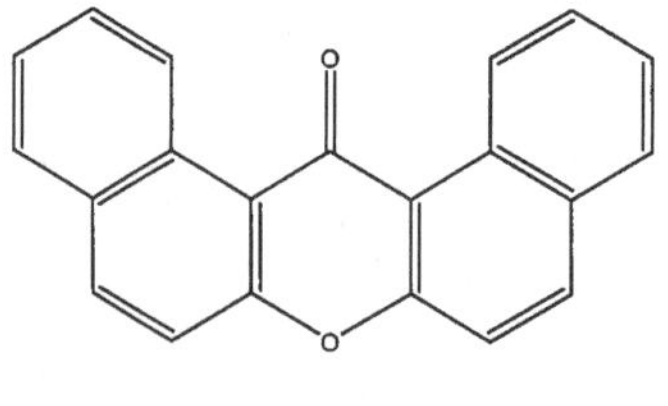

For simplicity, this compound is given the name DX-14-O. The maximum symmetry for DX-14-O is C_{2v}. The $vC=O$ mode belongs to the A_1 species if it has C_{2v} symmetry. The DX-14-O has two significant IR bands in the region expected for $vC=O$, and a solution study in $CHCl_3/CCl_4$ solution was used to help explain the presence of these two IR bands. In order for $vC=O$ to be in Fermi resonance in the case of DX-14-O both the combination (CT) or overtone (OT) and $vC=O$ must belong to the A_1 symmetry species. In addition, the CT or OT would have to occur in the range expected for $vC=O$. It is obvious from the ketone structure given here that the IR doublet could not be due to the presence of rotational isomers.

Figure 14.9 shows IR spectra of DX-14-O in the region $1550-1800\,cm^{-1}$. Spectrum (A) is for a saturated solution in hexane, spectrum (B) is for a saturated solution in carbon tetrachloride, and spectrum (C) is for a 0.5% solution in chloroform. In hexane the IR bands occur at 1651.9 and $1636.8\,cm^{-1}$, in carbon tetrachloride the bands occur at 1648.9 and $1634.9\,cm^{-1}$, and in chloroform the bands occur at 1645.4 and $1633.1\,cm^{-1}$. Inspection of their IR spectra shows that the absorbance ratio of the low frequency band to the high frequency band increases in the solvent order $n-C_6H_{14}$, CCl_4, $CHCl_3$. Figure 14.10 shows a plot of $vC=O$ and the OT or CT in Fermi resonance, and $vC=O$ and OT or CT corrected for Fermi resonance. The corrected data show that unperturbed $vC=O$ occurs at higher frequency than unperturbed OT or CT at mole % $CHCl_3/CCl_4$ below ~28%; at mole % $CHCl_3/CCl_4$ above ~28% unperturbed $vC=O$ occurs at a lower frequency than unperturbed OT or CT. Without FR correction, each IR band results from some combination of $vC=O$ and the OT or CT. At the ~28 mol% $CHCl_3/CCl_4$, both IR bands result from equal contributions of $vC=O$ or OT or CT.

Figure 14.11 shows plots of $vC=O$ and OT or CT and their corrected frequencies vs the solvent acceptor number (AN) for each of the eight solvents, numbered 1–8, and listed sequentially. These plots show that in general the two modes in Fermi resonance and unperturbed $vC=O$ decrease in frequency as the AN of the solvent is increased. The scattering of data points suggests that the AN values do not take into account the steric factor of the solvent, which causes variance in the solute-solvent interaction. It should be noted that unperturbed $vC=O$ occurs at lower frequency than the OC or OT in only chloroform and benzonitrile solutions.

SUBSTITUTED 1,4-BENZOQUINONES

The ketone 1,4-benzoquinone has the following planar structure:

It has two C=O groups, and these couple into an in-phase $(C=O)_2$ stretching vibration, $vip(C=O)_2$, and an out-of-phase $(C=O)_2$ stretching vibration, $v_{op}(C=O)_2$. In CCl_4 solution, 1,4-benzoquinone exhibits a strong IR band at $1670\,cm^{-1}$ and a medium strong band at $1656\,cm^{-1}$. Without consideration of the molecular symmetry of 1,4-benzoquinone, it would seem reasonable to assign the $1670\,cm^{-1}$ band to $v_{op}(C=O)_2$ and the $1656\,cm^{-1}$ band to $v_{ip}(C=O)_2$. However, 1,4-benzoquinone has a center of symmetry and it has V_h symmetry. The 30 fundamentals are distributed as $6A_g$, $1B_{1g}$, $3B_{2g}$, $5B_{1u}$, 5_{2u}, and $3B_u$. Only the u classes are IR active, and only the g classes are Raman active. The $v_{op}(C=O)_2$ mode belongs to the b_{1u} species, and the $v_{ip}(C=O)_2$ mode belong to the A_g species. Of course, 1,4-benzoquinone can not have rotational conformers. Therefore, one of the forementioned IR bands either results from the presence of an impurity, or else it must result from a B_{1u} combination tone in Fermi resonance with the $v_{op}(C=O)_2$, b_{1u} fundamental. It could not be in Fermi resonance from an overtone of a lower lying fundamental, because a first overtone would belong to the A_g species. The Raman band at 1661.4 cm^{-1} in CCl$_4$ solution is assigned to the $v_{ip}(C=O)_2$, A_g mode.

Figure 14.12 shows plots of $v_{op}(C=O)$, b_{1u} and the CT B_{1u} modes in Fermi resonance, and their unperturbed frequencies after correction for Fermi resonance for 1,4-benzonone in 0.5% wt./vol. or less in 0–100 mol% CHCl$_3$/CCl$_4$. The two observed IR band frequencies in Fermi resonance in this case increase in frequency as the mole % CHCl$_3$/CCl$_4$ is increased. However, unperturbed $v_{op}(C=O)_2$ decreases in frequency as the mole % CHCl$_3$/CCl$_4$ increases, and this is always the case for other carbonyl containing compounds as the mole % CHCl$_3$/CCl$_4$ is increased. It is noted that the unperturbed CTb$_{1u}$ mode increases in frequency as the mole % CHCl$_3$/CCl$_4$ is increased from 0–100%. In this case, at ~25 mol% CHCl$_3$/CCl$_4$ both IR bands result from equal contributions from $v_{op}(C=O)_2$ and the CT b_{1u} mode.

Table 14.12 lists IR and Raman data for several 1,4-benzoquinones in CCl$_4$ and CHCl$_3$ solutions (at 0.5% wt/vol. or less due to saturation). The point group pertaining to their molecular symmetry is given for each of these ketones. None of the other 1,4-benzoquinones show IR evidence for the $v_{op}(C=O)_2$ mode being in Fermi resonance. The $v_{op}(C=O)_2$ mode for these 1,4-benzoquines occurs in the range 1657–1702.7 cm^{-1} in CCl$_4$ solution and at slightly lower frequency in CHCl$_3$ solution. In CHCl$_3$ solution, $v_{ip}(C=O)_2$ occurs in the range 1666.9–1697.7 cm^{-1}. The increasing inductive effect of the halogen atoms (progressing in the order Br, Cl, F) together with their field effect increase both $v_{op}(C=O)_2$ and $v_{ip}(C=O)_2$ frequencies (22).

Tables 14.13–14.17 list IR data for tetrafluoro-1,4-benzoquinone, tetrachloro-1,4-benzoquinone, tetrabromo-1,4-benzoquinone, chloro-1,4-benzoquinone, and 2,5-dichlorobenzo-quinone in three different solvent systems.

In the IR, tetrafluoro-1,4-benzoquinone exhibits strong IR bands at 1702.7 and 1667.6 cm^{-1} in CCl$_4$ solution and at 1701.4 and 1668.4 cm^{-1} in CHCl$_3$ solution. In all cases the higher frequency band has more intensity than the lower frequency band. These two IR are assigned to $v_{op}(C=O)_2$ and $v_{op}(C=C)$, respectively. Figure 14.13 shows plots of $v_{op}(C=O)_2$ and $v_{op}(C=C)_2$ for 1,4-tetrafluorobenzoquinone vs mole% CHCl$_3$/CCl$_4$. The $v_{op}(C=O)_2$ mode decreases in frequency as expected as the mole% CHCl$_3$/CCl$_4$ is increased. The $v_{op}(C=C)_2$ ring mode increases in frequency as the mole% CHCl$_3$/CCl$_4$ is increased.

The frequency behavior of the other substituted 1,4-benzophenones is discussed in detail in Reference 21, and the reader is referred to this paper for further information on these interesting solute-solvent interactions.

Table 14.18 lists IR data for 3,3′,5,5′-tetraalkyl-1,4-diphenoquinones in CHCl$_3$ solution and in the solid phase (22). The 3,3′,5,5′-tetraalkyl-1,4-diphenoquinones have the following empirical structure:

When the 3,3',5,5' positions are substituted with identical atoms or groups, the compounds have D_{2h} symmetry (22). These molecules have a center of symmetry, and only the $v_{op}(C=O)$, B_{3u} fundamental is IR active. The $v_{op}(C=O)_2$, A_g fundamental is only Raman active. The $v_{op}(C=O)_2$ mode for the 3,3',5,5'-tetraalkyl-1,4-diphenoquinones occurs in the range 1586–1602 cm^{-1} in the solid phase and in the range 1588–1599 cm^{-1} in CHCl$_3$ solution. The compound 3,3'-dimethyl, 5,5'-di-tert-butyl-1,4-diphenoquinone has C_{2v} symmetry, and in this case both $v_{ip}(C=O)_2$ are IR active as well as Raman active. In this case the IR band at 1603 cm^{-1} is assigned to both $v_{op}(C=O)_2$ and $v_{ip}(C=O)_2$. The frequencies in brackets in Table 14.18 are calculated. All of these five 3,3',5,5'-tetraalkyl-1,4-diphenoquinones exhibit a weak IR band in the range 3168–3204 cm^{-1} in the solid phase and in the range 3175–3200 cm^{-1} in CHCl$_3$ solution. These bands are assigned to the combination tone $v_{op}(C=O)_2 + v_{ip}(C=O)_2$. Using the observed $v_{op}(C=O)_2$ and combination tone frequencies for these compounds, the $v_{ip}(C=O)_2$ frequencies are calculated to occur in the range 1579–1601 cm^{-1} in the solid phase and in the range 1583–1602 cm^{-1} CHCl$_3$ solution (22). These data confirm a previous conclusion that diphenoquinones exhibit a strong IR band near 1600 cm^{-1}, which must include stretching of the C=O bond (23). Thus, it is possible to distinguish between 4,4'-diphenoquinones and 1,4-benzoquinones, since the latter compounds exhibit carbonyl stretching modes 30 to 80 cm^{-1} higher in frequency (21).

CONCENTRATION EFFECTS

Table 14.19 lists data that show the dependence of the vC=O frequency of dialkyl ketones upon the wt./vol.% ketone in solution with CCl$_4$ or CHCl$_3$ (10). In CCl$_4$ solution, the vC=O mode for diisopropyl ketone decreases more in frequency in going from ∼0.8% to 5.25% than it does for di-tert-butyl ketone at comparable wt./vol. ketone in CCl$_4$ solution (−0.19 to −0.08 cm^{-1} at 5.25% wt./vol.). In CHCl$_3$ solution, the shift of vC=O is in the opposite direction to that noted for CCl$_4$ solutions. At 5.89% wt./vol. in CHCl$_3$ solution, vC=O increase 0.38 cm^{-1} for diisopropyl ketone and at 5.78 wt./vol. in CHCl$_3$ for di-tert-butyl ketone the increase is 0.1 cm^{-1}. The smaller vC=O frequency shifts in the case of the di-tert-butyl analog compared to the diisopropyl analog is attributed to steric factors of the alkyl group. The steric factor of the tert-butyl groups does not allow as much solute-solvent interaction between C=O and the solvent as it does in the case of the diisopropyl analog. With increase in the wt./vol. of ketone/CHCl$_3$ vC=O increases in frequency, indicating that the strength of the C=O···HCCl$_2$ Cl···(HCCl$_2$Cl)$_n$ intermolecular hydrogen bond becomes weaker as n becomes smaller.

REFERENCES

1. Schrader, B. (1989). *Raman/Infrared Atlas of Organic Compounds*, 2nd ed., Weinheim, Germany: VCH.
2. Nyquist, R. A. (1984). *The Interpretation of Vapor-Phase Infrared Spectra: Group Frequency Data*, Philadelphia: Sadtler Research Laboratories, a Division of Bio-Rad Labortories, Inc.
3. Hallam, H. E. (1963). *Infra-Red Spectroscopy and Molecular Structure*, p. 420, M. Davies, ed., New York: Elsevier.
4. Kirkwood, J. G., West, W., and Edwards, R. T. (1937). *J. Chem. Phys.*, 5, 14.
5. Bauer E. and Magot, M. (1938). *J. Phys. Radium*, 9, 319.

6. Josien M. L. and Fuson, N. (1954). *J. Chem. Phys.*, **22**, 1264.

7. Bellamy, L. J., Hallam, H. E., and Williams, R. L. (1959). *Trans. Farad. Soc.*, **55**, 1677.

8. Nyquist, R. A. (1989). *Appl. Spectrosc.*, **43**, 1208.

9. Nyquist, R. A. (1994). *Vib. Spectrosc.*, **7**, 1.

10. Nyquist, R. A., Putzig, C.L., and Yurga, L. (1989). *Appl. Spectrosc.*, **43**, 983.

11. Nyquist, R. A. (1986). *Appl. Spectrosc.*, **40**, 79.

12. Nyquist, R. A., Chrzan, V., and Houck, J. (1989). *Appl. Spectrosc.*, **43**, 981.

13. Lin-Vien, D., Colthup, N. B., Fatelely, W. G., and Grasselli, J. G. (1991). *The Handbook of Infrared and Raman Characteristic Group Frequencies of Organic Molecules*, San Diego: Academic Press, Inc.

14. Bowles, A. J., George, W. O., and Maddams, W. F. (1969) *J. Chem. Soc.*, B, 810.

15. Cottee, F. H., Straugham, B. P., Timmons, C. J., Forbes, W. F., and Shilton, R. (1967). *J. Chem. Soc.*, B, 1146.

16. (1982) *Sadtler Standard Infrared Vapor phase Spectra*, Philadelphia: Sadtler Research Laboratories, a Division of Bio-Rad, Inc.

17. Nyquist, R. A. (1963). *Spectrochim. Acta*, **19**, 1655.

18. Nyquist, R. A. (1990). *Appl. Spectrosc.*, **44**, 426.

19. Nyquist, R. A., Putzig, C. L., and Hasha, D. L. (1989). *Appl. Spectrosc.*, **43**, 1049.

20. Nyquist, R. A., Luoma, D. A., and Wilkening, D., (1991). *Vib. Spectrosc.*, **2**, 61.

21. Nyquist, R. A., Luoma, D. A., and Putzig, C. L. (1992). *Vib. Spectrosc.*, **3**, 181.

22. Nyquist, R. A. (1982). *Appl. Spectrosc.*, **36**, 533.

23. Gordon J. M. and Forbes, J. W. (1968). *Appl. Spectrosc.*, **15**, 19.

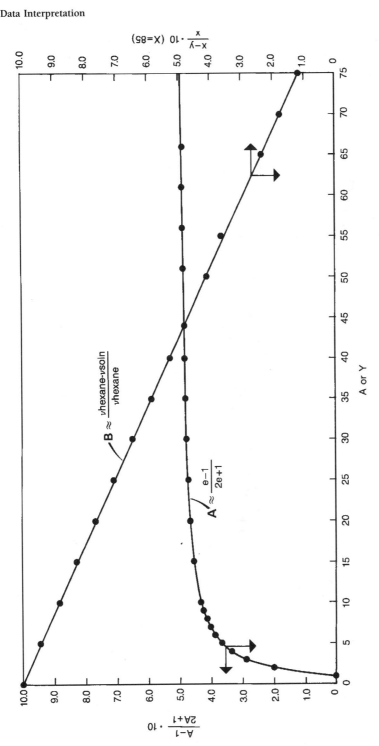

FIGURE 14.1 Plots of the number for A or Y vs the corresponding calculated value multiplied by a factor of 10 (see Table 14.2a).

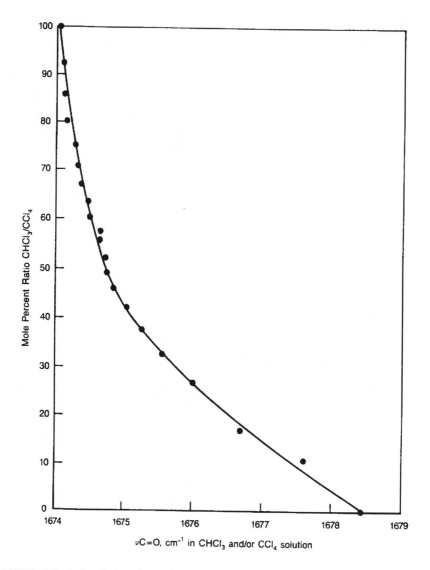

FIGURE 14.2 A plot of $\nu C{=}O$ for tert-butyrophenone (phenyl tert-butyl ketone) vs mol % $CHCl_3/CCl_4$.

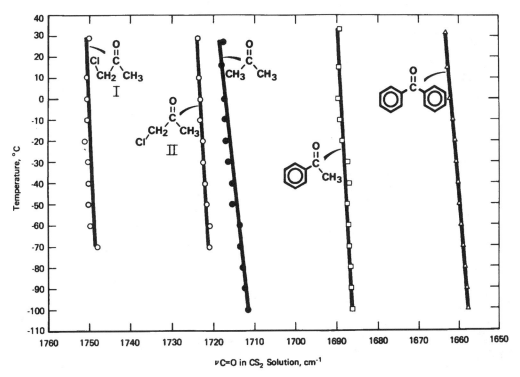

FIGURE 14.3 Plots of the νC=O frequencies for acetone, α-chloroacetone, acetophenone, and benzophenone in CS$_2$ solution between $\sim$29 and $-100\,^{\circ}$C.

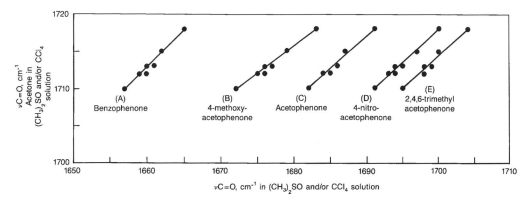

FIGURE 14.4 Plots of νC=O for 2% wt./vol. solutions of (A) benzophenone, (B) 4-methoxyacetophenone, (C) acetophenone, (D) 4-nitroacetophenone, (E) 2,4,6-trimethylacetophononone, and (F) acetone in mole % (CH$_3$)SO/CCl$_4$ solutions.

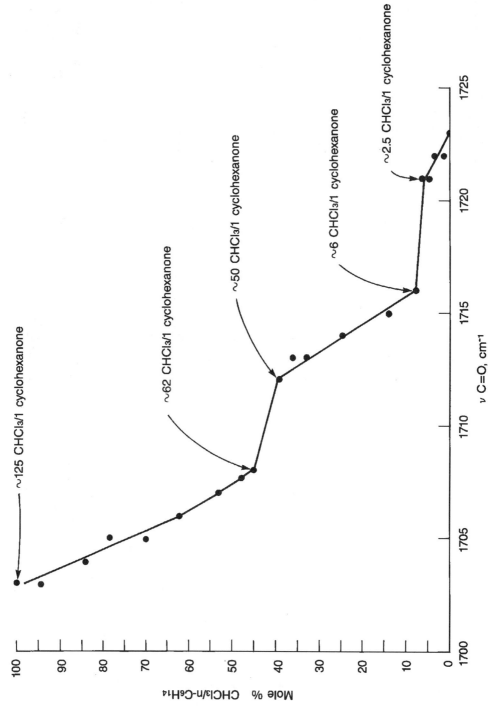

FIGURE 14.5   A plot of νC=O for 1% wt./vol. cyclohexanone vs mole % $CHCl_3/n\text{-}C_6H_{14}$ solutions.

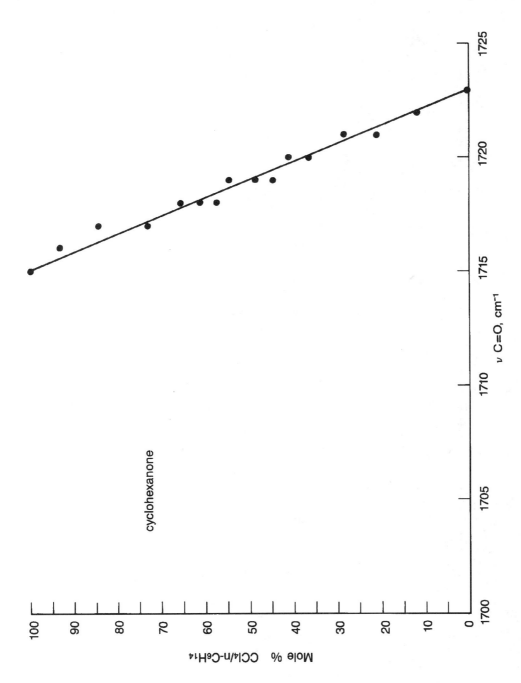

FIGURE 14.6   A plot of $\nu C=O$ for 1% wt./vol. cyclohexanone vs mole % $CCl_4/n\text{-}C_6H_{14}$.

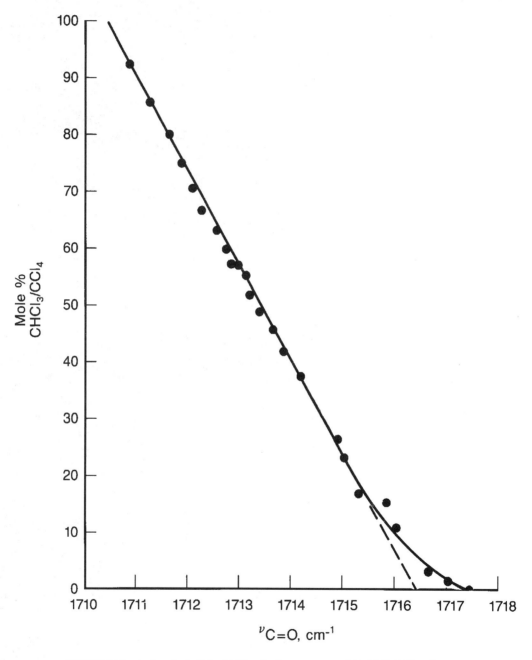

FIGURE 14.7   A plot of νC=O for 0.345 mole % acetone vs mol % CHCl$_3$/CCl$_4$ solutions.

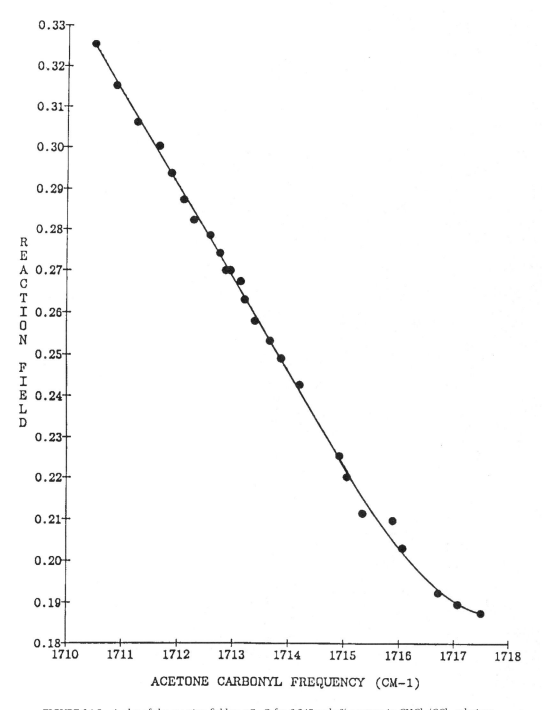

FIGURE 14.8   A plot of the reaction field vs νC=O for 0.345 mole % acetone in $CHCl_3/CCl_4$ solutions.

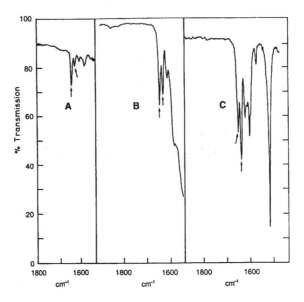

FIGURE 14.9 IR spectra for 14H-dibenzo [a,j] X anthen-14-one. (A) Saturated solution in hexane; (B) saturated solution in carbon tetrachloride; (C) 0.5% wt./vol. solution in chloroform.

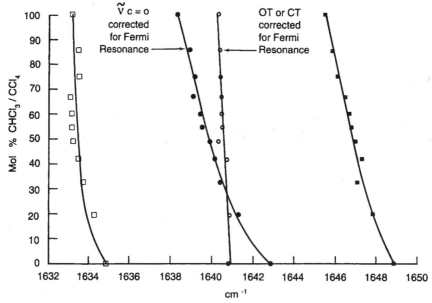

FIGURE 14.10 Plots of νC=O and OT or CT in Fermi resonance and their corrected unperturbed frequencies for 14H-dibenzo [a,j] xanthen-14-one vs mole % CHCl₃/CCl₄. The solid squares and open squares represent uncorrected frequency data, the solid circle represents νC=O corrected for Fermi resonance and the open circles represent OT or CT corrected for Fermi resonance.

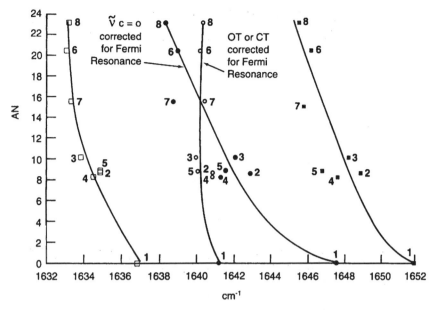

FIGURE 14.11 Plots of νC=O and OT or CT uncorrected and corrected for Fermi resonance for 14H-dibenzo [a,j]-xanthen- 14-one vs the solvent acceptor number (A) for (1) hexane; (2) carbon tetrachloride; (3) carbon disulfide; (4) benzene; (5) tetrahydrofuran; (6) methylene chloride; (7) nitrobenzene; and (8) chloroform.

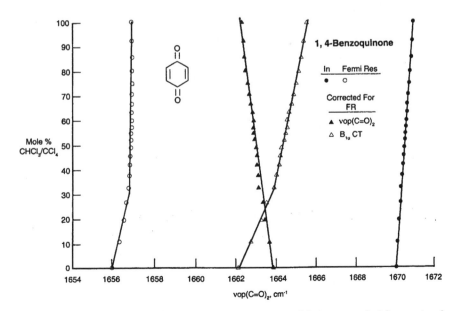

FIGURE 14.12 Plots of $\nu_{op}(C=O)_2$ and B_{1u} CT in Fermi resonance and their unperturbed frequencies after correction for Fermi resonance for 1,4-benzoquinone vs mole % $CHCl_3/CCl_4$.

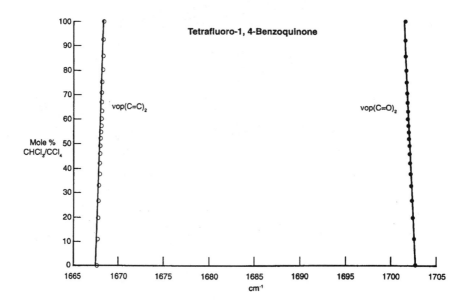

FIGURE 14.13   Plots of $\nu_{op}(C{=}O)_2$ and $\nu_{op}(C{=}C)_2$ for tetrafluoro-1, 4-benzoquinone vs mole % $CHCl_3/CCl_4$.

TABLE 14.1 IR vapor phase and Raman data for acetone and acetone-d₆

Compound	2(C=O str.) cm⁻¹(A)	a.CH₃ str. cm⁻¹(A)	s.CH₃ str. cm⁻¹(A)	C=O str. cm⁻¹(A)	? cm⁻¹(A)	a.CH₃ bend cm⁻¹(A)	s.CH₃ bend cm⁻¹(A)	a.CCC str. cm⁻¹(A)	CH₃ rock B2 cm⁻¹(A)	CH₃ rock B1 cm⁻¹(A)	delta C=O cm⁻¹	s.CCC str. Raman*1 liquid cm⁻¹	delta C=O Raman*1 liquid cm⁻¹	gamma C=O Raman*1 liquid cm⁻¹	gamma C=O IR2 vapor cm⁻¹	CCC bend IR*2 vapor
Acetone	3460 (0.03)	3000 (0.120) 2970 (0.150)	2941 (0.090)	1748 (1.110) 1735 (1.250) 1721 (1.050) 1745 (1.240)	1550 (0.050)	1435 (0.140)	1378 (0.500) 1362 (0.620) 1353 (0.550) 1070 (0.090)	1228 (0.530) 1212 (0.650) 1200 (0.510) 1250 (0.930)	1092 (0.040)	900 (0.050)	538 (0.080) 525 (0.135) 510 (0.120) 485 (0.110)	786.5 stg.,pol.	530 med.,pol.?	493 wk.,depol.	484 wk.type C	385 wk. type B
Acetone-db	3450 (0.030)	2250 (0.074)	2220 (0.090)	1738 (1.210) 1729 (1.240)	1415 (0.060)	1140 (0.060)	1050 (0.140) 1027 (0.090)	1239 (1.130)	986 (0.050)	795 (0.020)	472 (0.140) 458 (0.150)	695.5 stg.,pol.	478 wk.,depol.	393 wk.,depol.	405 wk.type C	321 wk. type B
Acetone/Acetone-db	1.003	1.32	1.325	0.998	1.095	1.38	1.297	0.978	1.108	1.132	1.112	1.131	1.109	1.254	1.195	1.199

*¹ Reference 1.
*² Reference 2.

TABLE 14.2 Application of the KBM equation using IR data for acetone C=O stretching frequencies in various solvents

Solvent	For acetone $[C=O(hexane)-C=O(soln.)]/ \times 10(3)$ $[C=O(hexane)]$	Calculated C=O cm^{-1}	delta C=O(obs.)−C=O(calc.) cm^{-1}
Hexane			
Diethyl ether	2.322	1715.9	−2.1
Benzene	4.645	1717.9	3.9
Toluene	2.904	1717.8	3.8
Carbon tetrachloride	2.322		
Pyridene	5.226	1714.2	1.2
Nitrobenzene	5.226	1713.7	0.7
Acetonitrile	5.226	1713.5	0.5
Dimethyl sulfoxide	6.969	1713.4	3.4
Methylene chloride	5.388	1714.5	2.5
Chloroform	6.388	1715.6	4.6
t-Butyl alcohol	6.969	1714.3	4.3
Isopropyl alcohol	7.549	1713.7	4.9
Ethyl alcohol	8.13	1713.7	5.7
Methyl alcohol	8.711	1814.7	7.7
Water	13.36	1713.3	14.3

TABLE 14.2A   The calculated values for A-1/2A + 1 and X-Y/X where A and Y equal 0 to 85 and X equals 85 [the KBM equation]

A or Y	A − 1/2A + 1	X−Y/X [X=85]
0	−1	1
1	0	0.988
2	0.22	0.976
3	0.286	0.964
4	0.333	0.953
5	0.364	0.941
6	0.385	0.929
7	0.4	0.918
8	0.412	0.906
9	0.421	0.894
10	0.428	0.882
15	0.452	0.824
20	0.463	0.765
25	0.47	0.706
30	0.475	0.647
35	0.478	0.588
40	0.481	0.529
45	0.483	0.471
50	0.485	0.412
55	0.486	0.353
60	0.488	0.294
65	0.489	0.235
70	0.4894	0.176
75	0.49	0.118
80	491	0.058
85	0.4912	0

TABLE 14.3   The C=O stretching frequencies for aliphatic ketones in the vapor phase and various solvents

Solvent	Dimethyl ketone cm^{-1}	Methyl ethyl ketone cm^{-1}	Diethyl ketone cm^{-1}	Ethyl isopropyl ketone cm^{-1}	Disopropyl ketone cm^{-1}	Di-t-butyl ketone cm^{-1}	AN
[Vapor]	1735	1742	1731	1730	1726	1699	0
Hexane	1722.4	1727.2	1725.1	1721.4	1720.3	1690.3	3.9
Diethyl ether	1719.6	1723	1721.4	1718.2	1717.8	1687.7	8.6
Carbon tetrachloride	1717.7	1721.3	1719.6	1716.5	1716	1685.9	
Carbon disulfide	1716.3	1720.1	1718.3	1715.2	1714.5	1684.9	18.9
Benzene	1715.8	1718.6	1717.4	1714	1713.5	1684.5	18.9
Acetonitrile	1713.3	1714	1713.7	1710.6	1707.8	1681.6	14.8
Nitrobenzene	1712.8	1714.3	1713.8	1710.6	1707.6	1681.7	15.5
Benzonitrile	1712.7	1713.9	1713.2	1710.2	1706.8	1682.2	20.4
Methyl chloride	1712	1712.6	1712.5	1709.6	1706.4	1680.5	
Nitromethane	1712.2	1712.4	1712.4	1709.5	1706.3	1680.6	
t-Butyl alcohol	1711.8	1711.2	1711.8	1709.2	1705.6	1678.5	29.1
t-Butyl alcohol*[1]		1722sh*[2]	1722.5sh	1718.8sh	1718sh	1687.0sh	39.1
Chloroform	1710.6	1710.2	1710.7	1708.2	1705.3	1680.8	23.1
Dimethyl sulfoxide	1709.2	1709.7	1710.1	1707.4	1704.6	1680.1	19.3
Isopropyl alcohol	1710.3	1709.8	1710.3	1707.7	1704.2	1677.9	33.5
Isopropyl alcohol*[1]		1720.5sh	1720.5sh	1716.2sh	1717sh	1687sh	33.5
Ethyl alcohol	1709	1708.5	1710.1	1706.1	1703.1	1676.8	37.1
Ethyl alcohol*[1]	1717sh	1719.4sh	1717.8sh	1715.4sh	1716sh	1686.4sh	37.1
Methyl alcohol	1708	1707.5	1707.7	1704.8	1701.1	1675.2	41.3
Methyl alcohol*[1]	1716.2sh	1718.6sh	1717.0sh	1715sh	1715sh	1684.8sh	41.3
$\Sigma 6^* \cdot \Sigma \alpha \cdot 10^{-2}$	0	0.7	2.8	15.7	35.7	184.8	
$\Sigma 6^*$	0	−0.1	−0.2	−0.29	−0.38	−0.6	
$\Sigma \alpha$	0	−0.07	−0.14	−0.54	−0.94	−3.08	

*[1] C=O not H bonded.

*[2] sh = shoulder.

TABLE 14.3A A comparison of the carbonyl stretching frequency difference (delta C=O str. in cm^{-1}) for dialkyl ketones in hexane and each of the other solvents

Solvent	Dimethyl ketone cm^{-1}	Methyl ethyl ketone cm^{-1}	Diethyl ketone cm^{-1}	Ethyl isopropyl ketone cm^{-1}	Diisopropyl ketone cm^{-1}	Di-t-butyl ketone cm^{-1}
Hexane	0	0	0	0	0	0
Diethyl ether	2.8	4.2	3.6	3.2	2.5	[2.6]
Carbon tetrachloride	4.7	5.9	5.4	4.9	4.3	[4.4]
Carbon disulfide	6.1	7.1	6.7	6.2	5.8	5.35
Benzene	6.5	8.6	7.65	7.4	6.8	5.8
Acetonitrile	9	13.25	11.4	10.8	[12.5]	8.7
Nitrobenzene	9.6	12.9	11.2	10.8	[12.7]	8.6
Benzonitrile	9.7	13.3	11.8	11.2	[13.5]	8.1
Methyl chloride	10.3	14.6	12.6	11.8	[13.9]	9.8
Nitromethane	10.15	14.8	12.7	11.9	[14.0]	9.7
t-Butyl alcohol	10.5	16	13.25	12.2	[14.7]	11.8
t-Butyl alcohol		5.2	2.6	2.6	2.3	[3.3]
Chloroform	11.75	17	14.35	13.2	[15.0]	9.5
Dimethyl sulfoxide	13.1	17.5	14.9	14	[15.7]	10.2
Isopropyl alcohol	12	17.4	14.7	13.7	[16.1]	12.4
Isopropyl alcohol		6.7	4.5	5.2	3.3	3.3
Ethyl alcohol	13.3	18.7	14.9	[15.3]	[17.2]	13.5
Ethyl alcohol	5.35	7.8	7.3	6	4.3	3.9
Methyl alcohol	14.3	19.7	17.4	16.6	[19.2]	15.1
Methyl alcohol	6.15	8.6	8.05	6.4	5.3	[5.5]
	0	0.7	2.8	15.7	35.7	184.8

TABLE 14.3B A comparison of the carbonyl stretching frequency difference (delta C=O str. in cm^{-1}) for dialkyl ketones in methyl alcohol and the other protic solvents

Protci solvent	Dimethyl ketone cm^{-1}	Methyl ethyl ketone cm^{-1}	Diethyl ketone cm^{-1}	Ethyl isopropyl ketone cm^{-1}	Diisopropyl ketone cm^{-1}	Di-t-butyl ketone cm^{-1}
Chloroform	2.6	2.7	3	3.4	4.2	5.7
t-Butyl alcohol	3.8	3.7	4.1	4.4	4.5	3.3
Isopropyl alcohol	2.3	2.3	2.7	2.9	3.1	2.7
Ethyl alcohol	1	1	2.5	1.3	2	1.7
Methyl alcohol	0	0	0	0	0	0

TABLE 14.3C A comparison of the differences in the carbonyl stretching frequencies (delta C=O str. in cm^{-1}) of dialkyl ketones in hexane solution and in alcohol solution [non-H-bonded C=O]

Alcohol	Dimethyl ketone cm^{-1}	Methyl ethyl ketone cm^{-1}	Diethyl ketone cm^{-1}	Ethyl isopropyl ketone cm^{-1}	Diisopropyl ketone cm^{-1}	Di-t-Butyl ketone cm^{-1}
Methyl	6.15	8.6	8.05	6.4	5.3	5.5
Ethyl	5.35	7.8	7.3	6	4.3	3.9
Isopropyl		6.7	4.5	5.2	3.3	3.3
t-Butyl		5.2	2.6	2.6	2.3	3.3

TABLE 14.4 The C=O stretching frequencies for *n*-butyrophenone and tert-butyrophenone in 0 to 100 mol % CHCl$_3$/CCl$_4$ I3ICCl4 solutions

Mole % CHCl$_3$/CCl$_4$	n-Butyro-phenone C=O str.	t-Butyro-phenone C=O str.	delta C=O
0	1691	1678.43	12.6
10.8	1689	1677.6	11.4
16.9	1687	1676.7	10.3
26.7	1687	1676	11
32.6	1686	1675.6	10.4
37.7	1686	1675.3	10.7
42.1	1685	1675.1	10.9
45.9	1685	1674.9	10.1
49.2	1685	1674.8	10.2
52.2	1685	1674.7	10.3
55.7	1685	1674.7	10.3
57.4	1685	1674.7	10.3
60.2	1685	1674.5	10.5
63.4	1685	1674.5	10.5
66.89	1684	1674.4	9.6
70.8	1684	1674.3	9.7
75.2	1684	1674.3	9.7
80.2	1684	1674.2	9.8
85.8	1684	1674.1	9.9
92.4	1683	1674.1	8.89
100	1682	1674.1	7.9

TABLE 14.5 IR data for acetone, α- chloroacetone, acetophenone, and benzophenone in CS₂ solution between ~29 and −100°C

°C	Acetone C=O str. [CS₂] cm⁻¹	°C	α-Chloro-acetone conformer 1 C=O str. [CS₂] cm⁻¹	α-Chloro-acetone conformer 2 C=O str. [CS₂] cm⁻¹	[conformer 1]− [conformer 2] cm⁻¹	°C	Acetophenone C=O str. [CS₂] cm⁻¹	°C	Benzophenone C=O str. [CS₂] cm⁻¹
27	1717.5	29	1750.2	1723.7	26.5	32	1689.4	31	1663.3
15	1717.3	11	1750.7	1723.7	27	15	1689.5	15	1662.7
0	1717.5	0	1750.5	1723.3	27.2	0	1689.7	0	1662.2
−10	1717.3	−10	1750.5	1723.3	27.2	−10	1689.2	−10	1661.5
−20	1717	−20	1751	1722.9	28.1	−20	1688.7	−20	1661
−30	1716.4	−30	1750.2	1722.4	27.8	−30	1687.4	−30	1660.7
−40	1715.4	−40	1749.9	1722	27.9	−40	1687	−40	1660.3
−50	1715.4	−50	1750	1721.6	28.4	−50	1687.4	−50	1660
−60	1713.6	−60	1749	1720.7	28.3	−60	1687.1	−60	1659.6
−70	1713.6	−70	1747.9	1721	26.9	−70	1687	−70	1659
−80	1713					−80	1686.5	−80	1658.5
−90	1712.7					−90	1686.4	−90	1658
−100	1711.7					−100	1686.2	−100	1657.8
delta C [−127]	delta C=O str. [−5.8]	delta C [−99]	delta str. C=O [−2.3]	delta C=O str. [−2.7]		delta C [−132]	delta C=O str. [−3.2]	delta C [−131]	delta C=O str. [−5.5]

TABLE 14.6 A comparison of carbonyl stretching frequencies for 2% wt./vol. ketone in dimethyl sulfoxide and/or carbon tetrachloride solution

Mole % $(CH_3)_2SO/CCl_4$	Acetone cm^{-1}	Aceto-phenone 2,4,6-Tri-methyl- cm^{-1}	Aceto-phenone 4-nitro- cm^{-1}	Aceto-phenone cm^{-1}	Aceto-phenone 4-methoxy- cm^{-1}	Aceto-phenone cm^{-1}
0	1718	1704	1700	1691	1683	1665
11.87	1715	1700	1697	1687	1679	1662
29.48	1713	1699	1695	1686	1677	1661
40.48	1713	1698	1694	1686	1676	1660
48.78	1712	1698	1694	1685	1676	1660
57.63	1712	1698	1693	1684	1675	1659
100	1710	1695	1691	1682	1672	1657

TABLE 14.7 IR vapor-phase data for conjugated ketones

Compound and Conformer	2(C=O str.)*	C=O str.* (see text)*	C=C. str. (see text)*	CH=CH twist	[C=O str.]−[C=C str.]*	[(A)C=C str.]/[(A)C=O str.]*	[C=C str.]/[HC=CH twist]	[(A)CH=CH twist]/[(A)C=C str.]*	C=CH₂ wag	CH=CH₂ wag
s-cis,trans CH=CH										
Chalcone	3358(0.005)	1680(0.690)	1620(1.230)	980(0.129)	60	1.78	640	0.11		
4,4'-Difluoro-		1680(0.372)	1619(0.614)	980(0.090)	61	1.65	639	0.15		
4'-Fluoro-methoxy	3350(0.005)	1678(0.365)	1605(1.220)	979(0.131)	73	3.34	626	0.11		
3,4-Dichloro-4'-methyl	3350(0.005)	1675(0.491)	1612(1.240)	979(0.180)	63	2.53	633	0.15		
2-Chloro-2',4'-dimethyl	3330(0.005)	1670(1.230)	1611(0.225)	979(0.225)	59	0.68	632	0.27		
4-Nitro-	3360(0.005)	1684(0385)	1619(0.393)	978(0.130)	65	1.02	641	0.33		
s-cis		1730(0.735)	1630(0.129)							
3-Buten-2-one	3400(0.011)	1715(1.245) 1695(1.239)	1627(0.120) 1616(0.112)	986(0.250) [C=CH₂ wag] 951(0.400)	88	0.09	641	2.08 [(A)C=CH₂ wag]/[CH=CH twist]-	951(0.400) [CH=CH twist]-	760(0.030)
s-cis										
3-Methyl-3-buten-2-one	3388(0.030)	1700(1.240)	1639(0.210)	929(0.520) [C=CH₂ wag]	61	0.17	710	2.48 [(A)C=C str.]	35 [C=CH₂ wag]	
s-cis 1-Penten-3-one	3399(0.020)	1712(1.240)	1624(0.210)	987(0.400)	88	0.17	637	1.9		
s-cis,trans HC=CH										
1-Phenyl-1-hepten-3-one	3395(0.010)	1690(0.920)	1620(1.210)	974(0.370)	70	1.32	646	0.31 [(Ao.p=C−H def.]/[(A)C=C str.]		
s-cis,cis(H,CH₃)		1710(1.140)								
Methyl 2-methyl-1-propenyl ketone	3400(0.011)	1705(0.940) 1699(1.150)	1634(1.240)	821(0.091)	71	1.32	813	0.07		
s-cis,cis(H,CH₃)?										
3-Methyl-4-phenyl-3-buten-2-one	3358(0.005)	1690(1.250)	1627(0.160)	860(0.030) [o.p.=CH def.]	63	0.13	767	0.19 [(A)o.p.=C−H def.]		

TABLE 14.7A Other fundamental vibrations for conjugated ketones

Compound	trans=C–H s. rock and a.CCC str.	i.p. Ring	i.p.o.p. 5H def.	o.p. Ring def.		
Chalcone	1320(0.600)	1015(0.500)	746(0.440)	691(0.381)		
4,4'-Difluoro-	1320(0.261)	1015(0.285)	[o.p.2H def] 821(0.285)			
4'-Fluoro-4-methoxy	1320(0.295)	1015(0.385)	821(0.300)			
3,4-Dichloro-4'-methyl	1322(0.370)	1030(0.600)	801(0.400)			
2-chloro-2',4'-dimethyl	1310(0.490)	1009(0.360)	[o.p.H def.] 873(0.445)	[o.p.2H def.] 817(0.180)	[o.p.4H def] 752(0.440)	
4-Nitro-	[CH2=bend] 1318(0.470)	1011(0.390)	851(0.232)	[a.NO2 str.] 1538(0.542)	[s.NO2 str.] 1348(1.230)	
3-Buten-2-one	1410(0.310)	[s.CH3 bend] 1380(0.359)	1240(0.365)	[s.CH3 str.] 1175(0.390)	760(0.030)	[CCC str.] 1145(1.030)
3-Methyl-3-buten-2-one	[a.CH2=str.] 3100(0.172)	[a.CH3 str.] 2988(0.540)	[s.CH3 str.] 2938(0.680)	[s.CH3 str.] 2880(0.285)	[a.CH3 bend] 1450(0.399); [CH2=twist?] 700(0.070)	[s.CH3 bend] 1369(0.830); [gamma C=O] 561(0.140)
1-Penten-3-one	[a.CH2=str.] 3102(0.103)	[CH2=bend] 1410(0.530)		[CCC str. ?] 1196(0.300)	[a.CH3 bend] 1460(0.191)	
Methyl-2-methyl-1-propenyl ketone	[CH=str.] 3020(0.190)	[a.CH3 str.] 2980(0.270)	[s.CH3 str.] 2870(0.140)	[a.CH3 bend] 1448(0.235)	[s.CH4 bend] 1390(0.370)	[s.CH3 bend] 1367(0.520)
1-Phenyl-1-hepten-3-one	[a.CH3 str.] 2970(0.940)	[a.CH2 str.] 2940(0.790)	[s.CH2 str.] 2885(0.400)	[CH2 bend] 1451(0.280)	[s.CH3 bend] 1385(0.100)	[o.p. Ring def.] 690(0.320)
3-Methyl-4-phenyl-3-buten-2-one	[CCC str.] 1325(0.430)			see text; 770(0.068)	[i.p.o.p.5H def.] 745(0.360); see text; 726(0.104)	695(0.146)
Cinnamaldehyde	[OC–H str. in F.R.] 2805(0.116)	[s(OC–H bend) in F.R.] 2737(0.109)	[OC–H bend] 1395(0.005)		[i.p.o.p.5H def.] 741(0.157)	[o.p. Ring def.] 688(0.111)
alpha-Hexyl cinnamaldehyde	2828(0.1180)	2705(0.115)	1400(0.035)		745(0.082)	695(0.131)

TABLE 14.8 Infrared data for 2-hydroxy-5-X-acetophenone in CCl_4 and CS_2 solutions

2-Hydroxy-5-X-acetophenones 5-X	C=O: H=O cm⁻¹ CCl_4 soln.	C=O: H=O cm⁻¹ NM	C=O: H=O cm⁻¹ $CHCl_3$	$C=OCH_3$ cm⁻¹ CCl_4 soln.	delta cm⁻¹	C-C(=)-C str. cm⁻¹ CS_2 soln.	C_6H_5-C(=)str. cm⁻¹ CS_2 soln.	C_6H_5-O str. cm⁻¹ CS_2 soln.	sym. CH_3 bend cm⁻¹ CCl_4 soln.
NO2	1658	1645			13	973	1295	1192	1360*1
$C=OCH_3$	1650			1687	37	968,958	1302	1212	1370;1359
H	1646.2		1642.3		3.9	957	1303	1212	1369
C_6H_5	1650					958	1301	1212	1368
I	1641					954	1283	1190	1365
CH_3	1654					960	1297	1213	1369
Br	1649					959	1318	1207	1363
Cl	1650					959	1321	1208	1367
F	1655					963	1322	1211	1371
CH_3O	1652					960	1370	1205	1370
NH_2	1648	1641			7	[969 solid]	1380	[1218 solid]	1380

NH_2 bend cm⁻¹

NH_2	1636

*1 and sym. NO_2 str.

TABLE 14.9   IR vapor-phase data and assignments for cyclobutanone and cyclopentanone

Compound	a.CH$_2$ str. cm^{-1}(A)	s.CH$_2$ str. cm^{-1}(A)	CH$_2$ bend cm^{-1}(A)	CH$_2$ wag cm^{-1}(A)	CH$_2$ twist cm^{-1}(A)	CH$_2$ rock cm^{-1}(A)	Ring def. cm^{-1}(A) breathing	Ring cm^{-1}(A)	2(C=O) str. cm^{-1}(A)	C=O str. cm^{-1}(A)
Cyclobutanone	3004 (0.240)	2930 (0.105)	1402 (0.070)	1254 (0.025)	1223 (0.050)	845 (0.010)	1072 (0.290)	945 (0.010)	3602 (0.021)	1828 (1.041)
	2980 (0.170)		1404 (0.060)		1210 (0.040)		1060 (0.232)			1815 (1.240)
		2904 (0.241)	1396 (0.070)		1195 (0.030)					1800 (1.141)
Cyclopentanone	2979 (0.841)	2895 (0.321)	1419 (0.175)	1275 (0.063)	1142		964		3510 (0.040)	1767 (1.250)
	2965 (0.690)									

TABLE 14.10　The C=O stretching frequencies for cyclopentanone and cyclohexanone in the vapor, neat, and solution phases

Cyclopentanone solvent	Corrected for fermi res. C=O str. 1% (wt./vol.) cm⁻¹	Corrected for fermi res. C.T cm⁻¹	AN	delta C=O [Hexane]-[soln.] cm⁻¹	Cyclohexanone	C=O str. 1% (wt./vol.) cm⁻¹	delta C=O str. [vap.]-[soln.] cm⁻¹	delta C=O [Hexane]-C=O str. [soln.] cm⁻¹
[Vapor]	1765					1732		
[Neat]	1739.2	1736.8						
Hexane	1750.6	1727.4	0			1723	0	27.6
Methyl t-Butyl ether	1745.8	1731.2	3.9	5.3		1718	5	27.3
Diethyl ether	1744.5	1730.5	8.6	6.1		1718	5	26.5
Carbon tetrachloride	1742.9	1731.1	8.2	7.7		1715	8	28.9
Benzene	1740.8	1731.2		9.8		1713	10	27.8
Carbon disulfide	1741.7	1727.3		8.9		1714	9	27.7
Benzonitrile	1736.9	1735.1	15.5	13.7		1706	10	30.9
Nitrobenzene	1736.1	1734.6	14.8	14.5		1707	18	29.8
Acetonitrile	1736.3	1736.7	18.9	14.3		1707	16	29.3
Methyl dichloride	1734.9	1735.1	20.4	15.7		1705	18	29.9
Nitromethane	1734.8	1737.2		15.8		1705	18	29.8
Dimethyl sulfoxide	1731.6	1735.4	19.3	19		1703(wet)	20	28.6
Chloroform						1703	20	30.6
Chloroform (4.0%)	1733.6	1735.4	23.1	17				
Chloroform (0.1%)	1733	1735	23.1	17.6				
Chloroform-d	1733.4	1735.6	23.1	17.2				
t-Butyl alcohol	1734.9	1737.1	29.1	15.7				
Isopropyl alcohol	1735.8	1738.2	33.5	14.8				
Ethyl alcohol	1735.2	1737.7	37.1	15.4		1702	21	33.2
Methyl alcohol	1733.5	1738.5	41.3	17.1		1701	22	32.5
Water	1728.8	1734.7	54.8	22.8				

TABLE 14.11 IR data for 14H-dibenzo[a,j] xanthen-14-one in CHCl$_3$/CCl$_4$ and various solvents

14H-dibenzo[a,j] xanthen-14-one Mole % CHCl$_3$/CCl$_4$	C=O str. in FR [1] cm^{-1}	C=O str. [2] in FR cm^{-1}	A[1]	A[2]	A[1]/A[2]	C=O str. corrected for FR cm^{-1}	OT or CT corrected for FR cm^{-1}	AN[1]
0	1648.85	1634.85	0.148	0.11	1.345	1642.9	1640.8	
54.62	1648.78	1633.14	0.094	0.108	0.87	1639.5	1640.3	
100	1645.4	1633.08	0.197	0.273	0.721	1638.2	1640.2	
Solvent								
Hexane	1651.89	1636.83	0.076	0.031	2.45	1647.5	1641.2	0
Carbon tetrachloride	1648.85	1634.85	0.148	0.031	1.345	1642.9	1640.8	8.6
Carbon disulfide	1648.18	1633.8	0.482	0.359	1.34	1642	1639.9	[10.1]
Benzene	1647.59	1634.49	0.178	0.166	1.07	1641.3	1640.8	8.2
Tetrahydrofuran	1646.74	1634.82	0.384	0.299	1.28	1641.5	1640	[8.8]
Methylene chloride	1646.06	1632.98	0.224	0.27	0.83	1638.9	1640.1	20.4
Benzonitrile	1645.79	1633.28	0.357	0.47	0.76	1638.7	1640.4	15.5
Chloroform	1645.4	1633.08	0.197	0.273	0.72	1638.2	1640.2	23.1

[1] AN = acceptor number.

TABLE 14.12   IR and Raman data for 1,4-benzoquinones in CCl_4 and $CHCl_3$ solutions

Substituted 1,4-benzoquinone	Point group	Species	IR data o.p.(C=O)$_2$ str. [CCl$_4$ soln.] cm^{-1}	IR data o.p.(C=O)$_2$ str. [CHCl$_3$ soln.] cm^{-1}	[CCl$_4$ soln.]–[CHCl$_3$ soln.] cm^{-1}	Species	Raman data i.p.(C=O)$_2$ str. [CCl$_4$ soln.] cm^{-1}	Raman data i.p.(C=O)$_2$ str. [CHCl$_3$ soln.]	[CCl$_4$ soln.]–[CHCl$_3$ soln.]
Tetrafluoro	Vh	B$_{1u}$	1702.71	1701.42	−1.29	A$_g$	1696.8	1697.7	0.9
Tetrachloro	Vh	B$_{1u}$	1694.68	1693.37	−1.31	A$_g$	[1675]	1685.1	
Trichloro	Cs	A′	[1694]*^{1}			A′			
Tetrabromo	Vh	B$_{1u}$	1687.48	1685.37	−2.11	A$_g$			
2,5-Dichloro	C2h	B$_u$	1682.18	1681.25	−0.09	A$_g$	1672.5	1673.6	1.1
Chloro	Cs	A′	1681.89	1680.95	−0.94	A′	1659.3	1659.3	0
Unsubstituted	Vh	B$_{1u}$	1663.51*^{2}	1662	−1.51	A$_g$	1661.4	1661.8	0.4
Methyl	Cs	A′	1662.47	1659.57	−2.9	A′	1665.7	1665.9	0.2
2,6-Dimethyl	C2v	A$_1$	1657			A$_1$		1666.9	2.5

*^{1} CS$_2$ soln.
*^{2} Corrected for FR.

TABLE 14.13   IR data for tetrafluoro-1,4-benzoquinone in CCl_4, $CHCl_3$, and C_6H_{14} solutions

Tetrafluoro-1,4-benzoquinone	o.p.(C=O)$_2$ str. cm^{-1}	o.p.(C=C)$_2$ str. cm^{-1}	A[1]	A[2]
Mole % $CHCl_3/CCl_4$				
0	1702.71	1667.63	1.279	1.164
26.53	1702.27	1667.88	1.308	1.157
52	1701.91	1668.05	1.506	1.314
75.06	1701.66	1668.21	1.614	1.416
100	1701.42	1668.35	1.805	1.525
delta cm^{-1}	−1.29	0.72		
Mole % $CHCl_3/C_6H_{14}$				
0	1705.03	1667.67	0.443	0.388
24.49	1704.07	1668.04	0.451	0.413
53.16	1702.82	1668.31	0.551	0.511
76.43	1702.09	1668.52	0.703	0.639
100	1701.21	1668.05	0.911	0.813
delta cm^{-1}	−3.82	0.38		
Mole % $CCl_4/C6H_{14}$				
0	1705.06	1667.67	0.466	0.408
28.26	1704.48	1667.72	0.495	0.439
51.96	1703.88	1667.78	0.539	0.49
73.01	1703.28	1667.79	0.563	0.517
100	1702.83	1667.79	0.78	0.713
delta cm^{-1}	−2.23	0.12		

TABLE 14.14 IR data for tetrachloro-1,4-benzoquinone in CHCl$_3$/n-C$_6$H$_{14}$, CCl$_4$/n-C$_6$H$_{14}$, and CHCl$_3$/CCl$_4$ solutions

Tetrachloro-1,4-benzoquinone	o.p.(C=O)$_2$ str. cm^{-1}	A
Mole % CHCl$_3$/n-C$_6$H$_{14}$		
0	1696.3	0.222
24.29	1695.27	0.267
53.16	1694.86	0.539
76.43	1693.76	0.726
100	1693.27	1.024
delta cm^{-1}	−3.03	
Mole % CCl$_4$/n-C$_6$H$_{14}$		
0	1696.4	0.217
28.26	1695.73	0.255
51.96	1695.19	0.32
73.01	1695.51	0.406
100	1694.66	0.593
delta cm^{-1}	−1.74	
Mole % CHCl$_3$/CCl$_4$		
0	1694.68	
26.563	1693.97	
52	1693.4	
75.06	1693.09	
100	1693.37	
delta cm^{-1}	−1.31	

TABLE 14.15   IR data for tetrabromo-1,4-benzoquinone in CCl$_4$/C$_6$H$_{14}$, CHCl$_3$/C$_6$H$_{14}$, and CHCl$_3$/CCl$_4$ solutions

Tetrabromo-1,4-benzoquinone	o.p.(C=O)$_2$ str. cm^{-1}	CT cm^{-1}
Mole % CCl$_4$/C$_6$H$_{14}$		
0	1689.87	
28.26	1688.52	1676.44
51.96	1688.11	1672.04
73.01	1687.52	1670.82
73.01	1687.52	1670.82
100	1687.59	1669.38
delta cm^{-1}	−2.28	
Mole % CHCl$_3$/C$_6$H$_{14}$		
0	1689.54	
24.49	1688.33	1670.82
49.31	1687.82	1670.02
72.99	1686.76	1669.45
100	1685.39	1669.33
delta cm^{-1}	−4.15	
Mole % CHCl$_3$/CCl$_4$		
0	1687.48	
26.53	1686.89	1669.88
52	1686.41	1669.84
75.06	1685.78	1669.9
100	1685.37	
delta cm^{-1}	−2.11	

TABLE 14.16   IR data for chloro-1,4-benzoquinone in CHCl$_3$/n-C$_6$H$_{14}$, CCl$_4$/n-C$_6$H$_{14}$, and CHCl$_3$/CCl$_4$ solutions

Chloro-1,4-benzoquinone	o.p.(C=O)$_2$ str.n cm^{-1}	i.p.(C=O)$_2$ str. cm^{-1}	A[1]	A[2]
Mole % CHCl$_3$/n-C$_6$H$_{14}$				
0	1683.76	1662.23	0.595	0.252
24.49	1683.32	1661.56	0.485	0.234
49.31	1682.23	1660.96	0.517	0.272
74.71	1681.85	1660.37	0.675	0.379
100	1680.9	1660.32	0.616	0.361
delta cm^{-1}	−2.86	−1.92		
Mole % CCl$_4$/n-C$_6$H$_{14}$				
0	1683.73	1662.19	0.599	0.256
28.26	1683.87	1661.71	0.526	0.235
51.96	1683.36	1661.33	0.565	0.254
77.17	1682.64	1660.86	0.613	0.288
100	1681.87	1660.37	0.689	0.34
delta cm^{-1}	−1.86	−1.82		
Mole % CHCl$_3$/CCl$_4$				
0	1681.89	1660.36	0.709	0.348
26.53	1682.01	1660.82	0.733	0.385
52	1681.59	1660.62	0.707	0.389
75.06	1681.31	1660.49	0.656	0.375
100	1680.95	1660.35	0.798	0.411
delta cm^{-1}	−0.94	−0.01		

TABLE 14.17 IR data for 2,5-dichlorobenzoquinone in $CHCl_3/n-C_6H_{14}$, $CCl_4/n-C_6H_{14}$, and $CHCl_3/CCl_4$ solutions

2,5-Dichloro-1,4-benzoquinone	o.p.(C=O)$_2$ cm^{-1}
Mole % $CHCl_3/n-C_6H_{14}$	
0	1684.86
24.29	1683.25
53.16	1682.76
76.43	1681.59
100	1681.59
delta cm^{-1}	−3.48
Mole % $CCl_4/n-C_6H_{14}$	
0	1684.74
28.26	1683.9
51.96	1683.2
73.01	1683.33
100	1682.15
delta cm^{-1}	−2.59
Mole % $CHCl_3/CCl_4$	
0	1682.18
26.53	1681.64
52	1681.87
75.06	1681.57
100	1681.25
delta cm^{-1}	−0.83

TABLE 14.18 IR data for 3,3′,5,5′-tetraalkyl-1,4- diphenoquinone in $CHCl_3$ solution and in the solid phase

3, 3′,5,5′-Tetraalkyl-1,4-diphenoquinone	Symmetry	o.p.(C=O)$_2$ str. solid cm^{-1}	o.p.(C=O)$_2$ str. CHCl$_3$ cm^{-1}	i.p.(C=O)$_2$ str. solid cm^{-1}	i.p.(C=O)$_2$ str. CHCl$_3$ cm^{-1}	o.p.(C=O)$_2$ str.+ i.p.(C=O)$_2$ str. solid cm^{-1}	o.p.(C=O)$_2$ str.+ i.p.(C=O)$_2$ str. CHCl$_3$ cm^{-1}
(CH$_3$)$_4$	D$_{2h}$	1589	1595	[1589]	[1597]	3178	3192
(iso C$_4$H$_9$)$_4$	D$_{2h}$	1589	1592	[1579]	[1583]	3168	3175
(tert C$_4$H$_9$)$_4$	D$_{2h}$	1602	1597	[1602]	[1583]	3204	3180
(cyclo C$_5$H$_{11}$)$_4$	D$_{2h}$	1598	1598	[1601]	[1602]	3199	3200
3,3′-(CH$_3$)2-5,5′	C$_{2v}$	1586	1588	[1582]	[1592]	3168	3180
(tert. C$_4$H$_9$)$_2$		1588	1589	1603	1603	3189	3195
						[3191]	[3192]

TABLE 14.19 The dependence of the C=O stretching frequency upon solute concentration of dialkyl ketones in CCl$_4$/CHCl$_3$ solutions

wt./vol. % solute [CCl$_4$]	Diisopropyl ketone C=O str.	wt./vol. % solute [CCl$_4$]	Di-t-butyl ketone C=O str.
0.75	1715.91	0.8	1685.9
1.5	1715.87	1.2	1685.88
2.63	1715.8	1.8	1685.86
3.75	1715.75	2.62	1685.87
5.25	1715.72	3.94	1685.83
7.5	1715.67	5.25	1685.82
		6.26	1685.81
delta C=O str.	−0.24		−0.09
[CHCl$_3$]		[CHCl$_3$]	
0.22	1704.94	0.4	1680.71
0.43	1704.95	0.8	1680.99
1.08	1704	1.2	1680.73
2.16	1705.05	1.6	1680.73
4.04	1705.2	2.3	1680.73
5.89	1705.32	3.47	1680.78
7.29	1705.38	4.63	1680.79
8.62	1705.47	5.78	1680.81
delta C=O str.	0.53		0.1

Carboxylic Acid Esters

*Numbers in parentheses indicate in-text page reference.

Table 15.1 lists IR data for the carbonyl stretching frequencies, $\nu C{=}O$, for some carboxylic acid esters. This table also shows how the $\nu C{=}O$ frequencies change or shift with change of phase or with change in the molecular structure (1).

In all cases, $\nu C{=}O$ occurs at higher frequency in the vapor phase than in the liquid or condensed phase. In the liquid phase the dipolar interaction of esters (dielectric effect) causes the C=O bond to weaken and occur at lower frequency. In any series of esters, the methyl analog will occur at the highest frequency, and the frequencies decrease with increased branching on the α-carbon atom. This is attributed to the inductive effect of the alkyl group, because the electron donation of the alkyl group increases with increased branching on the α-carbon atom, which causes $\nu C{=}O$ to weaken and vibrate at lower frequency.

The vinyl or phenyl esters of any series always occur at higher frequencies than their corresponding alkyl esters, and this is also due to the inductive effect. The vinyl or phenyl groups withdraw electrons, and this strengthens the C=O bond, which causes $\nu C{=}O$ to vibrate at higher frequency. For example, alkyl acetates exhibit $\nu C{=}O$ in the range 1755–1769 cm^{-1}, and $\nu C{=}O$ for vinyl acetate and phenyl acetates occur in the range 1781–1786 cm^{-1} in the vapor phase.

CONJUGATION EFFECT

Conjugation of C=O with a vinyl or phenyl group (C=C−C=O or C_6H_5-$\omega C{=}O$) causes $\nu C{=}O$ to decrease in frequency. For example, alkyl propionates exhibit $\nu C{=}O$ in the range 1751–1762 cm^{-1} and alkyl acrylates exhibit $\nu C{=}O$ in the range 1746–1751 cm^{-1} in the vapor phase.

INDUCTIVE EFFECT

The inductive effect of the phenyl group also causes $\nu C{=}O$ for phenyl benzoate to occur at higher frequency (1760 cm^{-1}) than $\nu C{=}O$ for alkyl benzoates (1737–1749 cm^{-1}). The $\nu C{=}O$ frequencies for dialkyl isophthalates, and dialkyl terephthalates occur in the range 1739–1753 cm^{-1}, and are essentially identical to those exhibited by the alkyl benzoates in the vapor phase.

INTRAMOLECULAR HYDROGEN BONDING

In the vapor phase at high temperature alkyl salicylates exhibit $vC=O$ in the range 1740–1755 and $v(C=O)\cdots H-O$ in the range 1687–1698 cm^{-1}, and phenyl salicylates exhibit $vC=O$ in the range 1750–1762 and $v(C=O)\cdots H-O$ in the range 1695–1706 cm^{-1}. The higher frequency $vC=O$ mode is attributed to C=O groups not intramolecularly hydrogen bonded and the lower frequency $v(C=O)\cdots H-O$ mode is attributed to intramolecularly hydrogen-bonded C=O groups.

DIESTERS

It should be noted that dialkyl malonates, dialkyl succinates, and dialkyl adipates which contain two ester groups exhibit $vC=O$ frequencies similar to those exhibited by the esters containing only one ester group in the vapor phase at elevated temperature.

VAPOR VS SOLUTION PHASES

Table 15.2 compares the $vC=O$ frequencies for alkyl alkanoates in the vapor phase and in various solvents at 0.5 % wt./vol. Here MA is methyl acetate, MP is methyl propionate, MIB is methyl isobutyrate, MTMA is methyl trimethylacetate, EA is ethyl acetate, EP is ethyl propionate, EIB is ethyl isobutyrate, and ETMA is ethyl trimethyl acetate (2).

A study of Table 15.2 shows that $vC=O$ for each of these esters occur at the highest frequency in the vapor phase, and in solution occur at the highest frequency in solution with n-hexane, a nonpolar solvent. In addition the $vC=O$ mode decreases in frequency as the branching is increased on the acetate α-carbon atom and as methyl is replaced by ethyl. From other studies, the $vC=O$ for the isopropyl and tert-butyl esters in each series would be expected to occur at subsequently lower frequency due to the increased inductive contribution of the isopropyl and tert-butyl groups.

It is interesting to compare the $vC=O$ frequency differences for these alkyl alkanoates in solution in n-hexane and in the neat phase. These frequency differences are 10.2, 9.1, 7.1, and 6.2 cm^{-1} and 8.3, 7.8, 6.3 and 6.1 cm^{-1} for the methyl and ethyl analogs of acetate, propionate, isobutyrate, and trimethyl isobutyrate, respectively. Moreover, the differences between the $vC=O$ frequencies of the methyl and ethyl analogs are 1.9, 1.3, 0.8 and 0.1 cm^{-1} progressing in the series acetate, propionate, isobutyrate, and trimethylacetate. These data indicate that as the steric factor of the R−C=O and O−R′ groups for R−C(=O)−OR′ is increased there is less dipolar interaction between ester molecules in the near phase; thus, there is less of a frequency difference between the $vC=O$ frequencies in n-hexane solution than in the near phase (2).

Figure 15.1 shows plots of $vC=O$ (solvent or neat) vs $vC=O$ (hexane) minus $vC=O$ (solvent or neat) for the alkyl alkanoates. These plots do not show data for the $v(C=O)\cdots H-O$ frequencies noted when in solution with the four alcohols included in this study. Methyl trimethylacetate exhibits two bands, and this is attributed to Fermi resonance (2). The plots demonstrate that the frequencies decrease in a progressive manner, as was discussed here. There is no theoretical significance to these linear plots because any set of data (numbers) treated mathematically in this manner yields a linear relationship.

Figure 15.2 shows plots of $v(C=O)\cdots HO(ROH)$ vs $vC=O$ (hexane) minus $v(C=O)\cdots HO(ROH)$ for alkyl alkanoates. These plots show that the $v(C=O)\cdots HO$ frequencies decrease in each series in the order tert-butyl alcohol through methyl alcohol, and this is the order of progressive decreasing ROH acidity and progressive refractive index of these four alcohols. However, the dielectric constant of each alcohol increases progressively in the solvent order tert-butyl alcohol through methyl alcohol. Using these data to calculate the reaction field $[(R) = (e-1)/(2e+n^2)]$, where e is the dielectric constant and n is the refractive index (3,4) shows that the reaction field increases in the order of butyl alcohol through methyl alcohol. Therefore, the reaction field is responsible for lowering the $vC=O$ frequencies in alkyl alkanoates in alkyl alcohols. The $v(C=O)\cdots HOR$ frequencies for these alkyl alkanoates are also progressively lowered in frequency by the increasing reaction field plus the additional lowering of the $vC=O$ mode due to intermolecular hydrogen bonding between the ROH proton and the carbonyl oxygen atom. Therefore, a quantitative measure of the strength of the intermolecular hydrogen bonds formed between $C=O\cdots HOR$ is the difference between the $vC=O$ and $vC=O\cdots HOR$ frequencies for each compound in each of the four alcohols. The larger the number, the stronger the intermolecular hydrogen bond. The strength of an intermolecular hydrogen bond between $C=O\cdots HOR$ is dependent upon three factors: 1) the acidity of the OH proton; 2) the basicity of the $C=O$ groups; and 3) the steric factors of both the alkyl group of each alcohol and the alkyl groups in each alkyl alkanote. The larger their steric factors the larger the $C=O\cdots HOR$ bond distance. The strength of the hydrogen bond decreases as this $C=O\cdots HOR$ bond distance increases, and causes the frequency separation between $vC=O$ and $vC=O\cdots HOR$ to decrease. The steric factor of the alkyl groups prevents the strongest intermolecular hydrogen bonds from being formed between the most acid OH proton (tert-butyl alcohol in this case) and the most basic carbonyl group (ethyl trimethylacetate in this case). The frequency differences between $vC=O$ and $vC=O\cdots HOR$ in the alkyl alcohol solution of ethyl trimethylacetate should be less than exhibited by methyl trimethylacetate. Table 15.3 lists the frequency difference between $vC=O$ and $vC=O\cdots HOR$ for the alkyl alkanoates. The data shows that the strongest intermolecular hydrogen bonds are formed between the carbonyl oxygen atom and the alcohol OH protons with minor exception in the case of methanol. With the exception of isopropyl alcohol, the strength of the intermolecular hydrogen bond formed is also greater in the case of methyl trimethylacetate than in the case of ethyl trimethyl acetate. These data support the supposition that steric factors of these molecules do affect the molecular association between solute and solvent. In addition, steric factors must then play a role in the dielectric or dipolar interaction between solute and solvent.

Figure 15.3 shows plots of $vC=O$ for alkyl alkanoates vs AN for the solvent or neat alkyl alkanoate(s). AN is the solvent acceptor number (6). These plots show that AN values are not a precise measure of solute-solvent interaction due to steric factors if the alkyl groups and basicity of the carbonyl group (2).

Figure 15.4 shows plots of $vC=O\cdots HO(ROH)$ vs AN for alkyl alkanoates in alkyl alcohols. With the exception of ethyl acetate in methyl alcohol, the $vC=O\cdots OH$ frequencies decrease in essentially a linear manner as the acidity of the OH proton decreases in the order tert-butyl alcohol through methyl alcohol. Reversal of that order of $vC=O\cdots HOR$ frequency decrease with decreasing strength of the intermolecular hydrogen bond indicates that dielectric properties of the solvent play a major role in affecting molecular vibrations.

Figure 15.5 shows plots of νC=O$\cdots$HOR for methyl acetate, ethyl acetate, isopropyl acetate, and tert-butyl acetate vs the solvent acceptor number (AN) for tert-butyl alcohol, isopropyl alcohol, ethyl alcohol, and methyl alcohol, numbered in solvent order 14–17 (5).

These alkyl acetates exhibit two bands in the carbonyl region of the IR spectrum in each of these four alkyl alcohols. Plots (A) through (D) are for the alkyl acetates in alcohol solution, but the carbonyl oxygen atoms are not intermolecularly hydrogen bonded. The lower frequency plots marked (A′) through (D′) are for the alkyl acetates in these alcohol solutions where the carbonyl oxygen atom is intermolecularly hydrogen bonded (C=O$\cdots$HOR). Plots (A′) through (D′) show a relationship with the AN values of both the nonprotic and protic solvents, and this indicates that the AN values include a value for the intermolecular hydrogen bond to a basic site in a solute molecule. This is understandable when considering how these AN values were determined. Since $(C_2H_5)_3P$=O was used as the solute molecule to determine the AN values of solvents by application of NMR(6), it is reasonable that in the case of the alkyl alcohols an intermolecular hydrogen bond was also formed between the P=O oxygen atom and the alcohol OH proton (P=O$\cdots$HOR).

Table 15.4 lists the frequency difference between νC=O and νC=O$\cdots$HO for alkyl acetates in solution with alkyl alcohols (5). In this case the basicity of the carbonyl group increases in the order methyl through tert-butyl acetate. With the exception of isopropyl acetate in solution in methanol, the strongest intermolecular hydrogen bonds are formed between the carbonyl oxygen atom of tert-butyl acetate and the OH proton of each of the four alcohols. Thus, the strongest intermolecular hydrogen bonds are formed in the molecule with the most basic carbonyl group. It then follows that branching on the O—Rα-carbon atom does not have as much effect on intermolecular hydrogen bonding as does branching on the acetate α-carbon atom. This is reasonable, because the alkyl group joined to an acetate α-carbon atom is much closer in space to the carbonyl oxygen atom than are the alkyl groups joined to the O—Rα-carbon atom.

Again it should be noted that both νC=O and νC=O$\cdots$HOR decrease in frequency as the acidity of the OH proton decreases, which is the opposite of what one might predict. The overriding factor is that the reaction field increases in the order tert-butyl alcohol through methyl alcohol, and this causes a general decrease in both vibrational modes. In other words there is more dipolar interaction between the solute and solvent as the reaction field increases, which subsequently weakens the C=O bond. As previously discussed, steric factors also must affect dipolar interaction between solute and solvent. It is reasonable to assume that the AN values of the νC=O frequencies for alkyl acetates in these alkyl alcohols which are not intermolecularly hydrogen bonded are comparable to those AN values for dialkyl ethers, because these are the values predicted by projection of these frequencies onto the (A′) through (D′) curves.

ROTATIONAL CONFORMERS

Table 15.5 lists the νC=O frequencies for alkyl 2,2-dichloroacetates and alkyl acetates in the vapor or net phases (1,7). The alkyl dichloroacetates in the vapor and neat phases exhibit rotational conformers 1 and 2, and both rotational conformer νC=O modes occur at lower frequency in the neat phase than the vapor phase by 11–16 cm^{-1}. The νC=O for rotational conformer 1 is 21–25 cm^{-1} higher in frequency than vC=O for conformer 2, and the corresponding alkyl analogs of alkyl acetate and conformer 2 of alkyl 2,2-dichloroacetates exhibit vC=O within ±4 cm^{-1} of each other. The inductive effect of the Cl atoms upon the C=O bond is independent of molecular orientation; therefore, another factor is needed to explain the relatively low frequency exhibited by vC=O of conformer 2 for alkyl dichloroacetates.

Comparison of the vC=O vapor-phase data for the methyl, ethyl, and butyl esters of 2-chloroacetates vs the corresponding alkyl acetates: (1770 vs 1769 cm^{-1}), (1764 vs 1761 cm^{-1}), and (1759 vs 1761 cm^{-1}), respectively, shows that substitution of a chlorine atom in the α-carbon atom has very little effect upon the vC=O mode. These vapor-phase data are recorded at high temperature. At ambient temperature in CCl$_4$ solution methyl chloroacetate exhibits vC=O for rotational conformer 1 at 1772 cm^{-1} and rotational conformer 2 at 1747 cm^{-1} (1). These data indicate that in the vapor phase at elevated temperature methyl chloroacetate exists in the form of rotational conformer 2 where the Cl atom is not cis of gauche with the carbonyl oxygen atom. It is suggested that the Cl atom is in a configuration in which the Cl atom is trans to the carbonyl oxygen atom.

In the case of the alkyl dichloroacetates, possible molecular rotational conformers are:

a. 2 Cl atoms gauche to the carbonyl oxygen atom;
b. 1 Cl atom cis and one Cl atom gauche to the carbonyl oxygen atom; and
c. 1 Cl atom trans and one Cl atom gauche to the carbonyl oxygen atom.

In a rotational conformer where one Cl atom is trans and one Cl atom is gauche to the carbonyl oxygen atom, the O−C(=O)−C bond angle is most likely increased due to electrostatic repulsion between the trans Cl atom and the R−O oxygen atom. The increase in the O−C(=O)−C bond angle would lower the vC=O frequency for rotational conformer 2, and this factor then offsets the inductive effect of the Cl atoms which raise the vC=O frequency. This would account for the fact that vC=O for conformer 2 for alkyl 2-chloroacetate and vC=O for comparable alkyl acetates vibrate at comparable frequency.

Even though vC=O frequencies are always observed at lower frequency in the neat liquid phase than in CCl$_4$ or CS$_2$ solution, it is of interest to compare the following data, which data illustrate the inductive effect. The IR data for ethyl 2-chloroacetate is for CS$_2$ solution data, and for ethyl 2,2-dichloroacetate is for liquid phase data (7,8):

	vC=O (cm^1) rotation comformer	vC=O (cm^1) rotation comformer 2
ethyl 2-chloroacetate (CS$_2$)	1765.0	1740.5
ethyl 2,2dichloroacetate (neat)	1771	1750
ethyl acetate (CS$_2$)		1740.8

The vC=O frequency for rotational conformer 2 increases with the addition of the second Cl atom when comparing data for ethyl 2-chloacetate and ethyl 2,2-dichloroacetate (Cl, 1740.5 cm^{-1} vs 2 Cl, 1750 cm^{-1}) while vC=O for rotational conformer 1 also increases with the addition of the second Cl atom (Cl, 1765.0 cm^{-1}, Cl$_2$, 1771 cm^{-1}). The addition of the second Cl atom shows the additional inductive effect attributed by the second Cl atom. The inductive effect of the first Cl atom is offset by the change in the O−C(=O)−C bond angle as already discussed here.

METHODS TO CONFIRM THE PRESENCE OF ROTATIONAL CONFORMERS

One method is to record IR spectra of a compound in CS_2 solution over a temperature range of 27°C to approximately −100 °C (8). The data in Table 15.5a is plotted in Fig. 15.6. The left plot shows νC=O rotational conformers 1 and 2 for ethyl 2-chloroacetate and νC=O for ethyl acetate in CS_2 solutions vs temperature in °C. The right figure shows a plot of the absorbance ratio for νC=O conformer 1/νC=O conformer 2 vs temperature in °C. These plots show that all of the νC=O modes decrease in frequency in a linear manner as the temperature of the CS_2 solution is lowered. In addition, the concentration of rotational conformer 1 increases while the concentration of rotational conformer 2 decreases with decrease in temperature.

Another method for determining the presence of rotational isomers is to record IR spectra of a compound in different solvents of mixtures of two solvents (8). Figure 15.7 illustrates this technique for methyl 2,4-dichlorophenoxyacetate in carbon tetrachloride and/or dimethyl sulfoxide solution. This figure shows a plot of the νC=O frequencies for rotational conformer 1 vs νC=O for rotational conformer 2 with change in the CCl_4 to $(CH_3)_2SO$ ratio. These data show that both frequencies decrease in a linear manner with change in the ratio of CCl_4 to $(CH_3)_2SO$. What it does not show is that both νC=O frequencies decrease in frequency as the concentration of CCl_4 is decreased or concentration of $(CH_3)_2SO$ is increased. This figure also shows that the absorbance ratio for (A)νC=O conformer 2/(A)νC=O conformer 1 decreases as the ratio of CCl_4 to $(CH_3)_2SO$ is decreased. Thus, the most polar isomer for methyl 2,4-dichlorophenoxyacetate is for rotational conformer 1. Rotational conformer 1 is where the phenoxy oxygen atom is near in space (cis or gauche) to the phenoxy oxygen atom, and rotational isomer 2 is where the phenoxy oxygen atom is away in space (trans or gauche) from the carbonyl oxygen atom (8).

LACTONES

Table 15.6 lists IR data for β-propiolactone in various solvents (1% wt./vol.) (9). Figure 15.8 shows IR spectra of β-propiolactone in CCl_4 solution (left) 52 mole % $CHCl_3/CCl_4$ solution (middle), and $CHCl_3$ solution (right). The β-propiolactone has the following empirical structure:

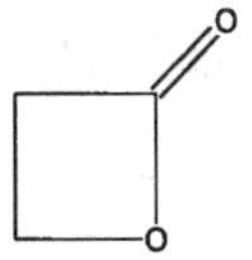

This structure is dsignated 4SR (4-membered saturated ring).

The C−C(=)−C bond angle is very small (~90°), and one would expect νC=O to occur at a very high frequency. Inspection of Fig. 15.8 shows that the two lower frequency bands increase in intensity while the higher frequency band decreases in intensity in progressing in the solvent series: CCl_4, 52 mol % $CHCl_3/CCl_4$, and $CHCl_3$. These bands are the result of $vC=O$ A' being in Fermi resonance with the combination $v_6 + v_{13}$, A' and the first overtone $2v_{10}$, A' (9). In this study of β-propiolactone an approximate correction for Fermi resonance based on perturbation theory for cases involving three modes was developed (9) (also see Chapter 1).

The data in Table 15.6 show that after correction for Fermi resonance, $vC=O$ for β-propiolactone occurs as high as 1850.2 cm^{-1} in n-hexane and as low as 1830.7 in methyl alcohol. Presumably the 1830.7 cm^{-1} band is for $vC=O \cdots HOCH_3$. The solvent acceptor numbers (AN) do not correlate with observed or unperturbed $vC=O$ β-propiolactone frequencies (9).

It is helpful to classify lactones as to the type of ring comprising the lactone group. For examples:

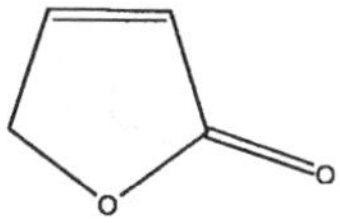

4 SR (see preceding text for β-propiolactone)

5 UR - 2,3(5-membered unsaturated ring where C=C group is in the 2,3-positions.

6 UR-2,3,4,5 (6-membered unsaturated ring where the C=C
    groups are in the 2,3- and 4,5-positions. The α-pyrone and
    coumarin are examples containing this group.)

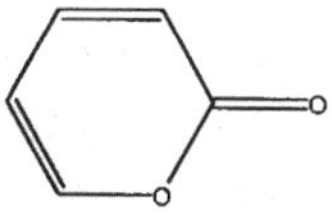

Lactones with 5 UR-2,3 structures are affected by alkyl substitution in the 4-position (1,10). In CCl_4 solution perturbed frequencies occur at 1785.4 and 1742 cm^{-1} for 4-H, 4-H analog, at 1782 and 1765 cm^{-1} for the 4-H, 4-CH_3 analog, and at 1776 and 1764 cm^{-1} for the 4-CH_3, 4-HC_3 analog (10) after correction for Fermi resonance, $vC=O$ occurs at 1781.4, 1776.2, and 1769.9 cm^{-1} for the 4H, 4H analog, the 4-H, 4-HC_3 analog, and the 4-HC_3, 4-HC_3 analog, respectively. The inductive effect of the CH_3 groups lowers the $vC=O$ frequencies. In addition, the $vC=O$ frequencies are lower in the case of 5 UR-2,3 compared to 5 SR lactones (1796 and 1784 cm^{-1} in CCl_4 solution and $vC=O$ is 1789.1 cm^{-1} after correction for Fermi resonance) due to conjugation of the C=C—C=O groups, which weakens the C=O bond and thus causes it to vibrate at a lower frequency (3).

The α-Pyrone has the following empirical structure:

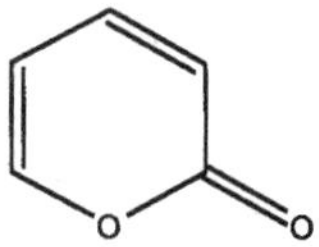

In CCl_4 solution, IR bands are noted at 1752 and 1716 cm^{-1}, and after correction for Fermi resonance unperturbed $vC=O$ is 1749.4 cm^{-1} (10).

Coumarin has the following empirical structure:

In CCl$_4$ solution at 1% wt./vol. coumarin exhibits IR bands at 1754.8 and 1741.3 cm^{-1}, and after correction for Fermi resonance νC=O is calculated to be 1747.2 cm^{-1}. In CHCl$_3$ solution at 1% wt./vol., coumarin exhibits IR bands at 1729.9 and 1754.9 cm^{-1}, and after correction for Fermi resonance νC=O is calculated to be 1734.9 cm^{-1} (10). Figure 15.9 show a plot of the unperturbed νC=O frequencies (corrected for F.R.) for coumarin vs the mole % CHCl$_3$/CCl$_4$. This plot is essentially linear over the mole % ~10–95 % CHCl$_3$/CCl$_4$ range. Deviation from linearity below 10 mole % CHCl$_3$/CCl$_4$ is caused by C=O . . . HCCl$_3$.

Figure 15.10 shows a plot of unperturbed νC=O (corrected for F.R.) for coumarin vs the mole % (CH$_3$)$_2$SO/CCl$_4$ and Fig. 15.7 shows a plot of unperturbed νC=O (corrected for F.R.) for coumarin vs the mole % (CH$_3$)$_2$SO/CHCl$_3$ (10). In Fig. 15.10 the plot is essentially linear over the ~10–95% (CH$_3$)$_2$SO/CCl$_4$ range. In only CCl$_4$ or (CH$_3$)$_2$SO solution, there is deviation from linearity (10). In Fig. 15.11 the plot is unique, as the νC=O frequencies increase up to ~35 mol % (CH$_3$)$_2$SO/CHCl$_3$ stays relatively constant in the 40–50 mol % range, and decrease in frequency in the ~55–100 mol % range. The unique plot in Fig. 15.11 most likely is the result of intermolecular hydrogen bonding between the CHCl$_3$ proton and the coumarin carbonyl group and between the CHCl$_3$ proton and the oxygen atom of dimethyl sulfoxide. In the case of CHCl$_3$/CCl$_4$ solutions, the strength of the intermolecular hydrogen bonding between Cl$_3$CH· · ·ClCCl$_3$ is much weaker than in the case of Cl$_3$CH· · ·OS(CH$_3$)$_2$. Therefore, in CHCl$_3$/CCl$_4$ solution formation between C=O· · ·HCCl$_3$ is apparent only after the first formation of (CH$_3$)$_2$SO· · ·HCCl$_3$ (10).

Figure 15.12 shows a plot of the unperturbed νC=O frequencies for coumarin (corrected for F.R.) vs the reaction field of the CHCl$_3$/CCl$_4$ solvent system (10). The reaction field is (R) = $(e - 1)/2e + n^2$), where e is the solvent dielectric constant and n is the refractive index of the solvent, and this equation is derived from bulk dielectric theory (3,4). Bulk dielectric theory predicts a single band (νC=O, for example) which shifts in frequency according to the composition of the solvent mixture. Figure 15.12 shows deviation from linearity, which is attributed to the formation of intermolecular hydrogen bonding between the CHCl$_3$ proton and the coumarin carbonyl group (10).

Table 15.7 lists IR group frequency data for coumarin and derivatives in the vapor and solution phases (1,10). Coumarin exhibits νC=O at 1776 cm^{-1} in the vapor phase, which is higher than it occurs in any of the solvent systems. Substitution of a 3-Cl or 6-CH$_3$ group apparently does not affect the νC=O frequency, as νC=O is observed at 1775 cm^{-1} in both cases (1). On the other hand, there is a noticeable increase in the νC=O frequency in the case of 3,4-dihydrocoumarin,

6UR-5, 6

In this case, the lactone has 6 UR-5,6 structure, and the C=O bond is not conjugated as it is in the case of coumarin; therefore, νC=O occurs at a higher frequency (1802 cm^{-1} vs 1776 cm^{-1} in the vapor phase (1)).

In summary, the unperturbed νC=O frequencies for lactones in CCl$_4$ solution occur at or in the range:

1748.0-1749.4 cm^{-1} with 6 UR-2,3,4,5 structure

1769.9-1781.4 cm^{-1} with 5 UR-2,3 structure

1789.1-1792.7 cm^{-1} with 5 SR structure

1841.8 cm^{-1} with 4 SR structure

Figure 15.13 shows plots of unperturbed νC=O for coumarin 1% wt./vol. In various solvents vs the solvent acceptor number (AN). The solvents are numbered sequentially 1–16 as they are listed in Table 15.6. Two linear plots are noted. Solvent 13–16 are for the four alcohols. In this case the OH group is intermolecularly hydrogen bonded with the coumarin carbonyl oxygen atom. Steric factors of solute and solvent cause variations in solute-solvent interaction, plus intermolecular hydrogen bonding prevent the AN values to be a precise predictor of νC=O values. However, they are useful in predicting approximate νC=O frequencies in many cases (10).

Table 15.8 lists IR data for 1% wt./vol. phenyl acetate in CHCl$_3$/CCl$_4$ solutions (11). In the case of phenyl acetate, the νC=O mode is in Fermi resonance with a combination tone (CT). Plots of the uncorrected and corrected νC=O and CT frequencies are shown in Fig. 15.14. After correction for Fermi resonance, the νC=O and CT frequencies converge as the mole % CHCl$_3$/CCl$_4$ increases. The νC=O mode decreases in frequency while the CT increases in frequency. It is apparent from these plots that intermolecularly hydrogen bonding occurs between the CHCl$_3$ proton and the phenyl acetate carbonyl oxygen atom at low mole % CHCl$_3$/CCl$_4$ concentration. Identical plots are obtained plotting the same IR data for phenyl acetate vs the reaction field (R) [see preceding text]. Therefore, νC=O is affected by bulk dielectric effects and refractive index of the solvent system as well as intramolecular hydrogen bonding (11).

Table 15.9 lists IR data for phenyl acetate 1% wt./vol. In various solvents, and Fig. 15.15 shows plots of the CT and νC=O modes in Fermi resonance and the νC=O and CT modes corrected for F.R. vs the solvent acceptor number (AN) (11). The observed uncorrected C=O and CT modes converge for solvents 1–11, 14, and 15 and diverge for solvents 12, 13, 16, 17. After correction for F. R., νC=O and CT converge for solvents 1–11, 14, 15, cross-over in the case of solvent 12, tert-butyl alcohol, and diverge even more for solvents 13, 16, 17, the other three alcohols in order of increasing acidity of the OH proton. Steric factors of the solvent and solute and intermolecular hydrogen bonding prevent a simple precise correlation to be developed between νC=O for any given compound in a particular solvent using its AN value (11).

ALKYL BENZOATES

Table 15.10 lists IR data for alkyl benzoates in CS_2 solution between 29 and -10 °C (11). In each case, the $vC=O$ mode(s) decreases in frequency as the temperature is decreased from ambient temperature to 90 or 100 °C. Both methyl 2-bromobenzoate and methyl 2-methoxy-benzoate exhibit two $vC=O$ IR bands. In each case, the lower frequency $vC=O$ mode decreases more in frequency than the higher frequency $vC=O$ mode. These two $vC=O$ modes result from the presence of rotational conformers. The higher frequency $vC=O$ band is assigned to conformer 1 and the lower frequency $vC=O$ band is assigned to conformer 2. Conformer 1 is assigned to the structure where the Br atom or methoxy oxygen atom is near in space to the carbonyl oxygen atom while conformer 2 is assigned to the structure where the Br atom or methoxy oxygen atom is near in space to the methyl ester oxygen atom. Methyl 2-methoxy-benzoate exhibits two linear segments above and below ~ 50 °C. The other five plots are all linear (see Fig. 15.16). The break in the plots for methyl 2-methoxybenzoate is attributed to change in the rotational configuration of the 2-methoxy group (rotation of the 2-methoxy group about the aryloxygen bond.

Figure 15.17 shows plots of the absorbance ratios A($vC=O$ conformer 1)/A($vC=O$ conformer 2). These plots show that conformer 1 increases and conformer 2 decreases in concentration as the temperature is lowered (11).

SALICYLATES AND 2-HYDROXYACETOPHENONE (A KETONE)

Table 15.11 lists IR data for methyl salicylate, phenyl salicylate, and 2-hydroxyacetophenone in various solvents at 1% wt./vol. The 19 solvents used in study this study are presented 1–19 in descending order (12). Figure 15.18 shows plots of these data vs the solvent acceptor number (AN) for each of the 19 solvents (12). These compounds are intramoleculairly hydrogen bonded as shown in this figure. Linear relationships are noted for $vC=O\cdots HO$ vs (AN) for each of these compounds in solution with the alkyl alcohols (solvents 14, 17–19). However, a linear relationship is not observed for $vC=O\cdots HO$ for vs (AN) for the other 14 solvents. However, there is a general increase in frequency as the (AN) value becomes larger. This behavior is attributed to steric factors of both the solute and solvent, which prevents the AN values from being precise predictors of vibrational frequencies.

Figure 15.19 shows plots of $vC=O\cdots HO$ frequencies for 2-hydroxyacetophenone, methyl salicylate, and phenyl salicylate vs the frequency difference between $vC=O\cdots HO$ (hexane) and $vC=O\cdots HO$ (solvent) for each of the other solvents. These linear plots are due to the mathematical treatment of the data. However, they do show that the number sequence in each of the three plots is not the same, showing that the (AN) values of these solvents are not precise predictors of vibrational frequencies.

It is noted that the $vC=O\cdots HO$ frequency for methyl salicylate and phenyl salicylate occur at 1680.4 and 1693.0 cm^{-1} in CS_2 solution, respectively, while $vC=O$ for methyl benzoate at ambient temperature in CS_2 solution occurs at 1727 cm^{-1}. The 46.6 cm^{-1} decrease in frequency between the $vC=O$ frequency for methyl benzoate and methyl salicylate is due to more than just intramolecular hydrogen bonding. The resonance effect of the OH group would weaken the C=O bond, causing a decrease in the frequency. This latter effect may be offset by the field effect between the hydroxyl oxygen atom and the carbonyl oxygen atom. The larger inductive effect of the phenyl group vs the methyl group causes νC=O$\cdots$HO for phenyl salicylate to occur at a higher frequency than νC=O$\cdots$HO for methyl benzoate.

In going from solution in CCl$_4$ to solution in CHCl$_3$ the νC=O$\cdots$HO frequency decreases by 4.3 and 4.8 cm^{-1} for methyl salicylate and phenyl salicylate, respectively. In the case of phenyl benzoate, the νC=O frequency decrease is 9 cm^{-1} in going from solution in CCl$_4$ to solution in CHCl$_3$. In going from solution in CCl$_4$ to solution in (CH$_3$)$_2$SO, the νC=O$\cdots$OH or νC=O frequency decrease is 7.2, 6.3, and 10 cm^{-1} for methyl salicylate, phenyl salicylate, and phenyl benzoate, respectively. It is thus readily apparent that the νC=O$\cdots$HO modes are not altered as much in frequency as νC=O modes in going from solution in CCl$_4$ to solution in CHCl$_3$ or (CH$_3$)$_2$SO. This is because the OH proton essentially neutralizes the basicity of the C=O oxygen atom and also acts as a site for OH$\cdots$ClCCl$_3$ or OH$\cdots$ClCHCl$_2$ interaction.

Figure 15.20 shows a plot of νC=O$\cdots$HO for phenyl salicylate vs mole % CHCl$_3$/CCl$_4$ (12). This plot shows essentially three linear segments with breaks at $\sim$23 and 69 mol % CHCl$_3$/CCl$_4$. The mole ratio of solvent to solute is 22.1 : 1 in CCl$_4$ solution and 26.7 : 1 in CHCl$_3$ solution. The mole ratio CHCl$_3$ to solute 6.1 : 18.4 at $\sim$23 and 69 mol % CHCl$_3$/CCl$_4$, respectively. These data indicate that different complexes are formed between phenyl salicylate and the CHCl$_3$/CCl$_4$ solvent system in the mole % CHCl$_3$/CCl$_4$ ranges 0–23, 23–69 and 69–100.

PHTHALATES

Table 15.12 shows Raman group frequency data for dialkyl pthalates in the neat phase. The relative intensity (RI) values are normalizes to the strongest Raman band in the spectrum occurring near 1045 cm^{-1}. The abbreviation denotes depolarization ratio. Figure 15.21 gives bar graphs of the Raman group frequencies for 21 dialkyl phthalates. The upper graph represents both the frequency ranges and intensities for each of the 12 group frequencies. The intensities are proportional to $(45\alpha')^2 + (4\beta')^2$ (α' and β' are the derivative of the mean polarizability and anisotropy, respectively, with respect to the normal coordinate of the polarizability tensor). The horizontal lines across the range of frequencies indicate minimum band intensity relative to the $\sim$1045 cm^{-1} group frequency. The lower graph of these Raman group frequencies is proportional to $(3\beta')^2$.

The Raman phthalate νC=O mode occurs in the range 1728–1738 cm^{-1}, whose relative intensity has a value in the range 40–51. The strongest Raman band occurs in the range 1042–1047 cm^{-1} and is assigned to the ortho phenylene ring breathing mode. It occurs at a slightly higher frequency than the breathing mode of the phenyl group for mono-substituted benzenes. Phthalates show a Raman band in the range 1277–1294 cm^{-1} having a relative intensity value in the range 24–40, and it is polarized. This band is assigned to a complex skeletal mode of the aryl [$-$C(=)O$-$R]$_2$ groups. The other Raman group frequencies have been correlated with in-plane ring mode of 1,2-dichlorobenzene (13).

ACRYLATES AND METHACRYLATES

Table 15.13 lists Raman data for the $\nu C=O$ and $\nu C=C$ stretching vibrations for acrylates (14). The $\nu C=O$ mode occurs in the range $1717-1774\,cm^{-1}$. The alkyl acrylates exhibit $\nu C=O$ in range $1717-1740\,cm^{-1}$ and aryl acrylates exhibit $\nu C=O$ in the range $1731-1774\,cm^{-1}$. Variations in frequency within each class can be attributed to the inductive effect of the alkyl or haloalkyl groups and branching on the $O-R\alpha$-carbon atom.

The s-trans $\nu C=C$ mode for alkyl and aryl acrylates occurs in the range $1623-1640\,cm^{-1}$ in the neat phase. The relative Raman band intensity (RI) for s-trans $\nu C=C$ is higher than it is for s-cis $\nu C=C$, and s-cis $\nu C=C$ occurs at lower frequency than s-trans $\nu C=C$ by $\sim 14-16\,cm^{-1}$. With the exceptions of hexafluoroisopropyl acrylate, pentabromophenyl acrylate, 1-naphthyl acrylate, and bis-phenol-A diacrylate, the ratio (RI) $\nu C=O$/(RI) s-trans $C=C$ is in the range 0.25 through 0.83.

Table 15.13a lists IR data for alkyl acrylates in $CHCl_3$ and/or CCl_4 solutions (1 % wt./vol. Of solute). In CCl_4 solution and in $CHCl_3$ solution, $\nu C=O$ occurs in the range $1722.9-1734.1\,cm^{-1}$ and $1713.8-1724.5\,cm^{-1}$, respectively (15,16). In CCl_4 solution and in $CHCl_3$ solution, s-trans $\nu C=C$ occurs in the range $1635.3-1637.0\,cm^{-1}$ and $1635.3-1637.9\,cm^{-1}$ respectively. In CCl_4 solution and in $CHCl_3$ solution, s-cis $\nu C=C$ occurs in the range $1619.2-1620.4\,cm^{-1}$ and $1618.5-1619.9\,cm^{-1}$, respectively. The $CH=CH_2$ twist and $C=CH_2$ wag modes occur in the range $982.6-985.6\,cm^{-1}$ and $966.3-983.4\,cm^{-1}$, respectively (15).

Figure 15.22 shows plots of $\nu C=O$ for seven alkyl acrylates vs mole % $CHCl_3/CCl_4$ (1% wt./vol.). The $\nu C=O$ mode for all of these acrylates decreases in frequency as the mole % $CHCl_3/CCl_4$ is increased. There is a distinct difference between the rate of $\nu C=O$ frequency in the $O \sim 35$ mol % $CHCl_3/CCl_4$ range than at higher mole % $CHCl_3/CCl_4$. At the lower mole % $CHCl_3/CCl_4$ range, the faster rate of $\nu C=O$ frequency decrease is attributed to the formation of $C=O\cdots HCCl_3$ bonds, $(Cl_3CH\cdots ClCCl_3)_x$ and $(ClCl_2CH\cdots)_n$. The general gradual linear decrease with change in the mole % $CHCl_3/CCl_4$ is attributed to an increasing value of the solvent reaction field (R). In the case of tert-butyl acrylate vs methyl acrylate the inductive contribution

$$CH_2=CH-\overset{O^-}{\underset{+}{C}}-O-C(CH_3)_3 \quad \text{is larger than the}$$

$$\text{contribution from} \quad CH_2=CH-\overset{O^-}{\underset{+}{C}}-O-CH_3, \quad \text{which}$$

causes $\nu C=O$ mode for the tert-butyl analog to occur at lower frequency than for the methyl analog (15).

Table 15.14 shows Raman data for alkyl and aryl methacrylates in the neat phase, and a summary of the IR CCl_4 and $CHCl_3$ solution data for $\nu C=O$ and $\nu C=C$ (14). In the neat phase, alkyl methacrylates exhibit Raman bands in the range $1732-1762\,cm^{-1}$ (the alkyl analogs in the range $1713-1742\,cm^{-1}$, and the aryl analogs in the range $1735-1762\,cm^{-1}$). These bands are assigned to the $\nu C=O$ mode. In all cases, the Raman band assigned as $\nu C=C$ occurs in the range $1635-1642\,cm^{-1}$, and with the exceptions of 2-phenoxyethyl methacrylate, 2-methallyl methacrylate, and pentachlorophenyl methacrylate the Raman band intensity ratio (RI) $vC=O/(RI)vC=C$ is always <1. In cases where the Raman spectra were recorded of the same alkyl or phenyl acrylates and methacrylates, the $vC=O$ mode always occurs at lower frequency in the case of the methacrylates than for that of the acrylates. The lower $vC=O$ mode frequencies in the case of the methacrylate analogs are due to the inductive contribution of the α-methyl group to the $C=O$ group. Moreover, with the exception of pentafluorophenyl methacrylate, the $vC=C$ mode occurs at higher frequency than both s-trans $vC=C$ and s-cis $vC=C$ for corresponding alkyl or aryl acrylates.

Figure 15.23 shows plots of $vC=O$ for corresponding alkyl acrylate and alkyl methacrylate analogs vs mole % $CHCl_3/CCl_4$ (15). The plots for the methacrylate and acrylate analogs appear to be comparable at ~ 40 mol % $CHCl_3/CCl_4$ and above, and are different below ~ 40 mol % $CHCCl_3/CCl_4$. That there are differences between the plots of these acrylate and methacrylate analogs warrants further discussion. In the case of the methacrylates the $C=O$ group is more basic than the $C=O$ group for the acrylates due to the inductive contribution of the α-methyl carbon atom to the $C=O$ bond. Thus, a stronger intermolecular hydrogen bond would be expected to be formed between $C=O\cdots HCCl_3$ for methacrylates than in the case of the corresponding alkyl acrylates. However, it is noted that the $vC=O$ frequency decrease for the allyl, butyl, and 2-ethylhexyl methacrylates in going from solution in CCl_4 to solution in $CHCl_3$ is 8.63, 9.43, and 9.99 cm^{-1}, respectively, while for the corresponding acrylate analogs the $vC=O$ frequency decrease is 9.43, 10.58, and 10.82 cm^{-1}, respectively. In both cases, the $vC=O$ frequency decreases in the order of increasing inductive contribution of the O–R alkyl group to the $C=O$ group. However, the decrease in the $vC=O$ frequency is larger in each case of the acrylates than in the case of the methacrylates. Thus, the $C=O\cdots HCCl_3$ intermolecular hydrogen bond appears to be stronger in the case of alkyl acrylates than for alkyl methacrylates. However, the contribution of hydrogen bonding of the $CHCl_3$ proton to $C=O$ group ($C=O\cdots HCCl_3$) can not be separated from the contribution of the reaction field R) for acrylates and methacrylates in $CHCl_3/CCl_4$ solution merely by subtracting the $vC=O$ frequencies in CCl_4 solution from those in $CHCl_3$ solution. As the reaction field (R) is a linear function of mole % $CHCl_3/CCl_4$ it is reasonable to extrapolate the linear portion of each plot to 0 mol % $CHCl_3/CCl_4$ to obtain the delta-$vC=O$ frequency decrease attributed to the $C=O\cdots HCCl_3$ intermolecular hydrogen bond. This delta-$vC=O$ frequency decrease divided by the total $vC=O$ frequency decrease between $vC=O(CCl_4)$ minus $vC=O(CHCl_3)$ multiplied by 100 yields the percentage of the $vC=O$ frequency decrease attributed to the intermolecular hydrogen bond. The results are as follows:

	acrylate	methacrylate
allyl	37.4%	45.1%
butyl	42.2%	49.4%
2-ethylhexyl	47.8%	54.6%

These calculations indicate that the percentage of $vC=O$ frequency decrease is larger in the case of the alkyl methacrylates than in the case of the alkyl acrylates. Thus, the reaction field (R) contributes more to the $vC=O$ frequency decrease in the case of alkyl acrylate than for alkyl methacrylates. The most reasonable explanation for the preceding data is that the alkyl methacrylate exist in the s-trans configuration. By analogy

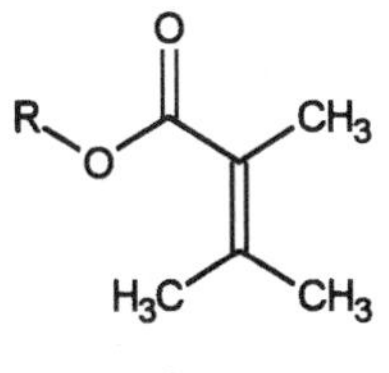

s - trans

with the alkyl α-methyl branched acetates discussed previously, the α-methyl group in alkyl methacrylates would affect the spatial distance between the carbonyl carbon atom and the solvent system (CHCl$_3$/CCl$_4$). Thus, the reaction field would be less in the case of alkyl methacrylates in the s-trans configuration than for the alkyl acrylates in either the s-cis or s-trans configurations. The methacylate C=O group is more basic than the acrylate C=O group, and apparently a stronger C=O⋯HCCl$_3$ bond is formed in the case of the methacrylates even though the spatial distance between the carbonyl oxygen atom and the CHCl$_3$ proton is greater in the case of alkyl methacrylates.

CINNAMATES

Poly (vinyl cinnamate) has the following empirical structure:

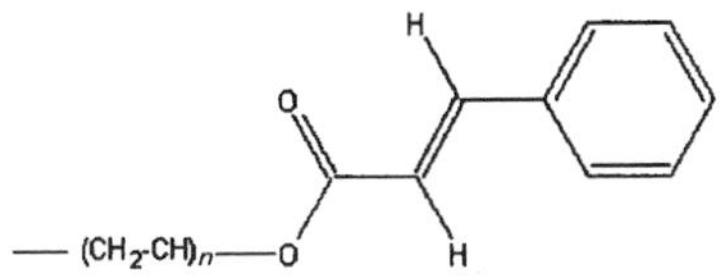

The Raman spectrum for poly (vinyl cinnamate) exhibits bands at 1701 and 1638 cm^{-1} whose relative band intensity ratio (RI) is 1:9 and these bands are assigned to νC=O and νC=C, respectively. Phenyl in-plane ring modes are assigned at 1601, 1030, 1002, and 620 cm^{-1} whose (RI) ratios are 6:9, 1:9, 4:9, and 1:9, respectively. In most cases the phenyl breathing mode for monosubstituted benzenes has the highest Raman band intensity. In this case it occurs at 1002 cm^{-1} with an (RI) of 4 compared to an (RI) of 9 for νC=C. Thus, the Raman band intensity of νC=C for this polymer is very strong. The Raman bands for other alkyl cinnamates are expected to occur at similar frequencies.

Table 15.15 lists IR vapor-phase data and assignments for alkyl cinnanates (1). In the vapor phase the alkyl cinnamates exhibit νC=O in the range 1727–1740 cm^{-1}. The highest and lowest νC=O frequencies are exhibited by the methyl and tert-butyl cinnamates, respectively. Again the νC=O frequencies decrease in the order of the increasing inductive contribution of the O–R alkyl group to the C=O bond. The νC=C mode occurs in the range 1649–1642 cm^{-1}. The IR band intensity ratio (A)νC=C/(A)νC=O varies between 0.33 and 0.41.

As noted in the empirical structure for poly (vinyl cinnamate), the CH=CH is presented in the trans configuration. In the IR, the band in the range 972–980 cm^{-1} results from the trans

CH=CH twisting vibration, and confirms the trans configuration for the C=C bond. The IR band intensity ratio (A) CH=CH twist/(A) vC=C varies between 0.42 and 0.67.

Table 15.15a lists IR vapor-phase data for alkyl cinnamates (1). In-plane phenyl ring modes are presented in columns [1] through [3] and these group frequencies occur in the ranges 1304–1314, 1245–1276 cm^{-1}, and 1198–1200 cm^{-1}, respectively. The phenyl ring in-phase out-of-plane 5-hydrogen deformation and the out-of-plane phenyl ring deformation occur in the range 750–765 and 685–696 cm^{-1}, respectively.

With the exception of the isopropyl analog, the strongest IR band for the alkyl cinnamates occurs in the range 1154–1171 cm^{-1}, it is attributed to a C—C(=)O stretching mode, and its intensity is used to normalize the 3 in-plane ring modes.

The ratio (A)[6]/(A)[7] is larger for the alkyl cinnamates than for benzyl cinnamate by a factor of 2 or more. The reason for this is that the comparable modes for the benzyl group are also absorbing at these frequencies. The ratio (A) vC=C/(A) [4] shows that the vC=C mode has much less IR absorbance than the absorbance for vC—C(=)O.

Table 15.15b lists IR data for the alkyl and phenyl groups of cinnamates.

The first overtone, $2v$C=O, occurs in the range 3435–3470 cm^{-1}. Phenyl ring carbon-hydrogen stretching modes occur in the ranges 3065–3075 cm^{-1} and 3030–3040 cm^{-1}. The $v$asym. CH$_3$ and $v$sym. CH$_3$ modes occur in the range 2961–2985 cm^{-1} and 2850–2945 cm^{-1}, respectively. The $v$asym. CH$_2$ and $v$sym. CH$_2$ modes occur in the range 2942–2960 cm^{-1} and 2868–2895 cm^{-1}, respectively. The δasym. CH$_3$ and δsym. CH$_3$ modes occur in the range 1452–1475 cm^{-1} and 1371–1400 cm^{-1}, respectively. An in-plane ring mode and/or δCH$_2$ occur in the range 1450–1453 cm^{-1}.

PHENOXARSINE DERIVATIVES

The compound 10-phenoxarsinyl acetate has the following empirical structure:

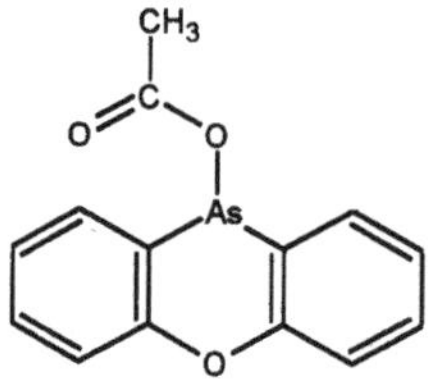

IR data for the 10-phenoxarsinyl esters are listed in Table 15.16. In this series 10-phenoxarsinyl acetate, the trichloroacetate, and the trifluoroacetate derivatives, the vC=O occurs at 1698, 1725, and 1738 cm^{-1} in CCl$_4$ solution, respectively, and this increase in the vC=O frequency is attributed to the inductive and field effects of the halogen atoms upon the C=O group (17). In CCl$_4$ solution, these vC=O frequencies occur ~ 50 cm^{-1} lower in frequency than vC=O occurs for the corresponding ethyl acetates. This frequency difference can be attributed to the fact that arsenic is less electronegative than oxygen, which would induce some ionic character to the functional group [e.g., R—C(=O)O$^-$—As$^+$]. This effect would reduce the C=O force constant because the C=O bond would contain less double character due to an increase of π overlap with the free pair of electrons on the O—As oxygen atom (17).

Figure 15.24 show IR spectra of 10-phenoxarsinyl chloroacetate in (A) CCl_4 (3800–1333 cm^{-1}) and CS_2 (1333–450 cm^{-1}) solutions and (B) as a split mull (Fluorolube 3800–1333 cm^{-1} and Nujol, 1333–450 cm^{-1}). In CCl_4 solution, $\nu C=O$ bands are noted at 1720 and 1696 cm^{-1} while in the Fluorolube mull $\nu C=O$ is noted at 1702 cm^{-1}. This occurs because in solution the chloroacetate derivative exists as rotational conformers while in the solid phase only one rotational conformer exists. The 1720 cm^{-1} rotational conformer in CCl_4 solution and the 1702 cm^{-1} band in the solid phase are assigned to the rotational conformer where the Cl atom is near in space to the carbonyl oxygen atom while the 1696 cm^{-1} CCl_4 solution band is assigned to the rotational conformer where the Cl atom is near in space to the O–As oxygen atom.

The 10-phenoxarsinyl thiol esters of form As–S–C(=O)R exhibit $\nu C=O$ in the range 1670–1681 cm^{-1}. Comparison of the As–S–C(=O)CH$_3$ vs the AsO–C(=O(CH)$_3$ (1681 cm^{-1} vs 1698 cm^{-1}) shows that there is a frequency decrease of ~ 17 cm^{-1} with the substitution of sulfur for oxygen; the reason for this is presented in a discussion on ordinary thiol esters [R–C(=O)–S–R'] that will follow.

The S-(10-phenoxarsinyl) thiol aryl esters of form As–S–C(=O)C$_6$H$_5$ exhibit $\nu C=O$ in the range 1638–1650 cm^{-1}, and the decrease in the $\nu C=O$ frequency compared to As–S–C(=O)R analogs is attributed to conjugation of the phenyl ring with the C=O group (17).

Compounds such as S-(10-phenoxarsinyl) α-(2,4,5-trichlorophenoxy) thiol acetate exhibit rotational conformer 1 at 1691 cm^{-1} and rotational conformer 2 at 1665 cm^{-1} in CCl_4 solution. In the solid phase, the $\nu C=O$ band at 1650 cm^{-1} is assigned to the more stable structure conformer 2. Rotational conformer 1 is where the phenoxy oxygen atom is near in space to the carbonyl oxygen atom and rotational conformer 2 is where the phenoxy oxygen atom is near in space to the As–S sulfur atom.

The compound, S-(10-phenoxarsinyl)-thiol-2-furoate exhibits rotational conformer 1 at 1646 cm^{-1} and rotational conformer 2 at 1631 cm^{-1} in CCl_4 solution, and the more stable conformer 1 at 1630 cm^{-1} in the solid state. Rotational conformer 1 is where the furan ring oxygen atom is near in space to the carbonyl oxygen atom while rotational conformer 2 is where the furan ring oxygen atom is near in space to A$_s$–S sulfur atom.

THIOL ESTERS

Thiol esters of form R–C(=O)–S–R' exhibit $\nu C=O$ near 1690 cm^{-1} while esters of form R–C(=O)–O–R' exhibit $\nu C=O$ at $\sim 1735 \pm 5$ cm^{-1} (18). The reason substitution of sulfur for oxygen causes $\nu C=O$ to vibrate at a lower frequency needs to be addressed. This change in the $\nu C=O$ frequency is attributed principally to the change in the C=O force constants. The resonance form

$$O^-$$
$$|$$
$$R-C=S^+$$

for thiol esters is more important than resonance form

$$O^-$$
$$|$$
$$R-C=O^+$$

for ordinary esters. In terms of electronic theory, there appears to be a greater tendency toward overlap of the carbonyl carbon atom π-electron with a nonbonding electron pair of the sulfur atom than with a nonbonding of the oxygen atom. This weakens the C=O bond, and causes νC=O to vibrate at a lower frequency than in the case of comparable ordinary esters (18). Otherwise, as will be discussed, the νC=O frequencies for thiol esters are affected by resonance, inductive, field, and hydrogen bonding.

Table 15.17 lists IR data for alkyl and phenyl thiol esters (18). The frequencies reported in the region $3800-1333\,\mathrm{cm}^{-1}$ are for CCl_4 solution, and in the region of $1333\,\mathrm{cm}^{-1}$ and below they are CS_2 solution data. In all cases, the thiol esters of form $R-C(=O)-S-C_6H_5$ exhibit higher νC=O frequencies than those for the corresponding $R-C(=O)-S-C_4H_9$ or $R-C(=O)-S-C_6H_{13}$ analogs. This occurs because the inductive effect of the phenyl ring is larger than that for the S—R alkyl group.

The inductive and field effects of halogen atoms on the α-carbon of thiol ester also increase the νC=O frequencies for compounds of forms: $CHCH_3C(=O)-S-C_4H_9$ $(1695\,\mathrm{cm}^{-1})$ $Cl_3C(=O)-S-C_4H_9$ $(1699\,\mathrm{cm}^{-1})$, and $Cl_3C(=O)-S-C_4H_9$ $(1710\,\mathrm{cm}^{-1})$, and for $CH_3C(=O)-S-C_6H_5$ $(1711\,\mathrm{cm}^{-1})$, $CCl_3C(=O)-S-C_6H_5$ $(1711\,\mathrm{cm}^{-1})$, and $CF_3C(=O)-S-C_6H_5$ $(1722\,\mathrm{cm}^{-1})$.

In the case of the mono- and dichloro thiol acetates in CCl_4 solution, both the S—R and S—C_6H_5 analogs exhibit rotational conformers (see Table 15.17a). The higher frequency νC=O band for the $ClCH_2-C(=O)-S-R$ or $-S-C_6H_5$ and $Cl_2CH-C(=O)-S-R$ or $-S-C_6H_5$ analogs results from the rotational conformer 1 where the Cl atom (s) is (are) near in space to the carbonyl oxygen atom. The lower frequency νC=O band is assigned to rotational conformer 2. In this case the Cl atom (s) is (are) near in space to S—R or S—C_6H_5 sulfur atom (18).

Thiol esters of oxalic ester have two C=O groups $[R-S-C(=O)-]_2$ and $[C_6H_5-S-C(=O)-]_2$; however, in each case only one νC=O band is observed (at 1680 and $1698\,\mathrm{cm}^{-1}$ for the S—C_4H_9 and S—C_6H_5 analogs, respectively). Thus, these S, S'-dialkyl or S, S'-diphenyl dithiol oxalates exist in a trans configuration where the molecular structure has a center of symmetry lying between the O=C—C=O groups. Only the out-of-phase $(C=O)_2$ stretching mode is IR active; the in-phase $(C=O)_2$ stretching mode is Raman active.

The S-butyl thiol formate and S-phenyl thiol formate exhibit νC=O at 1675 and $1693\,\mathrm{cm}^{-1}$, respectively. These compounds are readily identified by the formate νC—H and 2δCH frequencies, which occur in the region $2825-2835\,\mathrm{cm}^{-1}$ and $2660-2680\,\mathrm{cm}^{-1}$, respectively. The δCH mode occurs in the range $1340-1345\,\mathrm{cm}^{-1}$. In all cases these three C—H modes for the S—C_6H_5 analog occur at lower frequency than for the —S—C_4H_9 analog. Thus, the C—H bond is stronger for H—C(=O)S—R than for H—C(=O)S—C_6H_5.

Thiol esters of form $C_6H_5-C(=O)-S-C_6H_5$ exhibit νC=O at higher frequency than the correspondingly ring-substituted thiol benzoates of form $C_6H_5-C(=O)-S-R$ (see Table 15.17b). This is attributed to the inductive effect of the S—C_6H_5 being larger than the S—R group.

Both the S—R and S—C_6H_5 analogs of 2-fluoro-thiol benzoate and 2-methoxy thiol benzoate exist as rotational isomers in CCl_4 solution. The higher frequency band in each set is assigned to rotational conformer 1, the lower frequency band in each set is assigned rotational conformer 2. In these two cases, rotational isomers are presented here:

	RC1 $vC=O$, cm^{-1}	RC1 $vC=O$, cm^{-1}	RC2 $vC=O$, cm^{-1}	RC2 $vC=O$, cm^{-1}
X				
F	1675	1690	1648	1667
CH$_3$O	1672	1700	1640	1652

The phenyl 2-Cl, 2-Br, and 2-I thiol benzoates exhibit $vC=O$ at 1696, 1700, and 1698 cm^{-1} in CS$_2$ solution, respectively. In these cases, all exist as conformer 1, as the C=O group can not be coplanar with the phenyl group due to the bulky S atom of the S−R or S−C$_6$H$_5$ group.

Phenyl thiol salicylate exhibits $vC=O\cdots HO$ at 1640 cm^{-1}, and this relatively low frequency is mainly the result of the intramolecular hydrogen bond between the 2−OH proton and the free pair of electrons on the carbonyl oxygen atom (18).

THIOL ACIDS

Table 15.17c lists IR data for thiol acids, thiol anhydrides, and potassium thiol benzoate (18). Thiol acids exist as R−C(=O)−SH and C$_6$H$_5$−C(=O)−S−H, and this is in marked contrast to carboxyl acids (R−CO$_2$H and C$_6$H$_5$CO$_2$H), which exist as cyclic intermolecularly hydrogen-bonded dimers in the condensed phase (see Chapter 10).

Thiol acetic acid exhibits $vC=O$ and $vS-H$ at 1712 and 2565 cm^{-1} in CCl$_4$ solution, respectively, and thiol benzoic acid exhibits $vC=O$ and $vS-H$ at 1690 and 2585 cm^{-1} in CCl$_4$ solution, respectively. Conjugation lowers the $vC=O$ frequency as shown for thiol benzoic acid vs thiol acetic acid. The S−H proton for thiol acetic acid is more acidic than that for thiol benzoic acid, and consequently its $vS-H$ frequency occurs at lower frequency than it does for thiol benzoic acid. In CCl$_4$ solution, there is most likely intermolecular hydrogen bonding between the S−H proton and CCl$_4$ (e.g., SH$\cdots$ClCCl$_3$).

THIOL ACID ANHYDRIDES

Thiol benzoic anhydride or (dibenzoyl sulfide) exhibits IR bands at 1739, 1709, and 1680 cm^{-1}, and only two $vC=O$ modes are expected. An IR band assigned as =C−S stretching occurs at 860 cm^{-1}, and its first overtone would be expected at below 1720 cm^{-1}. Most likely, the third IR band in this region of the spectrum results from Fermi resonance of $2v$=C−S with v in-phase (C=O)$_2$ stretching (1680 cm^{-1}). The higher frequency IR band most likely results from the v out-of-phase (C=O)$_2$ mode (see Fig. 15.25, which compares the IR spectra of thiol benzoic anhydride (dibenzoyl sulfide) and benzoic anhydride (dibenzoyl oxide)).

Potassium thiol benzoate exhibits νasym. COS at 1525 cm^{-1} and its νC−S mode at 948 cm^{-1} (18).

Table 15.18 lists Raman data and probable assignments for propargyl acrylate and propargyl methacrylate. As noted in Table 15.18, νC≡C exhibits the most intense Raman band at 2132 cm^{-1} for both acrylate and methacrylate. The next most intense Raman band is assigned to νC=C and it is less intense than νC≡C by a factor of 7/9 and 5/9 for the acrylate and methacrylate, respectively. The νC=O modes occur at 1728 and 1724 cm^{-1} for the acrylate and methacrylate, respectively. The lower νC=O frequency for the methyacrylate is attributed to the inductive contribution of the α-methyl group, which weakens the C=O bond. Apparently this causes the depolarization of the electron cloud to be larger in the case of methacrylates compared to the acrylates, since the relative Raman band intensity is 3 for the methacrylate and only 2 for the acrylate.

OTHER ESTER VIBRATIONS

Table 15.19 lists IR vapor-phase data for carboxylic acid esters (1).

Alkyl alkanoates and dialkyl diesters exhibit a strong IR band in the range 1110–1250 cm^{-1}, which results from a complex mode involving R−C(=O)−OR′ stretching. In the case of alkyl formate, H−C(=O)−OR′, the mode most likely includes =C−O stretching, and it occurs in the range 1152–1180 cm^{-1}. In the case of alkyl acetates, the mode occurs in the range 1231–1250 cm^{-1} for compounds of form CH$_3$−C(=O)−OR′, and in the range 1201–1215 cm^{-1} when R′ is vinyl, isopropenyl, or phenyl. These data suggest that the stretching mode is complex and involves CH$_3$−C(=O)−OR′ skeletal stretching.

Study of α-substitution on the alkyl acetate series shows that the skeletal C−C(=O)−O−R′ stretching mode decreases as the 2-alkyl group increases in length from 2-methyl through 2-butyl. Moreover, the skeletal C−C(=O)−OR′ stretching mode decrease steadily in frequency as the substitution of α-methyl groups increases (e.g. (CH$_3$)$_2$CH−C(=O)−OR′, 1145–1159 cm^{-1} and (CH$_3$)$_3$C−C(=O)−OR′, 1110–1156. cm^{-1}). These data support the conclusion that the C−C(=O)−OR′ mode includes stretching of the C−C(=O)−OR′, (C−)$_2$C(=)O−R′, or (C−)$_3$C(=)−OR′ groups.

Table 15.20 lists IR vapor-phase data for conjugated esters (1). The alkyl aromatic esters skeletal aryl−C(=)−OR′ stretching modes in the range 1229–1311 cm^{-1} (strong) and in the range 1082–1145 cm^{-1} (medium). The alkyl crotonates exhibit skeletal C=C−C(=)−OR′ stretching modes in the range 1176–1190 cm^{-1} (strong) and 1021–1048 cm^{-1} (medium).

All esters show strong or medium IR bands in these general regions of the vibrational spectrum. It is always helpful to have a collection of IR and/or Raman standard reference spectra available for comparison and positive identification. The most comprehensive sets of IR and Raman spectra for all types of organic materials are available from Sadtler Research Laboratories, a Division of Bio-Rad Laboratories, Inc.

REFERENCES

1. Nyquist, R. A. (1984). *The Interpretation of Vapor-Phase Infrared Spectra: Group Frequency Data*, Philadelphia: Sadtler Research Laboratories, Division of Bio-Rad.

2. Nyquist, R. A. (1994). *Vib. Spectrosc.*, 7, 1.

3. Buckingham, A. D. (1960). *Can J. Chem.*, 308, 300.

4. Timmermans, J. (1959). *Physical-Chemical Constants of Binary Systems*, Vol. 1, pp. 308, 309, New York: Interscience Publishers.

5. Nyquist, R. A. (1991). *Vib. Spectrosc.*, 2, 221.

6. Gutman, V. (1978). *The Donor-acceptor Approach to Molecular Interactions*, New York: Plenum.

7. Nyquist, R. A. (1986). *Appl. Spectrosc.*, 40, 336.

8. Nyquist, R. A. (1986). *Appl. Spectrosc.*, 40, 79.

9. Nyquist, R. A., Fouchea, H. A, Hoffman, G. A., and Hasha, D. L. (1991). *Appl. Spectrosc.*, 45, 860.

10. Nyquist, R. A. and Settineri, S. E. (1990). *Appl. Spectrosc.*, 44, 791.

11. Nyquist, R. A. and Settineri, S. E. (1990). *Appl. Spectrosc.*, 44, 1629.

12. Nyquist, R. A., Putzig, C. L., Clark, T. L., and McDonald, A. T. (1996). *Vib. Spectrosc.*, 12, 93.

13. Nyquist, R. A. (1972). *Appl. Spectrosc.*, 26, 81.

14. (1996). *View Master-Raman Basic Monomers and Polymers*, Philadelphia: Sadtler Research Laboratories, A Division of Bio-Rad.

15. Nyquist, R. A. and Streck, R. (1994). *Vib. Spectrosc.*, 8, 71.

16. Nyquist, R. A. and Streck, R. (1995). *Spectrochim. Acta*, 51A, 475.

17. Nyquist, R. A., Sloane, H. J., Dunbar, J. E., and Strycker, S. J. (1966). *Appl. Spectrosc.*, 20, 90.

18. Nyquist, R. A. and Potts, W. J. Jr. (1959). *Spectrochim. Acta*, 15, 514.

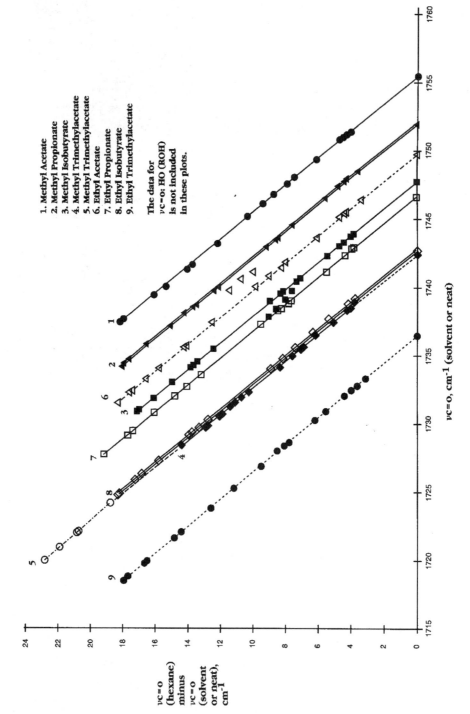

1. Methyl Acetate
2. Methyl Propionate
3. Methyl Isobutyrate
4. Methyl Trimethylacetate
5. Methyl Trimethylacetate
6. Ethyl Acetate
7. Ethyl Propionate
8. Ethyl Isobutyrate
9. Ethyl Trimethylacetate

The data for
$vc=o$: HO (ROH)
is not included
in these plots.

$vC=O$, cm^{-1} (solvent or neat)

$vC=O$ (hexane) minus $vC=O$ (solvent or neat), cm^{-1}

FIGURE 15.1   Plots of vCO (solvent or neat) vs vCO (hexane) minus vCO (solvent or neat) for alkyl alkanoate.

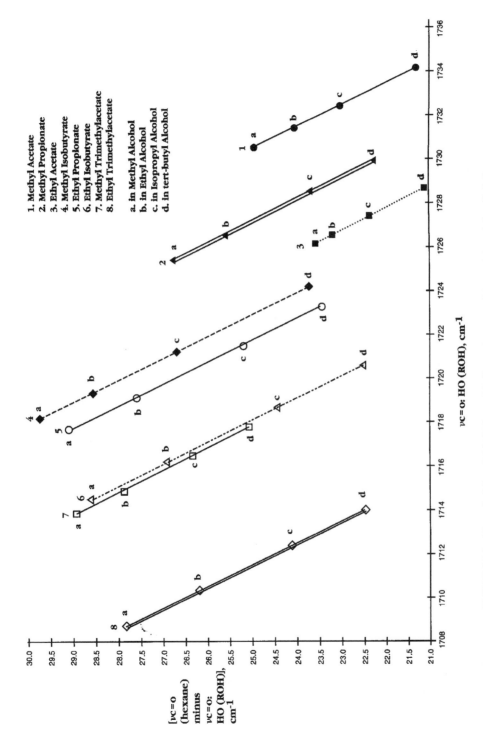

FIGURE 15.2   Plots of $\nu CO \cdots HO(ROH)$ vs νCO (hexane) minus $\nu CO \cdots HO(ROH)$ for alkyl alkanoates.

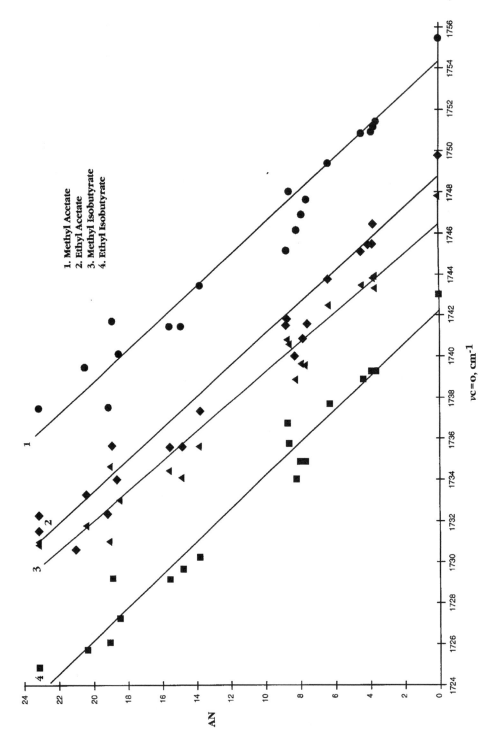

FIGURE 15.3   Plots of νCO for alkyl alkanoates vs the solvent acceptor number (AN) or neat alkyl alkanoate.

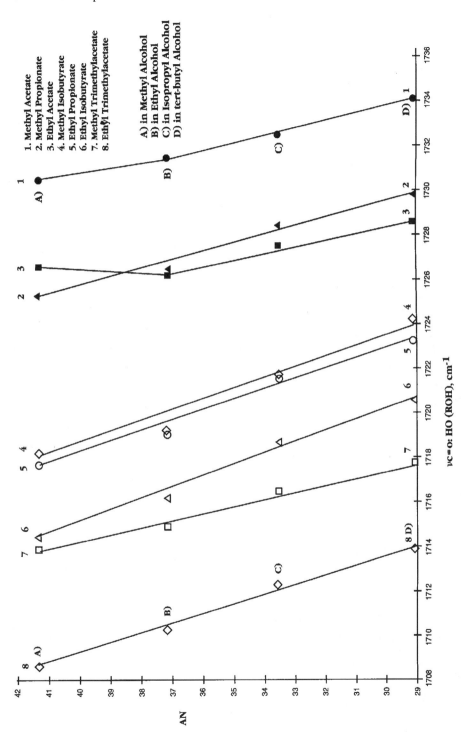

FIGURE 15.4 Plots of $\nu CO \cdots HO(ROH)$ vs the solvent acceptor number (AN) in alkyl alcohols.

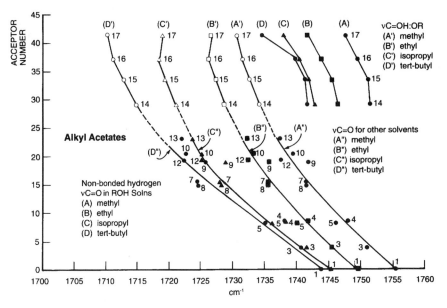

FIGURE 15.5 Plots of the νCO and νCO$\cdots$HO(ROH) frequencies vs the solvent acceptor number (AN). Extrapolation of the points on curve A to A, curve B to B, curve C to C, and curve D to D yields the postulated AN values of 9.15, 4.15, 3.00, and 2.06 for methyl alcohol through tert-butyl alcohol, respectively.

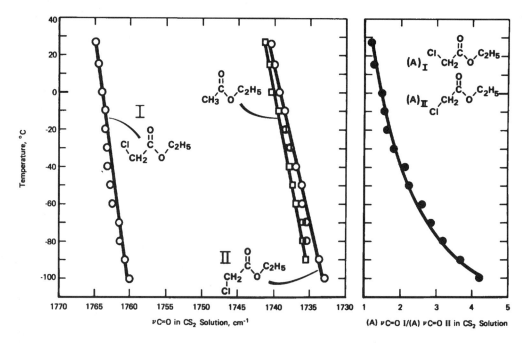

FIGURE 15.6 The left figure shows plots of νCO rotational conformers 1 and 2 for ethyl 2-chloroacetate and νCO for ethyl acetate frequencies in CS$_2$ solution vs temperature in $°$C. The right figure shows a plot of the absorbance ratio for νCO rotational conformer 1/rotational conformer 2 vs temperature in $°$C.

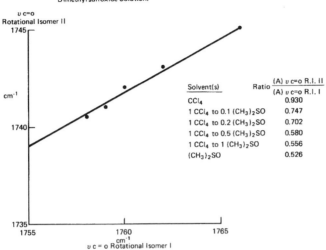

FIGURE 15.7   A plot of the νCO frequencies for rotational conformer 1 vs νCO for rotational conformer 2 for methyl 2,4-dichlorophenoxyacetate with change in the CCl₄ to (CH₃)₂SO ratio.

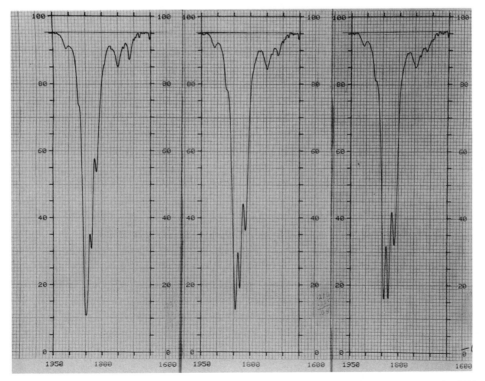

FIGURE 15.8   Infrared spectra of β-propiolactone in CCl₄ solution (left), 52 mol % CHCl₃/CCl₄ solution (middle), and CHCl₃ solution (right).

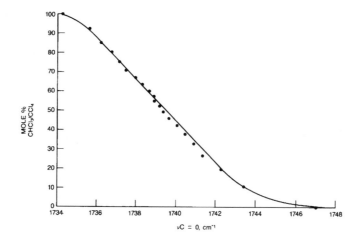

FIGURE 15.9 A plot of unperturbed νCO frequencies for coumarin in 1% wt./vol. $CHCl_3/CCl_4$ solutions vs mole % $CHCl_3/CCl_4$.

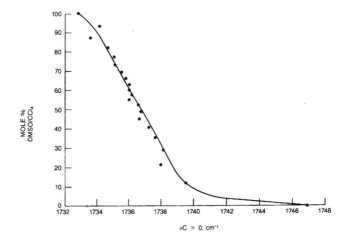

FIGURE 15.10 A plot of unperturbed νCO frequencies for coumarin in 1% wt./vol. $(CH_3)_2SO/CCl_4$.

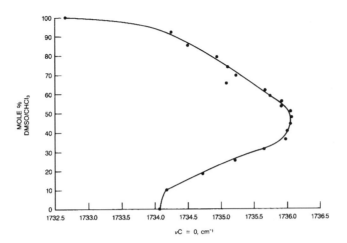

FIGURE 15.11   A plot of unperturbed νCO frequencies for coumarin 1% wt./vol. in $(CH_3)_2SO/CHCl_3$ solutions vs mole % $(CH_3)_2SO/CHCl_3$.

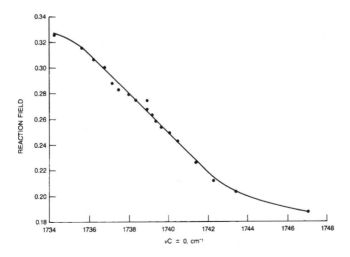

FIGURE 15.12   A plot of the unperturbed νCO frequencies for coumarin 1% wt./vol. in $CHCl_3/CCl_4$ solution vs the reaction field for $CHCl_3/CCl_4$ solutions.

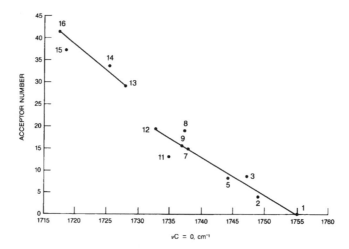

FIGURE 15.13 Plots of unperturbed νCO frequencies for coumarin 1% wt./vol. in various solvents vs the solvent acceptor number (AN).

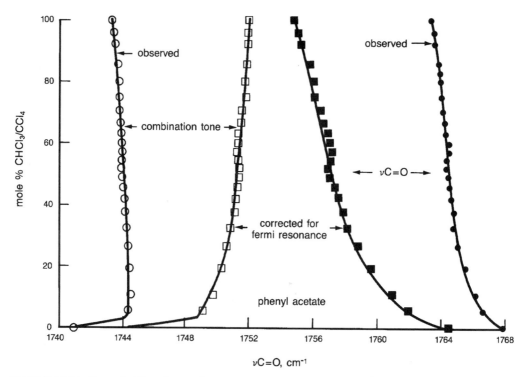

FIGURE 15.14 Plots of νCO and a combination tone in Fermi resonance and their unperturbed νCO and CT frequencies for phenyl acetate after correction for Fermi resonance vs mole % CHCl$_3$/CCl$_4$.

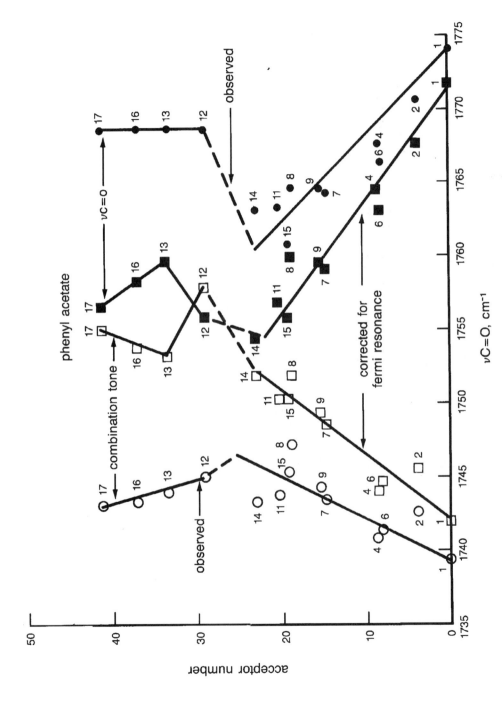

FIGURE 15.15 Plots of νCO and a combination tone in Fermi resonance and their unperturbed νCO and CT frequencies for phenyl acetate after correction for Fermi resonance vs the solvent acceptor number (AN).

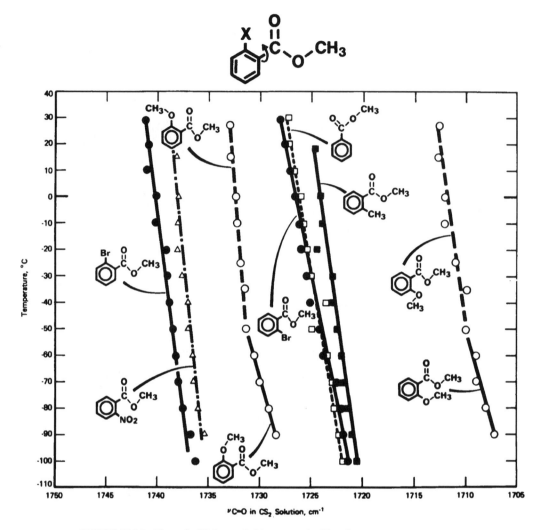

FIGURE 15.16   Plots of νCO for methyl benzoates in CS₂ solution vs temperature in °C.

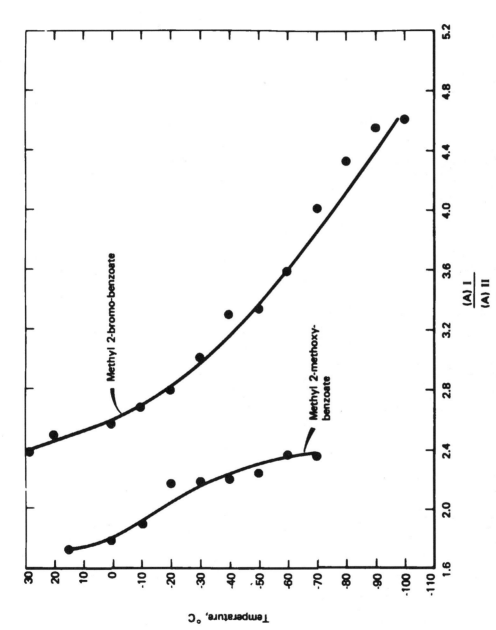

FIGURE 15.17 Plots of the absorbance ratio for A(νCO, rotational conformer 1)/A(νCO, rotational conformer 2) for methyl 2-methoxybenzoate and methyl 2-bromobenzoate in CS_2 solution vs temperature in °C.

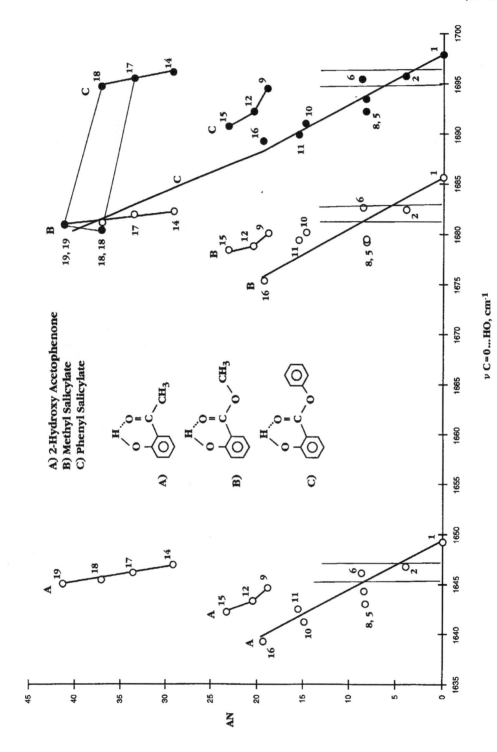

FIGURE 15.18   Plots of $\nu CO\cdots HO$ frequencies for 2-hydroxyacetophenone, methyl salicylate, and phenyl salicylate vs the solvent acceptor number (AN) for each of the solvents used in the study.

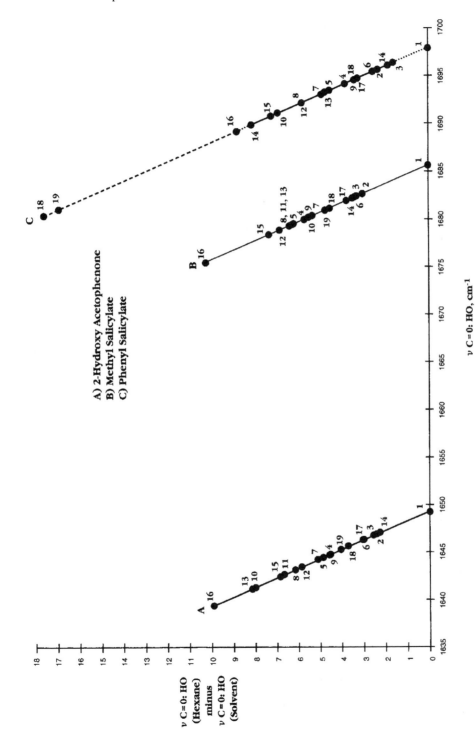

ν C=0: HO, cm^{-1}

A) 2-Hydroxy Acetophenone
B) Methyl Salicylate
C) Phenyl Salicylate

ν C=0: HO
(Hexane)
minus
ν C=0: HO
(Solvent)

FIGURE 15.19 Plots of νCO$\cdots$HO for 2-hydroxyacetophenone, methyl salicylate, phenyl salicylate vs the frequency difference between νCO$\cdots$HO (hexane) and νCO$\cdots$HO (solvent) for each of the other solvents.

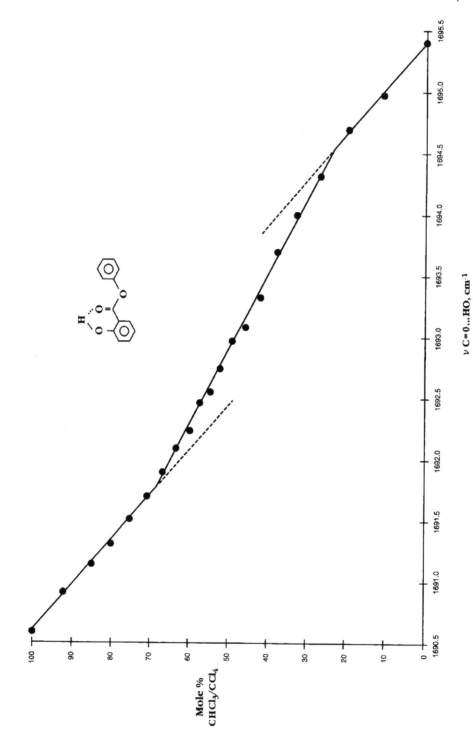

ν C=O...HO, cm^{-1}

FIGURE 15.20   A plot of νCO$\cdots$HO frequencies of phenyl salicylate vs mole % CHCl$_3$/CCl$_4$.

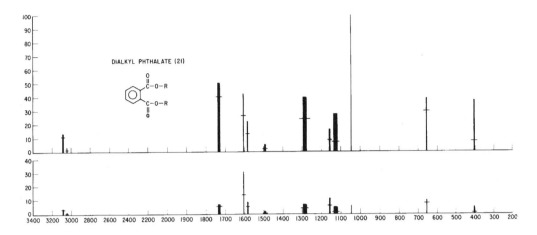

FIGURE 15.21 Bar graphs of the Raman group frequency data for 21 dialkyl phthalates.

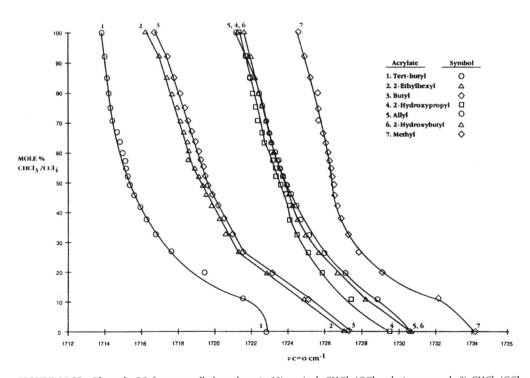

FIGURE 15.22 Plots of νCO for seven alkyl acrylates in 1% wt./vol. CHCl$_3$/CCl$_4$ solutions vs mole % CHCl$_3$/CCl$_4$.

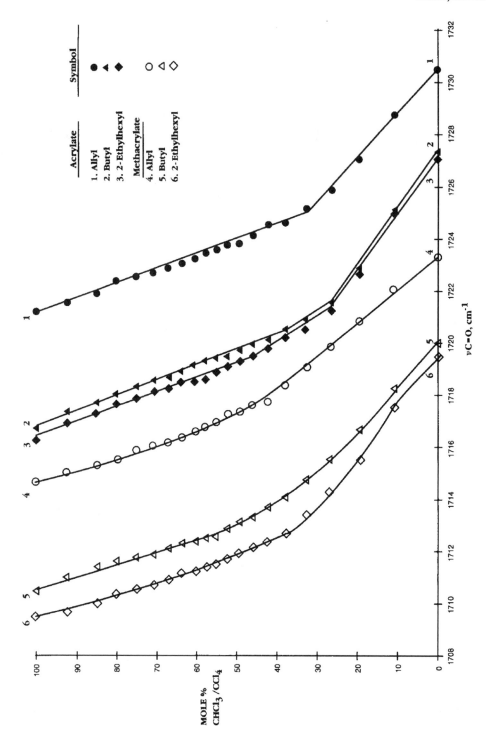

FIGURE 15.23   Plots of νCO for corresponding alkyl acrylates and alkyl methacrylates vs mole % CHCl$_3$/CCl$_4$ solutions.

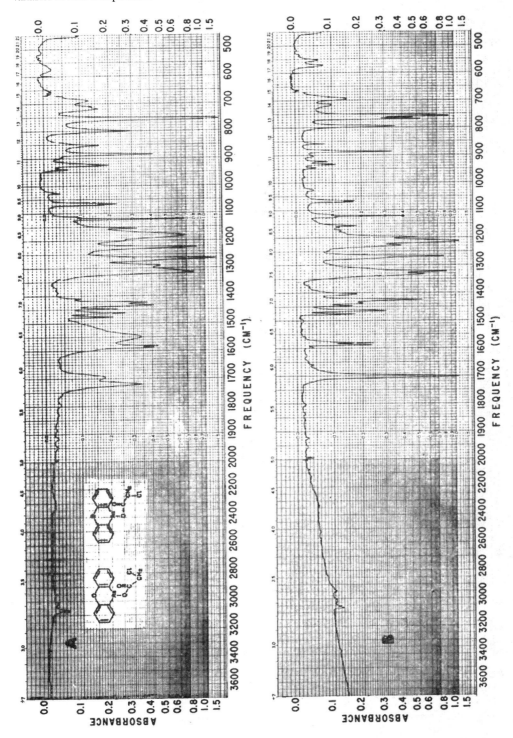

FIGURE 15.24 10-phenoxyarsinyl chloroacetate. (a) Saturated in CCl$_4$ and CS$_2$ solution, (b) split mull.

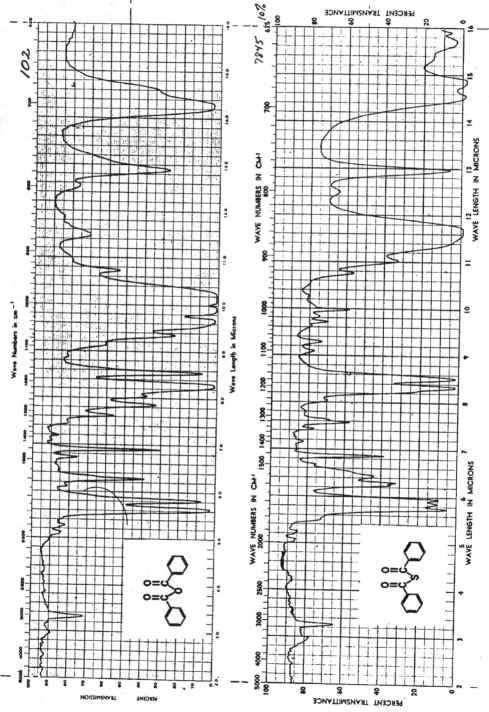

FIGURE 15.25   IR spectra of thiolbenzoic anhydride (dibenzoyl sulfide) and benzoic anhydride (dibenzoyl oxide).

TABLE 15.1 Carbonyl stretching frequencies for some carboxylic acid esters

Compound	C=O str. vapor cm^{-1}	C=O str. liquid cm^{-1}	Compound	C=O str. vapor cm^{-1}	C=O:H−O str liquid cm^{-1}
Alkyl formate	1741–1751	1715–1730			
Alkyl acetate	1755–1769	1735–1750			
Phenyl acetate	1781–1786				
Vinyl acetate	1784				
Alkyl propionate	1751–1762	1725–1740	Alkyl acrylate	1746–1751	
Phenyl propionate	1781				
Vinyl propionate	1777				
Alkyl isobutryate	1751–1760		Alkyl methacrylate	1739–1741	
Alkyl butyrate	1750–1760				
Alkyl valerate	1752–1761		Alkyl benzoate	1737–1749	
Alkyl hexanoate	1749–1759		Phenyl benzoate	1760	
Alkyl heptanoate	1752–1759		Methyl 2-nitrobenzoate	1751	
Alkyl octanoate	1751–1759				
Alkyl nonoate	1751–1759		Dialkyl phthalate	1735–1753	
Alkyl decanoate	1751–1760		Dialkyl isophthalate	1740–1746	
Vinyl decanoate	1780		Dialkyl terephthalate	1739–1749	
Alkyl undecanoate	1751–1760				
Alkyl octanoate	1740–1759		Alkyl isonicotinate	1750–1753	
Vinyl octanoate	1771				
Dialkyl malonate	1750–1770		Alkyl salicylate	1740–1755	1687–1698
Dialkyl succinate	1751–1761		Phenyl salicylate	1750–1762	1695–1705
Dialkyl adipate	1750–1760				

TABLE 15.2 The C=O stretching frequencies for alkyl alkanoates in the vapor phase and in various solvents

Solvent[*1]	MA C=O str. cm^{-1}	MP C=O str. cm^{-1}	MIB C=O str. cm^{-1}	MTMA C=O str. cm^{-1}	MTMA C=O str. cm^{-1}	EA C=O str. cm^{-1}	E C=O str. cm^{-1}	EIB C=O str. cm^{-1}	ETMA C=O str. cm^{-1}	AN
[vapor]	1769	1662	1760	1755	1755	1761	1755	1752	1750	
Hexane	1755.4	1752.1	1747.9	1742.8		1749.8	1746.7	1743	1736.4	0
Diethyl ether	1751	1747.8	1743.8	1739		1745.4	1742.8	1739.3	1732.4	3.9
Methyl t-butyl ether	1750.8	1747.3	1743.5	1738.7		1745.1	1742.4	1738.9	1732	[4.4]
Carbon tetrachloride	1748	1744.6	1740.6	1735.8		1741.8	1739.1	1735.7	1728.3	8.6
Benzene	1746.1	1742.9	1738.9	1734.4		1740	1737.2	1734	1726.9	8.2
1,2-Dichlorobenzene	1743.5	1739.7	1735.5	1732.5		1737.3	1733.6	1730.3	1723.8	[13.8]
Nitrobenzene	1741.5	1739.9	1734.1	1734.4		1735.5	171.9	1729.7	1725.3	14.8
Acetonitrile	1741.8	1738.7	1734.6	1731.6	1724.2	1735.7	1732.7	1729.3	1722	18.9
Benzonitrile	1741.4	1738.5	1734.3	1731.9	1722.1	1735.6	1732.7	1729.1	1721.6	1720
Nitromethane	1740.1	1737.1	1732.9	1738.8	1722.3	1734	1730.7	1727.2	1720	[18.5]
Methylene chloride	1739.4	1735.8	1731.8	1730.6	1721	1733.2	1729.4	1725.17	1718.7	20.4
Chloroform	1737.4	1734.1	1730.9	1729.9	1720	1732.2	1727.7	1724.9	1718.36	23.1
Chloroform-d	1737.5	1734.2	1730.8	1729.7	1720	1731.5	1727.7	1724.9	1718.9	23.1
t-Butyl alcohol	1751.4	1747.7	1743.3	1738.6		1746.4	1742.8	1739.3	1733.2	[3.7]
Isopropyl alcohol	1751.2	1748.6	1743.9	1738.6		1745.4	1742.9	1739.3	1732.8	[3.7]
Ethyl alcohol	1749.5	1746.6	1742.5	1738.6		1743.7	1741.2	1737.7	1731	[6.3]
Methyl alcohol	1747.6	1743.7	1739.5	1737.5		1741.5	1738.4	1734.9	1728.6	[7.6]
Dimethyl sulfoxide	1737.5	1734.6	1731	1728.5		1732.3	1729.1	1726.1	1719.9	19.1
[neat]	1745.2	1743	1740.8	1736.6	1720.9	1741.5	1738.9	1736.7	1730.3	[8.7]
Carbon disulfide	1746.9	1743.6	1739.6	1735.1		1740.8	1738.2	1734.9	1727.9	[7.9]

[*1] Abbreviations for the various solvents are spelled out in the text.

TABLE 15.3 The frequency difference between vC=O and vC=O:HOR for alkyl alkanoates in alkyl alcohols [0.5 wt./vol. solute in solvent]

Alcohol	MA cm^{-1}	MP cm^{-1}	EA cm^{-1}	MIB cm^{-1}	EP cm^{-1}	EIB cm^{-1}	MTMA cm^{-1}	ETMA cm^{-1}	[M + E]/2 cm^{-1}
tert-Butyl	17.32	17.82	17.55	19.16	19.55	18.77	20.81	19.26	18.78
Isopropyl	18.75	20.12	17.99	22.25	21.39	20.61	20.11	20.39	20.45
Ethyl	17.98	20.16	17.54	23.14	22.07	21.56	22.57	20.68	20.71
Methyl	17.05	18.33	14.97	21.36	20.71	20.42	21.81	20.01	19.33

TABLE 15.4 The frequency difference between vC=O and vC=O:HO for alkyl acetates in alkyl alcohols

Alcohol	Methyl acetate delta cm^{-1}	Ethyl acetate delta cm^{-1}	Isopropyl acetate delta cm^{-1}	tert-Butyl acetate delta cm^{-1}
Methyl	17.05	14.97	18.97	18.04
Ethyl	18.07	17.54	21.91	22.77
Isopropyl	17.75	18.02	22.43	23.41
tert-Butyl	17.32	17.75	21.67	23.13

TABLE 15.5 The C=O stretching frequencies for alkyl 2,2- dichloroacetates and alkyl acetates in the vapor and neat phases

Alkyl	[A] Alkyl 2,2-dichloroacetate [vapor phase] C=O str., Conformer 1 cm^{-1}	[B] Alkyl 2,2-dichloroacetate [vapor phase] C=O str., Conformer 2 cm^{-1}	[C] Alkyl acetate [vapor phase] C=O str. cm^{-1}	[A]–[C] cm^{-1}	[A]–[C] cm^{-1}	[B]–[C] cm^{-1}	[vapor]–[neat] cm^{-1}
Methyl	1790	1765	1769	25	21	−4	
Ethyl	1785	1762	1761	23	24	1	
Ethyl[neat]	[1771]	[1750]		[21]			[14]:[12]
Hexyl	1783	1761	1760	22	23	1	
Hexyl[neat]	[1767]	[1750]		[17]			[16]:[11]
Nonyl	1782	1761	1761	21	21	0	
Dodecyl	1782	1761	1761	21	21	0	
sec.Butyl	1779	1758	1755	21	24	3	
Methyl			1751				
t-Butyl			1740				
[delta C=O str.]			[11]				

TABLE 15.5A   IR data for ethyl acetate and ethyl 2-chloroacetate in CS_2 solution between 27 and $-100°C$

°C	Ethyl acetate acetate [CS$_2$] cm^{-1}	°C	Ethyl 2-chloroacetate conformer 1 [CS$_2$] cm^{-1}	Ethyl 2-chloroacetate conformer 2 [CS$_2$] cm^{-1}	A[conformer 1] A[conformer 2]	[conformer 1]– [conformer 1]– cm^{-1}
27	1741.3	26	1765	1740.5	1.19	24.5
15	1740.7	15	1764.5	1740	1.25	24.5
0	1740.5	0	1764.1	1739.4	1.47	24.7
−10	1739.5	−10	1763.6	1738.5	1.55	25.1
−20	1738.6	−20	1763.6	1738.5	1.6	25.1
−30	1738	−30	1763.5	1738	1.81	25.5
−40	1738	−40	1763.4	1737	2.14	26.4
−50	1737.6	−50	1762.8	1736.2	2.32	26.6
−60	1737	−60	1762.5	1736.1	2.62	26.4
−70	1736	−70	1761.6	1735.5	2.85	26.1
−80	1736	−80	1761.3	1735.4	3.18	26.2
−90	1735.5	−90	1760.6	1733.6	3.69	27
−100		−100	1759.8	1732.8	4.2	27
delta C [−117]	delta C=O str. [−5.8]	delta C [−126]	delta C=O str. [−5.2]	delta C=O str. [−7.7]	delta [A(conformer 1)]– [A(conformer 2)]– [3.01]	delta cm^{-1} [2.22]

TABLE 15.6   IR data for β-propiolactone in various solvents

β-Propiolactone 1% wt./vol. solutions in the solvents listed below	C=O str., A′ cm^{-1}	v6 + v13, A′ cm^{-1}	2v10, A′ cm^{-1}	C=O str., A′ corrected for F.R. cm^{-1}	v6 + V13A′ corrected for F.R. cm^{-1}	2v10, A′ corrected for F.R. cm^{-1}
Hexane	1857.18	1832.01	1817.52	1850.21	1830.69	1826.81
Diethyl ether	1852.24	1838.04	1814.14	1843.8	1834.65	1825.97
Carbon tetrachloride	1850.5	1833.08	1816.28	1841.78	1831.62	1831.62
Nitrobenzene	1830.93	1843.31	1808.44	1828.74	1830.62	1823.33
Acetonitrile	1832.18	1845.07	1810.23	1830.55	1832.56	1824.37
Benzonitrile	1830.77	1843.31	1829.78	1829.72	1831.92	1825.23
Methylene chloride	1831.46	1844.86	1812.36	1830.87	1832.43	1825.38
chloroform	1845.02	1831.7	1813.04	1834.7	1829.67	1825.39
Nitromethane	1831.8	1845.07	1809.8	1830.12	1832.48	1824.07
t-Butyl alcohol	1832.02	1845.66	1813.4	1830.79	1833.28	1821.01
Isopropyl alcohol	1833.07	1845.72	1810.82	1830.84	1832.72	1826.05
Ethyl alcohol	1833.81	1845.72	1810.37	1831.14	1832.78	1825.99
Methyl alcohol	1834.23	1843.23	1809.89	1830.72	1832.01	1825.35

TABLE 15.7 IR group frequency data for coumarin and derivatives in the vapor phase and in solution

Compound vapor phase	2(C=O) cm^{-1}(A)	C=O str. cm^{-1}(A)	CCOC str. cm^{-1}(A)	CCOC str. cm^{-1}(A)	o.p. Ring cm^{-1}(A)	o.p. Ring cm^{-1}(A)
Coumarin	3530(0.015)	1776(1.240)	1265(0.078)	1178(0.160)	751(0.131)	825(0.140)
3,4-dihydro	3590(0.021)	1802(1.250)	1235(0.710)	1140(1.040)	755(0.295)	
3-chloro	3545(0.020)	1775(1.210)	1242(0.160)	1120(0.199)	750(0.354)	
6-methyl	3525(0.010)	1755(1.240)	1260(0.118)	1163(0.210)	897(0.141)	819(0.172)

Coumarin 1 wt. % (0.1 mm KBr) solvent	C=O str. corrected for F.R. cm^{-1}	C=O: HO corrected for F.R.	AN	C=O str. not corrected for F.R. cm^{-1}	
Hexane	1755.16		0	1758.56	
Diethyl ether	1748.93		3.9	1755.16	
Carbon tetrachloride	1747.16		8.6	1741.25	
Carbon disulfide	1744.47			1739.49	
Benzene	1744.18		8.2	1738.64	
Toluene	1743.96			1739.3	
Nitrobenzene	1737.82		14.8	1733.47	
Acetonitrile	1737.31		18.9	1733.58	
Benzonitrile	1736.84		15.5	1733.5	
Nitromethane	1735.28			1732.06	
Chloroform	1734.89		23	1729.94	
Dimethylsulfoxide	1732.82		19.3	1729.22	
Tert-butyl alcohol		1727.9	29.1		
Isopropyl alcohol		1725.45	33.5		
Ethyl alcohol		1718.53	37.1	1739.74	
Methyl alcohol		1717.38	41.3	1738.18	

TABLE 15.8 IR data for phenyl acetate in CHCl$_3$/CCl$_4$

Phenyl acetate	C=O str. in F.R. cm^{-1}	CT in F.R. cm^{-1}	A[1]	A[2]	A[1]/A[2]	C=O str. corrected for F.R. cm^{-1}	CT corrected for F.R. cm^{-1}
Mole % CHCl$_3$/CCl$_4$							
0	1767.71	1740.99	0.433	0.059	0.133	1764.6	1744.1
52	1764.22	1743.94	0.612	0.35	0.572	1756.8	1751.3
100	1763.19	1743.32	0.287	0.217	0.756	1754.6	1751.9
delta C=O	−4.52	2.33				−9.9	7.8

TABLE 15.9   IR data for phenyl acetate in various solvents

Phenyl acetate 1 % wt./vol. solutions [.207 mm KBr cell]	C=O str. in F.R. cm⁻¹	CT*¹ in F.R. cm⁻¹	C=O str. corrected for F.R. cm⁻¹	CT corrected for F.R. cm⁻¹	AN*²
Solvent					
Hexane	1774.21	1736.5	1771.8	1741.9	0
Diethyl ether	1770.62	1742.62	1767.6	1745.6	3.9
Methyl t-butyl ether	1770.22	1743.3	1766.7	1746.8	
Carbon tetrachloride	1767.71	1740.99	1764.6	1744.1	8.6
Carbon disulfide	1766.46	1737.6	1763.9	1740.2	
Benzene	1766.39	1741.24	1763.1	1744.6	8.2
Nitrobenzene	1764.31	1743.48	1759.2	1748.6	14.8
Acetonitrile	1764.65	1747.24	1759.9	1751.9	18.9
Benzonitrile	1764.6	1744.34	1759.5	1749.4	15.5
Nitromethane	1763.88	1754.76	1758.1	1751.6	
Methylene chloride	1763.37	1743.79	1756.8	1750.3	20.4
t-Butyl alcohol	1768.72	1745.05	1755.9	1757.9	29.1
Isopropyl alcohol	1768.79	1744.07	1759.6	1753.3	33.5
Chloroform	1763.19	1743.32	1754.6	1751.9	23.1
Dimethyl sulfoxide	1760.83	1745.34	1755.8	1750.3	19.3
Ethyl alcohol	1768.82	1743.4	1758.3	1753.9	37.1
Methyl alcohol	1768.51	1743.17	1756.6	1755.1	41.3

*¹ CT = combination tone.

*² AN = acceptor number.

TABLE 15.10   IR data for alkyl 2-benzoates in CS_2 solution between 29 and $-100°C$

°C	Methyl 2-bromo-benzoate C=O str. conformer 1 $[CS_2]$ cm^{-1}	Methyl 2-bromo-benzoate C=O str. conformer 2 $[CS_2]$ cm^{-1}	[conformer 1]-[conformer 2] cm^{-1}	°C	Methyl 2-methoxy-benzoate C=O str. conformer 1 $[CS_2]$ cm^{-1}	Methyl 2-methoxy-benzoate C=O str. conformer 2 $[CS_2]$ cm^{-1}	[conformer 1]-[conformer 2] cm^{-1}
29	1741.2	1728	13.2	15	1732.8	1712.8	20
10	1741	1728	13				
0	1740	1727	13	0	1732.3	1712.2	20.1
−10	1740.2	1726.5	13.7	−10	1732.3	1712.1	20.2
−20	1739.2	1726	13.2	−24.5	1731.7	1711	20.7
−30	1739	1725.5	13.5	−35	1731.5	1709.8	21.7
−40	1738.8	1725	13.8	−40			
−50	1738.5	1724.2	14.3	−50	1731.5	1710.1	21.4
−60	1738.2	1723.8	14.4	−60	1730.5	1709	21.5
−70	1738	1722.5	15.5	−70	1730	1709	21
−80	1737.5	1722	15.5	−80	1729.1	1708	21.1
−90	1736.7	1722	14.7	−90	1728.5	1707.2	21.3
−100	1736.2	1721.5	14.7				
delta C [−129]	delta C=O str. [−5.0]	delta C=O str. [−6.5]	[1.5]	delta C [115]	delta C=O str. [−4.3]	delta C=O str. [−5.6]	[1.3]

°C	Ethyl 2-Nitro-benzoate C=O str. cm^{-1}	°C	Methyl benzoate C=O str. cm^{-1}	°C	Methyl 2-Methyl benzoate C=O str. cm^{-1}
15	1738.2	30	1727	18	1724.7
0	1738	0	1726	0	1724.3
−10	1738	−10	1725.8	−10	1724
−20	1738	−20	1725.4	−20	1723.5
−30	1737.6	−30	1725	−30	1723
−40	1737	−40	1724.5	−40	1723.3
−50	1737	−50	1724	−50	1722.6
−60	1737.5	−60	1723.5	−60	1722
−70	1736.5	−70	1723	−70	1722
−80	1736	−80	1722.8	−80	1721.5
−90	1735.5	−90	1722.3	−90	1721.1
		−100	1721.7	−100	1720.6
delta C [−115]	delta C=O str. [−2.7]	delta C [−130]	delta C=O str. [−5.3]	delta C [−118]	delta C=O str. [−4.1]

TABLE 15.11   The C=O stretching frequencies for methyl salicylate, phenyl salicylate, and 2-hydroxyacetophenone in various solvents

2-Hydroxy-acetophenone C=O:H−O cm^{-1}	Methyl salicylate C=O:H−O cm^{-1}	Phenyl salicylate C=O:H−O cm^{-1}	Phenyl salicylate C=O:(O−H)$_2$ cm^{-1}	Solvent	AN
1649.2	1685.6	1697.9		Hexane	0
1646.8	1682.3	1695.7		Diethyl ether	3.9
1646.7	1682.2	1696.3		Methyl t-butyl ether	
1644.7	1679.9	1694		Toluene	
1644.4	1679.4	1693.4		Benzene	8.2
1646.2	1682.5	1695.4		Carbon tetrachloride	8.6
1644.1	1680.4	1693		Carbon disulfide	
1643	1679.2	1692.1		1,2-Dichlorobenzene	8.2
1644.6	1680.1	1694.5		Acetonitrile	18.9
1641.2	1680.1	1691		Nitrobenzene	14.8
1642.5	1679.4	1689.8		Benzonitrile	15.5
1643.3	1678.7	1692.1		Methylene chloride	20.4
1641	1679.3	1693.2		Nitromethane	
1647	1682.1	1696.1		t-Butyl alcohol	29.1
1642.3	1678.3	1690.7		Chloroform	23.1
1639.3	1675.3	1689.1		Dimethyl sulfoxide	19.3
1646.2	1681.8	1695.5		Isopropyl alcohol	33.5
1645.5	1681.1	1694.7	1680.3	Ethyl alcohol	37.1
1645.1	1680.2		1680.9	Methyl alcohol	41.3

TABLE 15.12   Raman group frequency correlations for dialkyl phthalates in the neat phase

Dimethyl phthalate cm^{-1}	RI*1	DR*2	Di-(isooctyl) phthalate cm^{-1}	RI	DR
3082	11.1	0.31	3079	11.6	0.31
3042	2.1	0.64	3039	1.5	0.75
1731	45.3	0.14	1731	45.7	0.14
[1753 vapor, IR]					
1604	26.7	0.77	1603	31.9	1.76
1584	14	0.43	1583	15.9	0.45
1494	3.8	0.31	1492	2.2	0.33
1287	26.2	0.16	1278	30.8	0.19
1170	9	0.83	1167	14.2	0.73
1129	17	0.1	1135	9.4	0.15
1044	100	0.05	1043	100	0.06
652	30.2	1.3	652	32.6	0.27
405	37	0.11	405	13.1	0.14

*1RI = relative intensity.

*2DR = depolarization ratio.

TABLE 15.13 Raman data for C=O and C=C stretching for acrylates

Group	[R]C=O str. cm^{-1}	RI	C=C str. cm^{-1}	RI	C=O str.–C=C str. cm^{-1}	RI C=O/ RI C=C
Methyl	1726	4	1636	9	90	0.44
Propyl	1723	5	1639	9	84	0.56
Butyl	1723	4	1639	9	84	0.44
Heptyl	1725	4	1639	9	86	0.44
Nonyl	1726	3	1639	8	87	0.38
Undecyl	1726	3	1640	9	86	0.33
2-Ethylhexyl	1725	4	1638	7	87	0.57
Isopropyl	1721	3	1640	9	81	0.33
Cyclohexyl	1720	3	1639	6	81	0.51
isoButyl	1725	3	1638	9	87	0.33
isoAmyl	1725	5	1639	9	86	0.56
Benzyl	1723	1	1636	2	87	0.51
2-Phenylethyl	1722	1	1637	4	85	0.25
2-(2-Ethyoxyethoxy) ethyl	1724	4	1639	9	85	0.44
2-Hydroxyethyl	1722	3	1639	9	83	0.33
2-Hydroxypropyl	1722	3	1639	9	83	0.33
2-Hydroxybutyl	1721	3	1638	9	83	0.33
2-Methoxyethyl	1723	4	1639	9	84	0.44
3-Methoxybutyl	1724	5	1639	9	85	0.56
Triethyleneglycol,di-	1721	4	1638	9	83	0.44
1,4-Tetramethylene,di	1720	4	1636	9	82	0.44
Ethylene,di	1723	4	1639	9	85	0.44
1,2-Propanediol,di	1724	5	1638	9	86	0.56
1,3-Propanediol,di	1721	4	1637	7	84	0.57
1,6-Hexamethylene,di	1721	4	1638	9	83	0.44
1,10-Decanediol,di	1722	5	1638	9	84	0.56
2,-Butene-1,4-diol,di	1722	4	1637	9	85	0.44
Cinnamyl	1722	1	1637	3	85	0.33
2-Bromoethyl	1726	2	1637	5	89	0.41
2,3-Dibromopropyl	1717	1	1636	3	81	0.33
Tribromoneopentyl	1730	1	1637	3	93	0.33
2,2,2,Trifluoroethyl	1748	4	1638	9	110	0.44
Hexafluorobutyl	1746	4	1639	9	107	0.44
Pentafluorooctyl	1752	4	1639	9	113	0.44
1,H,1H,11H-Eicos-fluorodecanyl	1747	1	1639	2	108	0.51
Hexafluorisopropyl	1757	3	1640	4	117	0.75
Phenyl	1739	1	1635	3	104	0.33
p-Chlorophenyl	1742	3	1637	9	105	0.33
p-Nitrophenyl	1748	~ 0.51	1632	2	116	~ 0.25
2,4,6,Tribromophenyl	1737	2	1623	~ 0.50	114	~ 0.13
Pentabromophenyl	1738	2	1625	~ 0.25	113	~ 4.0
Pentachlorophenyl	1752	1	1638	3	114	0.33
Pentaflourorphenyl	1774	2	1636	5	138	0.41
p-Phenylene,di-	1726	5	1637	6	89	0.83
Bisphenol A,di-	1726	6	1626	5	100	1.21
1-Naphthyl	1731	1	1637	1	94	1
2-Naphthyl	1735	1	1634	2	101	0.51

(continued)

TABLE 15.13 (*continued*)

Group	[R]C=O str. cm⁻¹	RI	C=C str. cm⁻¹	RI	C=O str.–C=C str. cm⁻¹	RI C=O/ RI C=C
Vinyl	1740	4	1630	9	110	0.44
Propargyl	1728	2	1637	7	91	0.29
N,N-dimethylaminoethyl	1724	4	1638	9	86	0.44
2-(N-morpholino)ethyl	1723	4	1637	4	86	0.57
3-Dimethylamino-neopentyl	1725	4	1636	7	89	0.57
3-Sulfopropyl potassium salt	1719	1	1638	4	81	0.25
Range	1717–1774		1623–1640		81–138	

TABLE 15.13A IR C=O stretching frequency data and other group frequency for alkyl acrylates [CHCl$_3$ and CCl$_4$ solutions]

Mole % CHCl$_3$/CCl$_4$	Methyl acrylate C=O str.	2-Hydroxy-butyl acrylate C=O str.	Allyl acrylate C=O str.	2-Hydroxy-propyl acrylate C=O str.	Butyl acrylate C=O str.	2-Ethyl-hexyl acrylate C=O str.	tert-Butyl acrylate
[vapor]	1751				1741		
0	1734.1	1730.6	1730.5	1729.4	1727.3	1727.1	1722.9
100	1724.5	1721.6	1727.1	1721.3	1716.7	1716.3	1713.8
[delta cm⁻¹]	16.9;26.5;9.6	9	3.4	8.1	13.8;24.3;10.6	10.8	9.1
	s-trans C=C str.	s-trans C=C str.	s-trans C=C str.	s-trans C=C str.	s-trans C=C str.	s-trans C=C str.	s-trans C=C str.
0	1635.3	1636.5	1635.4	1637	1637.2	1635.7	1635.6
100	1635.1	1636.1	1635.4	1636.6	1635.8	1636.3	1635.8
[delta cm⁻¹]	−0.2	−0.4	0	−0.4	−1.4	0.6	0.2
	s-cis	s-cis	s-cis	s-cis	s-cis	s-cis	s-cis
0	1620.4	1619.3			1619.7		1619.2
100	1619.9				1619.2		1618.5
[delta cm⁻¹]	−0.5				−0.5		−0.7
	CH$_2$=bend	CH$_2$=bend	CH$_2$=bend	CH$_2$=bend	CH$_2$=bend	CH$_2$=bend	CH$_2$=bend
0		1406.4	1404.6	1406.6	1407.2	1406.5	1401.6
100		1409.4	1407	1409.5	1410.3	1409.9	1404
[delta cm⁻¹]		3	2.4	2.9	3.1	3.4	2.4
	HC=CH twist	HC=CH twist	HC=CH twist	HC=CH twist	HC=CH twist	HC=CH twist	HC=CH twist
0	984.9	982.6	984.7	983.3	983.5	983.6	984.45
100	985	983.4	984.4	983.9	984.3	984.5	985.6
[delta cm⁻¹]	0.1	0.8	−0.3	0.6	0.8	0.9	1.1
	C=CH$_2$ wag	C=CH$_2$ wag	C=CH$_2$ wag	C=CH$_2$ wag	C=CH$_2$ wag	C=CH$_2$ wag	C=CH$_2$ wag
0	968.1	982.6			968.4		966.3
100	970.3	983.4			969		966.5
[delta cm⁻¹]	2.2	0.8			0.6		0.2

TABLE 15.14 Raman data for methacrylates in the neat phase and summary of IR data in CHCl₃ and CCl₄ solutions

	C=O str.	RI	C=C str.	RI	C=O str. minus C=C str.	RI C=O str./ RI C=C str.	C=O str.Acr.- C=O str. Methacr.	C=C str.Methacr.- C=C str.Acr.
Methyl [IR vapor]	1741							2
Propyl	1719	5	1641	7	78	0.71	4	
Pentyl [IR vapor]	1739							
Pentyl	1719	4	1641	6	78	0.67		
Hexyl	1720	3	1640	5	80	0.6		
Heptyl	1720	3	1640	4	80	0.75	5	1
Octyl	1720	3	1641	4	79	0.75	2	
Nonyl	1720	3	1641	79	0.75	6		
Decyl [IR vapor]	1739							
Decyl	1720	3	1641	4	79	0.75	6	1
Undecyl	1720	3	1641	9	79	0.33	6	2
Dodecyl	1721	2	1641	3	80	0.67	7	3
Hexadecyl	1721	2	1641	2	80	1		
Octadecyl	1721	1	1640	2	81	0.5		
2-Ethylhexyl	1720	3	1641	4	79	0.75	5	3
Oleyl	1720	2	1641	3	79	0.67	8	2
Isopropyl	1716	4	1641	6	75	0.67	5	1
sec-Butyl	1715	4	1640	5	75	0.8		
isoBornyl	1716	2	1640	3	76	0.67	6	2
isoButyl	1719	5	1641	7	78	0.71	6	3
isoAmyl	1719	4	1640	6	79	0.67	6	1
isoDecyl	1720	3	1641	4	79	0.75		
2-Phenylethyl	1717	2	1641	2	77	1	5	3
2-Phenoxyethyl	1720	1	1639	2	81	0.5		
2-Ethylethyl	1719	4	1640	79	0.67			
Ethoxytriethylene glycol	1719	4	1640	5	79	0.8		
2-Hydroxyethyl	1718	5	1640	9	78	0.56	4	1
2-Hydroxybutyl	1716	4	1639	9	77	0.44	5	1
2-Methoxypropyl	1717	5	1640	7	77	0.71		
2-Methoxybutyl	1718	4	1640	6	78	0.67	6	1
Glyceryl tri-	1721	7	1639	9	82	0.78	3	1
Ethylene, di-	1720	6	1640	9	80	0.67		
1,9-Nonanediol,di	1717	4	1640	6	77	0.67		
1,10-Decanediol,di	1717	4	1640	6	77	0.67	5	2

(continued)

TABLE 15.14 (continued)

	C=O str.	RI	C=C str.	RI	C=O str. minus C=C str.	RI C=O str./ RI C=C str.	C=O str.Acr.− C=O str. Methacr.	C=C str.Methacr.− C=C str.Acr.
2-Bromoethyl	1720	1	1639	2	81	0.5	6	2
tribromoneopentyl	1723	1	1639	2	84	0.5	7	2
Trichloroethyl	1742	2	1637	1	105		2	
2,2,2-Trifluoroethyl	1738	6	1640	9	98	0.67	10	2
1,H,1H,3H-Tetra-fluoropropyl	1735	6	1640	9	95	0.67		
Hexafluorobutyl	1736	6	1641	9	95	0.67	10	2
Dodecafluoro-1-heptyl	1739	4	1641	7	98	0.57		
Pentadecylfluoro-octyl	1743	5	1641	7	102	0.71	9	2
1H,1H,2H-Hepta-decylfluorodecyl	1727	3	1642	4	85	0.75		
Phenyl	1735	3	1639	4	96	0.75	4	4
p-Nonylphenyl	1738	4	1639	5	99	0.8		
p-Nitrophenyl	1742	2	1636	1	106	2	6	4
Pentabromophenyl	1737	1	1635	1	102	1	1	10
Pentachlorophenyl	1745	4	1637	1	108	1.33	7	−1
Pentafluorophenyl	1762	2	1639	2	123	1	12	3
Bisphenol A, di-	1718	5	1639	7	79	0.79	8	13
4-Hydroxy-benzophenone	1737	2	1629	7	108	0.29		
2-Naphthyl	1729	2	1639	2	90	1	6	5
Methyallyl	1720	5	1640	8	80	0.63		
Allyl	1720	5	1641	8	79	0.63		
Propargyl	1724	3	1640	5	84	0.6		
N,N-Dimethylamino-ethyl	1719	6	1640	8	79	0.75	5	2
2-(1-Aziridinyl) ethyl	1718	6	1640	8	78	0.75		
2-Aminoethyl hydrochloride	1719	4	1639	8	78	0.5		
Trimethylammonium ethyl methosulfate	1717	1	1640	3	77	0.33		
3-Sulfopropyl	1713	2	1639	3	74	0.67	6	1
Range	1713–62		1635–42		77–123			
Infrared Range in CCl$_4$ soln.	1719–26		1637.3–38					
Infrared Range in CHCl$_3$ soln.	1709.5–18		1635.7–37.3					

TABLE 15.15 IR vapor-phase data and assignments for alkyl cinnamates [C=C stretching, CH=CH twisting, and C=O stretching]

Cinnamate	C=O str.	C=C str.	[(A)C=C]/ [(A)C=O]	HC=CH twist	[(A)HC=CH twist]/ [(A)C=C str.]	[(A)HC=CH twist]/ [(A)C=O str.]
Methyl	1740(1.141)	1640(0.431)	0.38	975(0.181)	0.42	0.16
Ethyl	1735(1.050)	1640(0.370)	0.35	975(0.169)	0.46	0.16
Butyl	1731(1.141)	1641(0.379)	0.33	978(0.205)	0.54	0.18
Isobutyl	1735(0.806)	1641(0.310)	0.38	975(0.130)	0.42	0.16
Isopentyl	1737(0.830)	1642(0.282)	0.34	980(0.171)	0.61	0.21
Isopropyl	1731(1.250)	1641(0.500)	0.41	982(0.310)	0.62	0.25
Tert-butyl	1727(1.030)	1640(0.370)	0.36	972(0.247)	0.67	0.24
Cyclohexyl	1731(0.654)	1642(0.214)	0.33	980(0.129)	0.6	0.2
Benzyl	1739(0.959)	1640(0.371)	0.39	980(0.246)	0.66	0.26

TABLE 15.15A IR vapor-phase data and assignments for alkyl cinnamates [in-plane and out-of-plane phenyl ring vibrations]

Cinnamate	[1] cm^{-1}(A)	[2] cm^{-1}(A)	[3] cm^{-1}(A)	[4] CCO str. cm^{-1}(A)	[5] COC str. cm^{-1}(A)	[6] i.p.o.p.5H Ring def. cm^{-1}(A)	[7] o.p.Ring def. cm^{-1}(A)
Methyl	1314(0.750)	1269(0.850)	1199(0.700)	1166(1.240)	1045(0.180)	765(0.199)	695(0.159)
Ethyl	1309(0.599)	1259(0.690)	1199(0.490)	1165(1.240)	1042(0.325)	761(0.158)	690(0.120)
Butyl	1309(0.490)	1265(0.605)	1199(0.465)	1170(1.240)		764(0.136)	688(0.110)
Isobutyl	1310(0.440)	1250(0.500)	1199(0.460)	1161(1.240)		763(0.115)	690(0.080)
Isopentyl	110(0.470)	1252(0.450)	1200(0.370)	1165(1.210)		764(0.110)	688(0.072)
Isopropyl	1309(0.690)	1268(0.770)	1199(1.250)	1171(1.250)		763(0.162)	688(0.130)
Tert-butyl	1314(0.589)	1276(0.559)	1199(0.580)	1155(1.250)		761(0.166)	685(0.094)
Cyclohexyl	1307(0.280)	1268(0.365)	1199(0.292)	1170(1.240)	1042(0.180)	764(0.081)	690(0.061)
Benzyl	1304(0.511)	1245(0.750)	1198(0.480)	1154(1.234)	1011(0.331)	750(0.192)	696(0.345)

	[(A)[1]]/[(A)[4]	[(A)[2]]/[(A)[4]]	[(A)[3]]/[(A)[4]]	[(A)[6]]/[(A)[7]]	[(A)C=C str.]/[(A)[4]]
Methyl	0.61	0.68	0.56	1.25	0.35
Ethyl	0.48	0.56	0.41	1.32	0.31
Butyl	0.41	0.48	0.38	1.24	0.31
Isobutyl	0.35	0.4	0.37	1.44	0.25
Isopentyl	0.39	0.37	0.31	1.53	0.23
Isopropyl	0.55	0.62	0.48	1.25	0.41
Tert-butyl	0.47	0.45	0.46	1.75	0.31
Cyclohexyl	0.23	0.29	0.24	1.32	0.17
Benzyl	0.41	0.61	0.39	0.56	0.131

TABLE 15.15B IR data for alkyl cinnamates

Cinnamate	2(C=O) cm⁻¹(A)	Ring C–H cm⁻¹(A)	Ring C–H cm⁻¹	a.CH₃ str. cm⁻¹(A)	a.CH₂ str. cm⁻¹(A)	s.CH₃ str. cm⁻¹(A)	s.CH₂ str. cm⁻¹(A)	a.CH₃ def. cm⁻¹(A)	CH₂ bend+ Ring cm⁻¹(A)	s.CH₃ def. cm⁻¹(A)
Methyl	3470(0.010)	3075(0.200)	3040(0.150)	2961(0.281)		2850(0.051)		1442(0.290)		
Ethyl	3459(0.010)	3070(0.160)	3035(0.100)	2985(0.211)	2942(0.119)	2910(0.070)	2885(0.050)	1470(0.090)	1450(0.110)	1400(0.070)
Butyl	3458(0.005)	3075(0.149)	3040(0.090)	2984(0.392)	2943(0.199)		2895(0.112)	1452(0.132)		1385(0.122)
Isobutyl	3458(0.005)	3075(0.080)	3040(0.081)	2975(0.350)	2960(0.200)		2885(0.129)	1469(0.080)	1451(0.100)	1379(0.150)
Isopentyl	3458(0.005)	3075(0.120)	3040(0.070)	2969(0.440)	2950(0.190)	2920(0.170)	2882(0.120)	1470(0.101)	1451(0.110)	1390(0.090)
Isopropyl	3446(0.011)	3075(0.175)	3040(0.115)	2985(0.487)		2945(0.142)	2888(0.060)	1470(0.090)	1451(0.140)	1380(0.169)
Cyclohexyl	3446(0.005)	3075(0.0830)	3040(0.052)		2942(0.652)		2868(0.159)		1453(0.095)	
Tert-butyl	3435(0.008)	3065(0.150)	3030(0.090)	2984(0.420)		2939(0.161)		1475(0.090)	1451(0.120) 1371(0.300)	1395(0.170)
Benzyl	3469(0.005)	3070(0.289)	3039(0.080)		2960(0.099)	2899(0.029)			1450(0.143)	

TABLE 15.16 The C=O stretching frequencies for phenoxarsine derivatives in the solid state and in CCl₄ solution

Phenoxarsine X=O−(C=O)−R	C=O str. [CCl₄ soln.] cm⁻¹	C=O str. [Nujol mull] cm⁻¹
R		
methyl	1698	
chloromethyl	1720	1702
	1696	
trichloromethyl	1725	1717
trifluoromethyl	1738	1737
X=S−(C=O)−R		
R		
methyl	1681	
ethyl	1680	1671
propyl	1671	1661
isopropyl	1675	1669
isobutyl	1675	
octyl	1677	
cyclohexyl	1669	1660
2-cyclohexylethyl	1671	1661
benzyl	1670[CS2]	1671
phenoxymethyl	1694;1665	
2,4,5-trichloro-phenoxymethyl	1691;1665	1650
alpha-(2,4,5-tri-chlorophenoxy)ethyl	1682;1665	
carbethoxy	1759;1675	1762;1670
	1742	
2-furyl	1646;1631	1630
phenyl	1645	
4-t-butylphenyl	1650	
4-methoxyphenyl	1638[CS2]	1628
4-n-butoxyphenyl	1641	
4-n-pentoxyphenyl	1643	1635
3,4,5-triethoxyphenyl		1625
diethylamino	1624	
piperidino	1631	

TABLE 15.17 IR data for alkyl thiol esters and phenyl thiol esters

Alkyl thiol ester R–C(=O)–S–R' R' is C4H9 or C6H13	C=O str. cm^{-1}	C=O str. cm^{-1}	Phenyl thiol ester R–C(=O)–S–C6H5	C–C=str. R' analog cm^{-1}	C–C=str. C6H5 analog cm^{-1}	S–C=str. R' analog cm^{-1}	S–C=str. C6H5 analog cm^{-1}
R			R				
Formate*	1675	1693	Formate			755	730
Acetate	1695	1711	Acetate	1137	1111	955	947
chloro-	1671(s)	1691(s)	chloro-	1089	1065	1000	986
	1699(m)	1725(m)					
dichloro-	1682(s)	1700(s)	dichloro-	1085	1070	990	976
	1703(m)	1736(m)					
trichloro-	1699	1711	trichloro-	?	?	1032	1018
trifluoro-	1710	1722	trifluoro-	?	955	?	930
Propionate	1691	1710	Proprionate	1090	1088	937	925
Butyrate	1693	1710	Butyrate	1111	1111	989	975
Dialkyl dithiol esters							
Oxalate	1680	1698	Oxalate			790	770
Succinate	1690	1705	Succinate	~1050	~1050	985	970
Adipate		1710	Adipate		~1010		~940
	C–H str. cm^{-1}	C–H str. cm^{-1}		C–H bend cm^{-1}	C–H bend cm^{-1}	2(C–H bend) cm^{-1}	2(C–H bend) cm^{-1}
Formate*	2835	2825		1345	1340	2680	2660

TABLE 15.17A IR data for alkyl thiolesters and phenyl thiol esters

Alkyl thiol ester R–C(=O)–S–R' R' is C4H9 or C6H13 R	C=O str. cm^{-1}	C=O str. cm^{-1}	Phenyl thiol ester R–C(=O)–S–C6H5 R
Acetate	1695	1711	Acetate
chloro-	1671(s)	1691(s)	chloro-
	1699(m)	1725(m)	
dichloro-	1682(s)	1700(s)	dichloro-
	1703(m)	1736(m)	
trichloro-	1699	1711	trichloro-
trifluoro-	1710	1722	trifluoro-
Propionate	1691	1710	Propionate
Butyrate	1693	1710	Butyrate

TABLE 15.17B   IR data for Thiolbenzoates

Thiol benzoate ring substitution	Alkyl thiol benzoate C=O str. cm^{-1}	Phenyl thiol benzoate C=O str. cm^{-1}	Alkyl analog C—C str. cm^{-1}	Phenyl analog C—C str. cm^{-1}	Alkyl analog S—C str. cm^{-1}	Phenyl analog S—C str. cm^{-1}
Thiol benzoate	1665	1685	1203	1205	915	898
2-F	1648	1667	1205	1196	920	906
	1675	1690				
	1701(w)	1701(w)				
2-Cl		1696		1198		898
2-Br		1700		1198		898
2-I	1679	1698	1205	1200	910	897
2-CH$_3$O	1640	1652	1193	1190	904	888
	1672	1700				
2-HO		1640		1192		918
3-I	1670	1689	1195	1191	938	913
3-NO$_2$		1689		1203		939
4-Br	1669	1681	1204	1199	913	898
4-NO$_2$		1683		1197		907
Dithiol phthalate		1675		1193		913
diphenyl		1690		1211		

TABLE 15.17C   IR data for thiol acids, thiol anhydrides, and potassium thiol benzoate

Compound	C=O str.	C—C str.	S—C str.	S—H str.	S—H bend
Thiol acetic acid	1712	1122	988	2565	828
Thiol benzoic acid	1690	1210	950	2585	835
2-chloro-	1700	1207	945	2580	837
	[O=C]2—S str.				
Thiol benzoic anhydride	1739	1202	860		
	1709	1178			
	1680				
	asym.COS str.	1203	948		
Potassium thiol benzoate	1525				

TABLE 15.18 Raman data and assignments for propargyl acrylate and propargyl methacrylate

Propargyl acrylate	Assignment	Propargyl methacrylate
3111(1)	a.CH_2=str.	3110(1)
3042(2)	s.CH_2=str.	
2994(1)	a.CH_2 str.	2998(2)
2953(1)	s.CH_2 str.	
	s.CH_3 str.	2933(2)
2132(9)	CC str.	2132(9)
1728(2)	C=O str.	1724(3)
1637(7)	C=C str.	1640(5)
1439(1)	CH_2 bend	1439(1)
1410(2)	CH_2=bend	1405(3)
1368(0)	CH_2 wag	1372(1)
	CH=rock	1319(0)
1293(2)	CH=rock	
1073(0)	CH_2=rock	
	s.COC str.	1016(1)
992(2)	CH_2 rock	
958(2)	s.COC=str.	962(2)
	CH_3 rock?	946(2)
935(1)	C—C str.	
	C—C str.	922(0)
	C—C str.	852(1)
842(2)	C—C str.	
	CH_2=twist	820(3)
		605(1)
		592(1)
		564(0)
408(1)		
		377(1)
	C—O—O_3 bend,a′	350(2)
309(5)	C—CC bend,a″	306(5)
238(6)	C—CC bend,a′	227(5)

TABLE 15.19   IR vapor-phase data for R—C(=)—OR' skeletal stretching

R—C(=O)—O—R'	R' is Alkyl cm^{-1}	R' is Allyl cm^{-1}	R' is Benzyl cm^{-1}	R' is Vinyl cm^{-1}	R' is Isopropenyl cm^{-1}	R' is Phenyl cm^{-1}
Formate	1158–1180		1152–1159			
Acetate	1231–1250	1231		1215	1202	1201–1203
Propionate	1182–1194	1180		1162		1200
Butyrate	1176–1185		1170	1155		
Valerate	1171–1178	1165				
Hexanoate	1168–1171					
Heptanoate	1165–1170					
Octanoate	1165–1169	1160				
Nonanoate	1162–1169	1160				
Decanoate	1160–1169			1151		
Tetradecanoate	1170–1177					
Octadecanoate	1165–1178			1150		
Acetate	1231–1250					
2-Methylacetate	1182–1194					
2-Ethylacetate	1176–1181					
2-Propylacetate	1173–1178					
2-Butylacetate	1168–1171					
2-Alkylacetates	1165–1172					
2,2-Dimethylacetate	1145–1159					
2,2,2-Trimethylacetate	1110–1156					
2,-Cyanoacetate	1160–1172					
2-Chloroacetate	1159–1171					
2,2-Dichloroacetate	1160					1101
2,2,2-Trichloroacetate	1235–1241					1181
Diesters						
Oxalate	1152–1164					
Malonate	1139–1153					
Succinate	1159–1166					
Glutarate	1170–1175					
Adipate	1151–1179					
Sebacate	1161–1171					

TABLE 15.20   IR vapor-phase data for conjugated esters [aryl—C(=)—OR or C=C—C(=)—OR skeletal stretching

Compound	Alkyl or Dialkyl cm^{-1}	Phenyl cm^{-1}	Alkyl or Dialkyl cm^{-1}	Phenyl cm^{-1}
Benzoate	1235–1294	1260	1089–1145	1200
Phthalate	1260–1281		1111–1128	
Isophthalate	1229–1240		1092–1095	
Terephthalate	1263–1270		1100–1107	
Nicotinate	1272–1280		1105–1111	
Isonicotinate	1272–1281		1115–1120	
Picolinate	1305–1311		1130–1131	
Salicylate	1300–1305	1300	1082–1111	1061
Crotonate	1176–1190		1021–1048	

Organic Carbonates, Thiol Carbonates, Chloroformates, Thiol Chloroformates, Acetyl Chloride, Benzoyl Chloride, Carbamates, and an Overview of Solute-Solvent Effects upon Carbonyl Stretching Frequencies

Table 16-4a	413 (392)	Table 16-5	415 (393)
Table 16-4b	413 (392)	Table 16-6	415 (392, 293)
Table 16-4c	414 (393)	Table 16-7	416 (393, 397)
Table 16-4d	414 (393)	Table 16-8	417

*Numbers in parentheses indicate in-text page reference.

In developing spectra-structure correlations it is helpful to know the molecular vibrations of some relatively simple model compounds. In this case, these include dihalocarbonyl compounds such as $F_2C=O$, $FClC=O$, and $Cl_2C=O$. The normal modes as obtained by Overend and Scherer (1) for these three carbonyl halides are depicted in Figs. 16.1a–c. The normal skeletal stretching modes for $F_2C=O$ and $Cl_2C=O$ are best described as symmetric and asymmetric X_2C stretching, while for $FClC=O$ they are best described as C–X and C–Y stretching. These model compounds are useful in predicting where the similar molecular vibrations for compounds of forms $(R-O)_2C=O$, $(R-S-)_2C=O$, $(R-S-)(R-O-)C=O$, $(R-S)C(=O)Cl$, $(R-O-)C(=O)Cl$, $(R-O-)C(=O)NH_2$, etc. are expected to occur in the IR region of the electromagnetic spectrum.

In addition, Overend and Evans (2) have shown that the force constant of the out-of-plane skeletal deformation is similar to the sum of Taft σ_R and σ_I parameters, therefore, it is expected that this out-of-plane skeletal deformation is sensitive to the mass as well as the resonance and inductive parameters of the X and Y substituents for compounds containing the $XYC=O$ skeletal structure.

Table 16.1 summarizes the IR group frequency data for organic carbonates, thiol carbonates, chloroformates, thiol chloroformates, carbamates and related compounds (3). In all of the compounds listed in Table 16.1 it is noted that when compounds whose carbonyl substituents are phenyl–O– or phenyl–S– are compared to analogous compounds whose substituents are alkyl–O– or alkyl–S–, consistent frequency differences are noted: (1) the carbonyl stretching frequency is always higher for the aromatic containing compounds than for the aliphatic containing compounds; (2) the frequency of the asymmetric $X_2C=$ or $Y_2C=$stretch, or the X–C= stretch in unsymmetrical $XYC=O$ compounds, is always lower for the aromatic compounds than for the analogous aliphatic compounds; and (3) the carbonyl stretching frequency for the alkyl–O– or phenyl–O– always occurs at higher frequency than the analogous alkyl–S– or phenyl–S–$X_2C=O$ or $XYC=O$ compounds.

A possible explanation for these frequency shifts is that they are caused by resonance competition between the π-electron of the phenyl ring and the π-electron of the carbonyl atom for overlap with the nonbonding electron pair of oxygen or sulfur: increase of electron overlap between the phenyl π electron and a nonbonding pair on oxygen or sulfur takes place at the expense of overlap between the carbonyl carbon π-electron and oxygen or sulfur non-bonding pairs; this results in a reduced force constant and stretching frequency for $X_2C=$, $Y_2C=$ and $XYC=$ bonds, but increased force constant and stretching frequency of the C=O bond (3).

CARBONATES $(-O-)_2C=O$

The carbonyl stretching frequency for compounds of type $(R-O-)(R'-O)C=O$ is $\sim 1739\,cm^{-1}$, for compounds of type $(\phi-O-)(R-O-C=O)$, the range is $1754–1787\,cm^{-1}$, and it is 1775–

1819 cm^{-1} for compounds of type $(\phi-O-)_2C=O$. These carbonates exhibit asymmetric $C(-O-)_2$ stretching in the range 1205–1280 cm^{-1}, and the out-of-plane skeletal deformation occurs in the range 785–800 cm^{-1} (3).

MONOTHIOL CARBONATES $-O-C(=O)-S-$

Compounds of types $(R-O-)C(=O)-S-R)$ and $(\phi-O-)C(=O(-S-R)$ exhibit their carbonyl stretching frequencies in the range 1702–1710 cm^{-1} and 1730–1739 cm^{-1}, respectively, while for compounds of type $(R-O-)C(=O(-S-\phi)$ in the range 1719–1731 cm^{-1}. The band in the range 1056–1162 cm^{-1} is assigned to the C$-$O stretching vibration. The out-of-plane skeletal deformation occurs in the range 650–670 cm^{-1}.

CHLOROFORMATES $R-O-C(=O)Cl$ AND $\phi-O-C(=O)Cl$

The alkyl chloroformates exhibit the carbonyl stretching frequency in the range 1775–1780 cm^{-1}, and it occurs at 1784 cm^{-1} in the case of phenyl chloroformate. The C$-$O stretching vibration is assigned in the range 1139–1169 cm^{-1} and at 1113 cm^{-1} for alkyl chloroformates and phenyl chloroformate, respectively.

DITHIOL CARBONATES $(-S-C(=O)-S-)$

The carbonyl stretching frequencies for the dithiol carbonates occur in the range 1640–1655, 1649, and 1714–1718 cm^{-1} for compounds of $(R-S)_2C=O$, $(R-S-)(\phi-S-)C=O$, and $(\phi-S-)_2C=O$, respectively. The strong band in the range 827–880 cm^{-1} is assigned to the asymmetric $S_2C=$stretching vibration (see Fig. 16.2), which shows the IR spectrum of diallyl dithiol carbonate (upper) and dipropyl dithiol carbonate (lower), respectively. The weak band in the range 554–595 cm^{-1} is assigned to the out-of-plane carbonyl skeletal deformation. A weak band in the region 700–750 cm^{-1} is assigned to C$-$S stretching.

The asymmetric $S_2C=$ mode exhibits a strong first overtone that is sometimes higher or lower in frequency than the carbonyl stretching absorption band and it is in Fermi resonance with $\nu C=O$. The $\nu C=O$ frequencies are corrected in Table 16.1.

RING STRAIN

Ethylene carbonate, ethylene monothiol carbonate, and ethylene dithiol carbonate exhibit $\nu C=O$ at 1818.3, 1757, and ~ 1687 cm^{-1}, respectively. These $\nu C=O$ frequencies occur at higher frequency than their analogous open chain analogs by approximately 70, 50, and 40 cm^{-1}, respectively, after correction for Fermi resonance. This is the order of increasing ring size due to the fact that sulfur is larger than oxygen. As the ring size becomes smaller, there is a decrease in the X$-$C$-$X bond angle, which makes it more difficult for the carbonyl carbon atom to vibrate in and out of the 5-membered ring. Thus, as the ring strain is increased, the carbonyl stretching mode increases in frequency.

THIOL CHLOROFORMATES Cl—C(=O)—S

Alkyl thiol chloroformates and aryl thiol chloroformates exhibit $vC=O$ in the range 1766–1772 cm^{-1} and 1769–1775 cm^{-1}, respectively, which is not as much difference as shown between alkyl chloroformates and phenyl chloroformate (1775–1780 cm^{-1} and 1784 cm^{-1}). However, the $vC=O$ mode for alkyl thiol chloroformate is ~ 1760 cm^{-1} CCl$_4$ solution after correction for Fermi resonance vs 1769–1775 cm^{-1} for aryl thiol chloroformate (see Fig. 16.3 for a comparison of the IR spectrum of methyl thiol chloroformate vs phenyl chlorothiol chloroformate). The strong IR band at ~ 845 cm^{-1} for the methyl ester and the strong IR band at ~ 815 cm^{-1} for the phenyl ester are assigned to asymmetric S—C—Cl stretching. A more comprehensive vibrational assignment for methyl thiol chloroformate will be presented later in Table 16.6.

Ethylene carbonate has C$_{2v}$ symmetry, its vibrational spectrum has been assigned, and its carbonyl stretching mode has been reported to be in Fermi resonance with the first overtone of the skeletal breathing mode. The skeletal breathing mode occurs at 897 cm^{-1} in the liquid phase, at 894 cm^{-1} in CHCl$_3$ solution, at 880 cm^{-1} in the vapor phase, and at 900 cm^{-1} in water solution, while the perturbed C=O stretching mode is located in the region 1810–1870 cm^{-1}, depending upon the physical state of the sample (4).

Table 16.2 lists the carbonyl stretching frequency for ethylene carbonate in various solvents. The IR band in the ranges 1770–1780 cm^{-1} and 1790–1815 cm^{-1} is in Fermi resonance. In each solvent system, the $vC=O$ mode has been corrected for Fermi resonance, and it occurs as low as 1791.7 cm^{-1} in solution in methanol and as high as 1812.5 cm^{-1} in solution in carbon tetrachloride (5).

Figure 16.4 shows plots of the $vC=O$ frequencies of ethylene carbonate corrected for Fermi resonance vs the solvent acceptor number (AN). Two separate plots are apparent. The upper plot includes methylene chloride, chloroform, tert-butyl alcohol, isopropyl alcohol, ethyl alcohol, and methyl alcohol while the lower plot includes diethyl ether, carbon tetrachloride, benzene, nitrobenzene, acetonitrile, benzonitrile, and dimethyl sulfoxide. Thus, it is apparent that the protic solvents correlate in a different manner than the so-called aprotic solvents. Thus, it again shows that the AN values do not take into account the factors determining the strength of the intermolecular hydrogen bond formed between the carbonyl oxygen atom of ethylene carbonate and the OH or C—H proton of the solvent system (5).

Figure 16.5 shows a plot of the carbonyl stretching frequencies of ethylene carbonate corrected for Fermi resonance vs the reaction field of the CHCl$_3$/CCl$_4$ solutions. The linear plot demonstrates a good correlation between $vC=O$ and its surrounding reaction field.

Table 16.3 compares the vibrational assignments for methyl chloroformate, 3-propynyl chloroformate, and chlorofluoro carbonyl (6). This comparison shows the value of utilizing the vibrational modes of a model compound such as FCl C=O in assigning the six OCl C=O skeletal vibrations for the alkyl chloroformates (6). It was shown that 3-propynyl chloroformate exists as rotational conformers.

Tables 16.4, 4a, and 4b list IR and Raman data and assignments for acetyl chloride, acetyl—d$_3$ chloride, and acetyl—d$_1$ chloride (7). These data serve as model compounds in the development of spectra structure correlations for compounds of forms CH$_3$—C(=O)(—S—R) and CH$_3$—C(=O)(—S—aryl). In the case of the d$_1$ analog, the presence of trans and gauche conformers is apparent (7). The $vC=O$ mode occurs at 1807, 1812, and 1802 cm^{-1} for the

CH$_3$, CD$_3$, and CDH$_2$ analogs, respectively. Data such as these show that the C=O stretching mode is not a "pure" molecular vibration.

Table 16.4c lists IR spectra-structure correlations for carboxylic acid halides in the vapor phase. Comparison of the νC=O frequency for acetyl fluoride (1869 cm^{-1}) to νC=O for benzoyl fluoride (1832 cm^{-1}) shows that the benzoyl analog exhibits νC=O 37 cm^{-1} lower in frequency than the acetyl analog. This lower νC=O frequency is the result of conjugation of the phenyl group with the CO group, which weakens the C=O bond.

The νC=Cl mode for compounds of form R−C(=O)Cl occurs in the range 570–601 cm^{-1}, and in the range 821–889 cm^{-1} for compounds of form ϕ−C(=O)Cl. The νC−F mode for acetyl fluoride is assigned at 827 cm^{-1} and for benzoyl fluoride it is assigned at 1022 cm^{-1}.

Table 16.4d lists IR spectra-structure correlations for benzoyl halides in the neat phase. Most of these compounds exhibit IR bands in the region 1750–1812 cm^{-1} and in the region 1693–1815 cm^{-1}. The presence of two IR bands is the result of νC=O being in Fermi resonance with an overtone or a combination tone of a lower lying fundamental (s). The νC=O and the overtone have been corrected for Fermi resonance (see Table 16.4d).

An lR band in the region 1172–1245 cm^{-1} is assigned to a complex mode involving aryl−C stretching. The νC−X mode is assigned in the region 840–1000 cm^{-1} in the neat phase.

Table 16.5 lists IR data for benzoyl chloride in CS$_2$ solution at temperatures between 31 and −70 °C. Figure 16.6 shows plots of νC=O, 2νC−C(=)Cl, and $\nu\phi$−C(=)Cl frequencies for benzoyl chloride in CS$_2$ vs temperature in °C. The carbonyl stretching frequency is in Fermi resonance with the first overtone of the C$_6$H$_5$−C−Cl stretching mode in the case of benzoyl chloride. Neither νC=O nor $2\nu\phi$−C−Cl has been corrected for Fermi resonance in this case (8). Perturbed νC=O decreases in frequency from 1774.6 cm^{-1} at 31 °C to 1771.2 cm^{-1} at −70 °C while perturbed 2νC−C−Cl decreases in frequency from 1732.5 cm^{-1} at 31 °C to 1731.3 cm^{-1} at −70 °C. Moreover, νC−C−Cl increases in frequency from 871.5 cm^{-1} at 31 °C to 874.1 cm^{-1} at −70 °C. In addition, the absorbance ratio A[νC=O]/A[2νC−C−Cl] decreases from 4.44 at 31 °C to 3.03 at −70 °C. Thus, the absorbance of 2νC−C−Cl becomes larger as the νC−C−Cl frequency increases with a decrease in temperature. These experimental data prove conclusively that the extent of Fermi interaction between these two vibrational modes increases with decreases in temperature. Of course, the combination or overtone must belong to the same symmetry species as the fundamental vibration. In this case νC=O and 2νC−C−Cl belong to the A′ symmetry species.

Table 16.6 compares the vibrational data for S-methyl thiol chloroformate with those for Cl$_2$C=O and s-methyl phosphoro-dichloridothioate. These comparisons show the value of model compounds in making the vibrational assignments for the S−C(=O)Cl group based on the molecular vibrations for Cl$_2$C=O. Moreover, the value of comparing vibrational assignments for a compound containing the CH$_3$−S− group is also demonstrated (9).

Table 16.7 lists IR spectra-structure correlations for carbamic acid: aryl-, alkyl esters (10). These carbamates have the following empirical structure:

In CCl_4 solution these compounds exhibit the C=O stretching mode in the range 1730–1755 cm^{-1} and the intermolecularly hydrogen-bonded vC=OH−N frequency in the range 1705–1734 cm^{-1}. In the solid state, vC=O occurs at lower frequency at ~1690 cm^{-1}. The N−H stretching mode occurs in the range 3409–3461 cm^{-1} and vN−H$\cdots$O=C occurs in the range 3295–3460 cm^{-1}. These IR bands are no longer present in dilute CCl_4 solution because in dilute solution these carbamates are not intermolecularly hydrogen bonded.

The in-plane bending and out-of-plane N−H bending modes occur in the ranges 1504–1546 cm^{-1} and 503–570 cm^{-1}, respectively. In the solid phase the out-of-phase N−H bending mode occurs at even higher frequency, 624–680 cm^{-1}. A complex mode involving aryl-N stretching is assigned in the range 1237–1282 cm^{-1}.

In solution, an IR band occurs in the range 1195–1225 cm^{-1} and shifts to the range 1219–1257 cm^{-1} in the solid phase. This complex mode most likely is a mixture of N−C−O stretching and N−H in-plane bending (10).

INTRAMOLECULAR HYDROGEN BONDING

These carbamates have the following empirical structure:

where X is a halogen atom or a phenyl group.

The vN−HX mode for 2-substituted carbamates occurs at 3345, 3419, 3409, and 3425 cm^{-1} in 10 % wt./vol. In CCl_4 solution where X is F, Cl, Br, and phenyl, respectively. Comparable 3- and 4-substituted carbamates show IR evidence for intermolecular hydrogen bonding, but only the 2−F analog shows IR evidence for a small amount of intermolecular hydrogen bonding. In this case, F forms the weakest hydrogen bond in the series, which is most likely due to the relatively small size of the F atom. The strength of the intramolecular hydrogen N−H$\cdots$X bond increases in the order F, CI, and Br, and this is in the order of increasing size of the halogen atom. The larger size of Cl and Br also sterically interferes with intermolecular hydrogen bonding, and most likely contributes to the stabilization of the intramolecular hydrogen bond (10).

In the case of the 2-phenyl analog, the vN−H$\cdots\phi$ mode occurs at 3425 cm^{-1}. The 2-phenyl group in this case has to be perpendicular to the carbamate phenyl group in order for the N−H group to intramolecularly hydrogen bond to its π system.

The 1-naphthyl alkyl carbamates differ from the 2-naphthyl alkyl carbamates in that they exhibit vN−H in the ranges 3441–3461 cm^{-1} and 3425-3442 cm^{-1} while the 2-naphthyl analog only exhibits a band near 3445 cm^{-1}. The two vN−H bands in the case of the 1-naphthyl analogs are attributed to the existence of rotational conformers [rotation about the napthyl-N bond (10)].

The vN−H frequencies within each series vary by as much as 20 cm^{-1} depending on the nature of the O−R group in compounds of form ϕ−N−H−C(=O)−O−R. Compounds of form ϕ−NH−C(=O)−O−CH$_2$−CH$_2$−N(−CH$_3$)$_2$ exhibit the lowest N−H frequencies in each series studied, and occur in the region 3411–3440 cm^{-1}. These data suggest that there is a weak intramolecular hydrogen bond formed between the N—H proton and the β-nitrogen atom of the O—R group (10).

AN OVERVIEW OF THE SOLUTE-SOLVENT EFFECTS UPON CARBONYL STRETCHING FREQUENCIES

In CHCl$_3$/CCl$_4$ solutions the carbonyl stretching mode for a variety of compounds decreases in frequency as the mole % solvent is increased from 0 to 100. Mole % solvent is directly equivalent to the Reaction field surrounding the solute molecules. The Reaction field (R) involves both the dielectric constant and the refractive index of the solvent system.

$$|R| = (\varepsilon - 1)/(2\varepsilon + n^2)$$

where ε is the dielectric constant and n is the refractive index of the solvent.

$$\varepsilon = \frac{Q_1 Q_2}{f r^2} \quad \text{(Reference 11)}$$

Q_1 and Q_2 are the two charges of the solvent.
f is the force between the two charges.
r is the distance between the two charges.

Thus, the carbonyl stretching mode decreases in frequency as the Reaction field (R) increases; this is due mainly to the increasing electrically charged solvent molecules surrounding the solute molecules. In an aprotic solvent system such as CCl$_4$/C$_6$H$_{14}$ the carbonyl stretching frequency for a compound such as 1,1,3,3-tetramethylurea decreases in a linear manner as the mole % CCl$_4$ increases from 0 to 100. However, in the CHCl$_3$/CCl$_4$ solvent system the carbonyl stretching frequency for 1,1,3,3-tetramethylurea decreases in frequency in linear segments A, B, and C. It is suggested that segment A represents TMU(CHCl$_3$)$_1$, segment B represents TMU(CHCl$_3$)$_2$, and segment C represents TMU(CHCl$_3$)$_3$ complexes within the CHCl$_3$/CCl$_4$ solvent system (10). In contrast, acetone in CHCl$_3$/CCl$_4$ solvent system exhibits only one point of deviation from linearity as the mole % is increased from 0 to 100. In this case, the decrease due to intermolecular C=OHCCl$_3$ occurs within 17 mol % CHCl$_3$/CCl$_4$ (12).

The degree of carbonyl stretching frequency decrease in going from solution in CCl$_4$ to solution depends upon the basicity of the carbonyl group. For example, the carbonyl stretching frequency for acetone occurs at 1717.5 cm^{-1} in CCl$_4$ and at 1710.5 cm^{-1} in CHCl$_3$, a decrease of 7 cm^{-1} (12). In the cases of 1,1,3,3-tetramethylurea, 1,1,3,3-tetraethylurea, and 1,1,3,3-tetra-butylurea the carbonyl stretching frequency in going from CCl$_4$ to CHCl$_3$ solution decreases by 25.6, 26.1, and 27.0, respectively (13). The carbonyl group becomes more basic as the four alkyl groups become larger as reflected by larger frequency decrease in going from CCl$_4$ to CHCl$_3$ solution.

The carbonyl stretching frequency for acetone occurs at 1735 cm^{-1} in the vapor phase and at 1713 cm^{-1} in the liquid phase while for 1,1,3,3-tetramethylurea it occurs at 1685 cm^{-1} in the vapor phase and at 1649.6 cm^{-1} in the liquid phase (10, 12). Thus, acetone shows a decrease of 22 cm^{-1} and 1,1,3,3-tetramethylurea shows a decrease of 35.4 cm^{-1} in going from the vapor to the neat phase. These data indicate that the dipolar interaction between carbonyl groups $[(+C-O^-)_n]$ is larger in the case of 1,1,3,3-tetramethylurea than in the case of acetone (12). This is what is expected, as the carbonyl group of 1,1,3,3-tetramethylurea is more basic than the carbonyl group for acetone.

It has been shown that the carbonyl stretching frequency for ketones and esters is affected by steric factors of the $R-(C=O)-R'$ groups and for the R_1 group of $R_1 - (C=O)-OR$. The steric factor of these alkyl groups increases as the dipolar distance increases between the dipolar sites in the solute and the dipolar sites of the solvent.

In summary, factors affecting solute-solvent interactions include the following:

a. basicity of the $C=O$ group (dipolarity);
b. acidity of the solvent group;
c. the dielectric constant of the solvent (dipolarity);
d. the refractive index of the solvent;
e. steric factors of the solvent;
f. steric factors of the solute;
g. other basic sites in the solute; and
h. concentration of the solute in the solvent.

Thus, the dipolar interaction between sites in the solute and solvent which causes shifts in the carbonyl stretching mode is complex and the magnitude of the carbonyl stretching shift can not be determined by a simple equation.

SOLVENT ACCEPTOR NUMBERS (AN)

Gutman developed solvent acceptor numbers utilizing NMR spectroscopy (14). The AN is defined as a dimensionless number related to the relative chemical shift of ^{31}P in $(C_2H_5)_3PO$ in that particular solvent with hexane as the reference solvent on one hand, and $(C_2H_5)_3PO \cdot SbCl_5$ in 1,2-dichloroethane on the other—to which the acceptor numbers 0 and 100 have been assigned, respectively:

$$AN = \frac{\delta \text{ corr} - 100}{\delta \text{ corr } (C_2H_5)_3PO \cdot SbCl_5} = \text{Corr} - 2.348$$

Studies included in this book show that these AN values for the alcohols also included intermolecular hydrogen bonding between $C=O \cdot HOR$. In cases where both $C=O$ and $C=O\cdots HOR$ were determined, the carbonyl frequencies for the ketones or esters not intermolecularly hydrogen bonded, but surrounded by intermolecularly hydrogen-bonded alkyl alcohols, exhibit their carbonyl stretching mode at frequencies comparable to those recorded in dialkyl ethers. Thus, a large portion of the AN value for each alkyl alcohol is due to intermolecular hydrogen bonding, and a smaller portion of the AN value results from the $R-O$ portion of the alcohol complex.

Steric factors of the solute and solvent also appear to affect a linear correlation of carbonyl stretching frequencies vs AN. However, the AN values do help in spectra-structure identification of unknown chemical compounds.

CONCENTRATION EFFECTS

In an aprotic solvent such as CCl_4 dialkyl ketones would tend to cluster due to dipolar interaction between molecules $[(+C-O^-)_n]$, where n becomes larger as the wt. % solute/volume solvent increases. The carbonyl stretching frequency for diisopropyl ketone decreases $0.24 \, cm^{-1}$ in going from 0.75 to 7.5 % wt./vol. In CCl_4 solution and for di-tert-butyl ketone it decreases $0.05 \, cm^{-1}$ in going from 0.8 to 6.56 % wt./vol. In CCl_4; and although this carbonyl frequency decrease is attributed to clustering of ketone molecules, steric factors of the tert-butyl group prevent close interaction of the carbonyl groups compared to that for diisopropyl ketone (15).

In the case of $CHCl_3$, the carbonyl stretching frequency for diisopropyl ketone increases $0.53 \, cm^{-1}$ in going from 0.22 % wt./vol. To 8.6 % wt./vol. in $CHCl_3$ solution, and for di-tert-butyl ketone $0.1 \, cm^{-1}$ in going from 0.40 % wt./vol. to 5.78 % : wt,/vol. in $CHCl_3$. This suggests that clustering of the diisopropyl ketone molecules at higher % wt./vol. solutions decreases the effect of intermolecular hydrogen bonding by a small amount. The effect appears to be small in the case of the tert-butyl analog due to steric effects of the tert-butyl groups compared to the isopropyl groups (14).

DIPOLARITY-POLARIZABILITY EFFECT

Recently, it has been reported that there is a coupled dipolarity-polarizability influence of the solvent upon the carbonyl stretching mode for 1,1,3,3-tetramethylurea (16). This observation is based in part on an SCRF-MO model, which assumes that the solute is located in a spherical cavity within an unstructured dielectric continuum. It is stated that the solvent dipolarity appears to exert the larger effect, along with making solvent polarizability detectable. However, it does not predict the effect of hydrogen bonding upon the carbonyl stretching frequency for 1,1,3,3-tetramethylurea (15).

It is interesting to compare the IR data used to develop the SCRF-MO model developed by Kolling (16) and the data reported by Nyquist (10, 13) and Wohar (17) as shown in Table 16.8. The AN values are those developed by Gutman (14) with the exception of the AN values in brackets, which are estimated from IR spectra-structure correlations.

Comparison of columns A, B, and C shows that there are some serious discrepancies between the three sets of data generated in three different laboratories.

It should be pointed out that with the exception of chloroform, methyl alcohol, ethyl alcohol, isopropyl alcohol, and tert-butyl alcohol, the carbonyl stretching frequencies correlate in a linear manner in the case of 1,1,3,3-tetramethylurea (10). The four alkyl alcohols also show two linear relationships vs AN. The linear relationship for the $\nu C=O\cdots HOR$ frequencies includes intermolecular hydrogen bonding in the Gutman AN value. The linear relationship for the $\nu C=O$ frequency (molecules surrounded by intermolecular hydrogen-bonded alcohols, but where C=O is not hydrogen bonded) vs AN occurs at a significantly lower frequency. Projecting the points on the linear segment of $\nu C=O$ vs AN onto points on the plot for the aprotic solvents indicates that the AN values for intermolecularly hydrogen-bonded tert-butyl alcohol, isopropyl alcohol, and ethyl alcohol are approximately 5, 7, and 10, respectively (see Fig. 16.7). Thus, approximately 26–27 of the Gutman AN values for the four alkyl alcohols appear to be the result of intermolecular hydrogen bonding (e.g., C=O $\cdots$ HOR). The AN values of 5, 7, and 10 for tert-butyl alcohol, isopropyl alcohol, and ethyl alcohol for intermolecularly hydrogen-bonded alcohols, respectively, are comparable to the AN values for diethyl ether [3.9], methyl tert-butyl ether [5.0], and tetrahydrofuran [8.9] (10).

In the writer's opinion there are two reasons why the Gutman AN values are not a precise predictor of vC=O frequencies in solvent systems. The first reason is intermolecular hydrogen bonding. The second is the steric factor of the solute and of the solvent, which alters the distance between the carbonyl group and the interactive site of the solvent molecules. Otherwise the AN values are useful in predicting the general direction of vC=O frequency shift in a particular or similar type solvent.

The Nyquist values do correlate well with the AN values. However, there are serious discrepancies in the three sets of IR data for vC=O for 1,1,3,3-tetramethylurea in hexane, diethyl ether, tetrahydrofuran, and hexamethyl phosphoramide (see Table 16.8). In these four cases the data are significantly lower than the Nyquist (12) or Wohar data (17).

In the writer's opinion, the Kolling model is not correct in assuming that the solute is located in a spherical cavity within an unstructured dielectric continuum because steric factors of both the solute and solvent alter the spatial distance between the dipolar inter-active sites between the solute and solvent. Furthermore, the accuracy of the experimental data is in question. The presence of a more polar solvent not flushed from the IR cell, or from the presence of water in the solute-solvent system would lower the vC=O frequencies. Further experimental data are suggested to help clarify the theoretical aspects of solute-solvent interaction.

Finally, carbonyl stretching vibrations are often perturbed by Fermi interaction. Solvents either increase or decrease the amount of Fermi resonance between vC=O and a combination or overtone of the same symmetry species as vC=O. Therefore, in order to obtain the exact frequency for vC=O in this case in any solvent or solvent system, it is necessary to correct for Fermi resonance (vC=O for 1,1,3,3-tetramethylurea is not in Fermi resonance).

REFERENCES

1. Overend, J. and Scherer, J. R. (1960). *J. Chem. Phys.*, **32**, 1296.

2. Overend, J. and Evans, J. C. (1959). *Trans. Faraday Soc.*, **551**, 1817.

3. Nyquist, R. A. and Potts, W. J. Jr. (1961). *Spectrochim. Acta*, **17**, 679.

4. Angell, C. L. (1956). *Trans. Faraday Soc.*, **52**, 1178.

5. Nyquist, R. A. and Settineri, S. E. (1991). *Appl. Spectrosc.*, **45**, 1075.

6. Nyquist, R. A. (1972). *Spectrochim. Acta*, **28A**, 285.

7. Overend, J., Nyquist, R. A., Evans, J. C., and Potts, W. J. Jr. (1961). *Spectrochim. Acta*, **17**, 1205.

8. Nyquist, R. A. (1986). *Appl. Spectrosc.*, **40**, 79.

9. Nyquist, R. A. (1967–68). *J. Mol. Structure*, **1**, 1.

10. Nyquist, R. A. (1973). *Spectrochim. Acta*, **29A**, 1635.

11. Lange, N. A. (ed.) (1961). *Handbook of Chemistry* p. 1728, 10th ed., New York: McGraw-Hill.

12. Nyquist, R. A., Putzig, C. L., and Hasha, D. L. (1989). *Appl. Spectrosc.*, **43**, 1049.

13. Nyquist, R. A. and Luoma, D. A. (1991). *Appl. Spectrosc*, **45**, 1491.

14. Gutman, V. (1978). *The Donor-Acceptor Approach to Molecular Interactions*, p. 29, New York: Plenum Press.

15. Nyquist, R. A. and Putzig, C. L. (1989). *Appl. Spectrosc.*, **43**, 983.

16. Kolling, O. W. (1999). *Appl. Spectrosc.*, **53**, 29.

17. Wohar, M., Seehra, J., and Jagodzinski, P. (1988). *Spectrochim. Acta*, **44A**, 999.

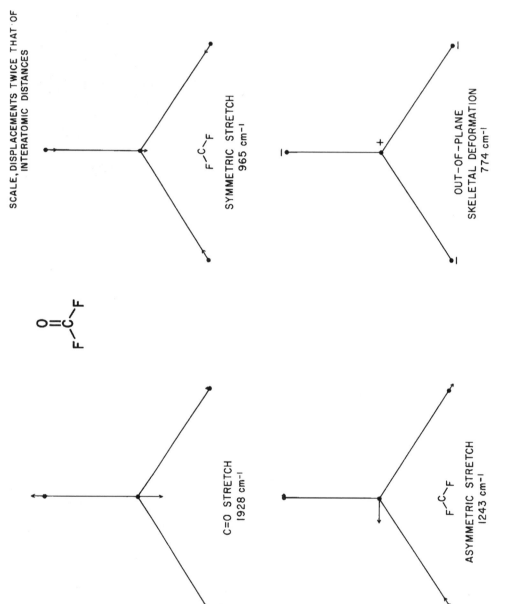

FIGURE 16.1a   Normal modes for COF₂.

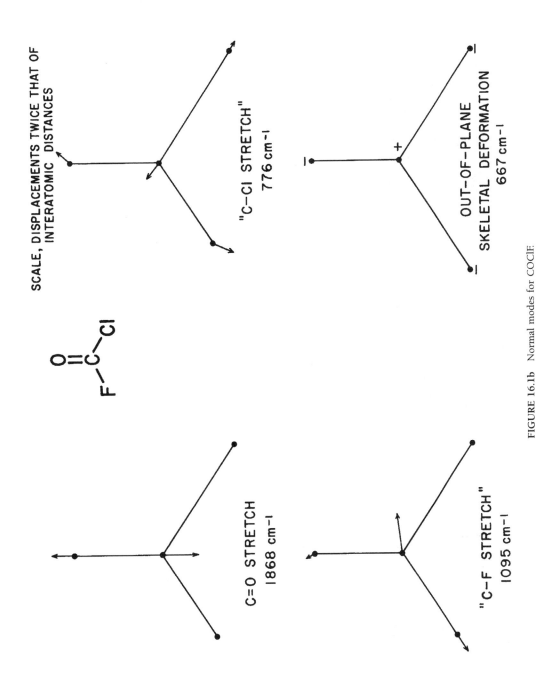

FIGURE 16.1b Normal modes for COClF.

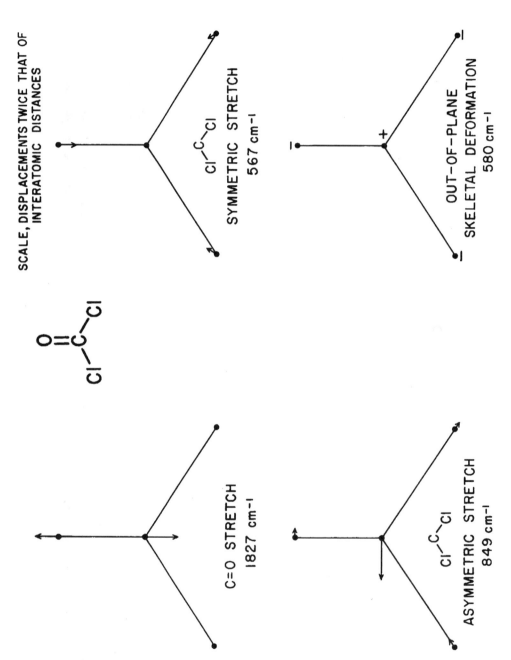

FIGURE 16.1c Normal modes for $COCl_2$.

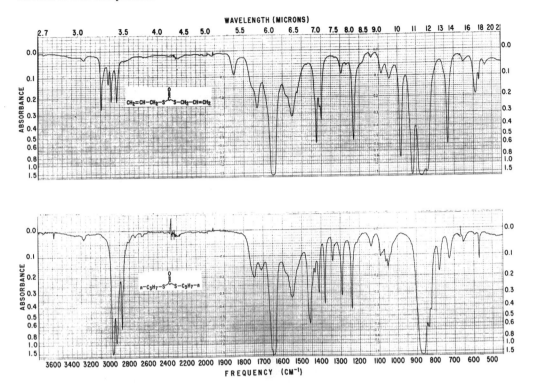

FIGURE 16.2 (Upper)—Infrared spectrum of diallyl dithiol carbonate. (Lower)—Infrared spectrum of dipropyl dithiol carbonate.

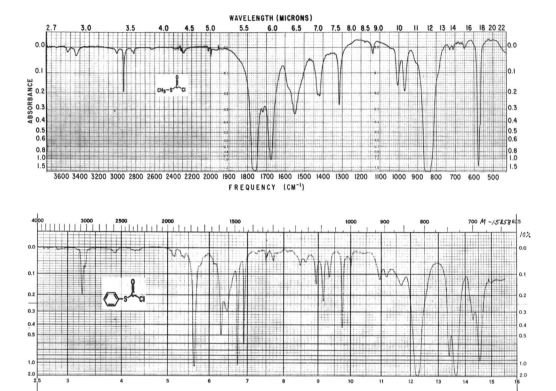

FIGURE 16.3 (Upper)—Infrared spectrum of methyl thiol chloroformate. (Lower)—Infrared spectrum of phenyl thiol chloroformate.

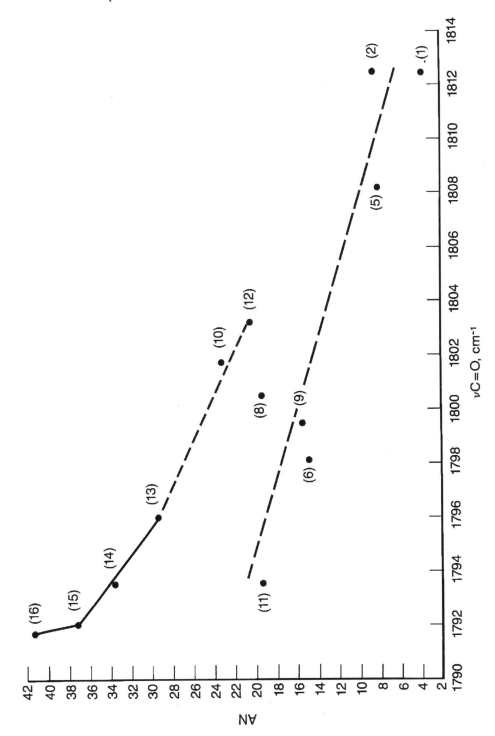

FIGURE 16.4   Plots of $\nu C{=}O$ frequencies of ethylene carbonate corrected for Fermi resonance vs the solvent acceptor number (AN).

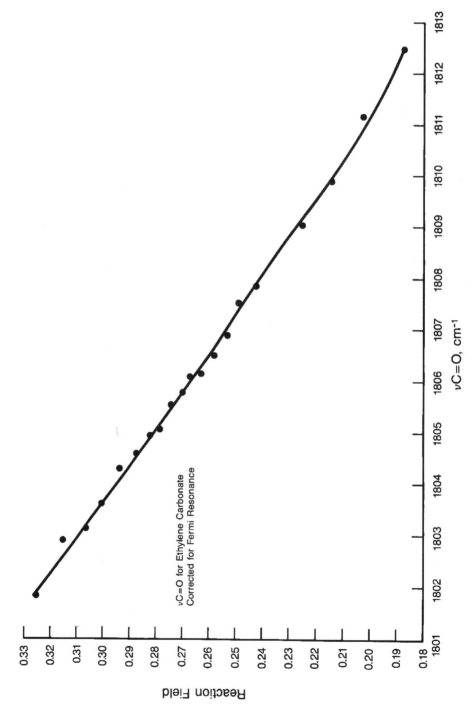

FIGURE 16.5 A plot of the $\nu C=O$ frequencies for ethylene carbonate corrected for Fermi resonance vs the reaction field of the $CHCl_3/CCl_4$ solutions.

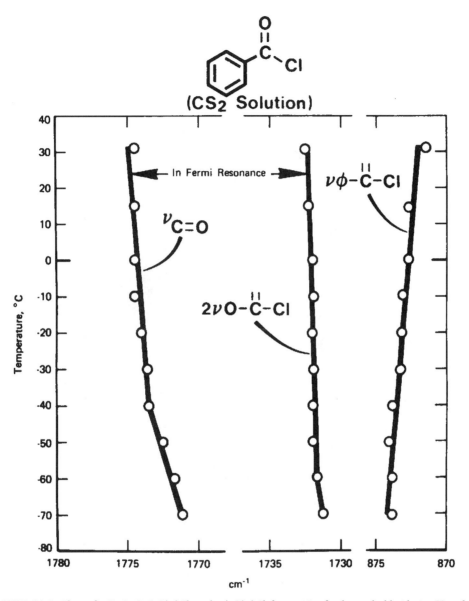

FIGURE 16.6 Plots of $\nu C{=}O$, $2\nu C\text{-}C({=})Cl$, and $\nu\phi\text{-}C({=})Cl$ frequencies for benzoyl chloride in CS_2 solution vs temperature in °C.

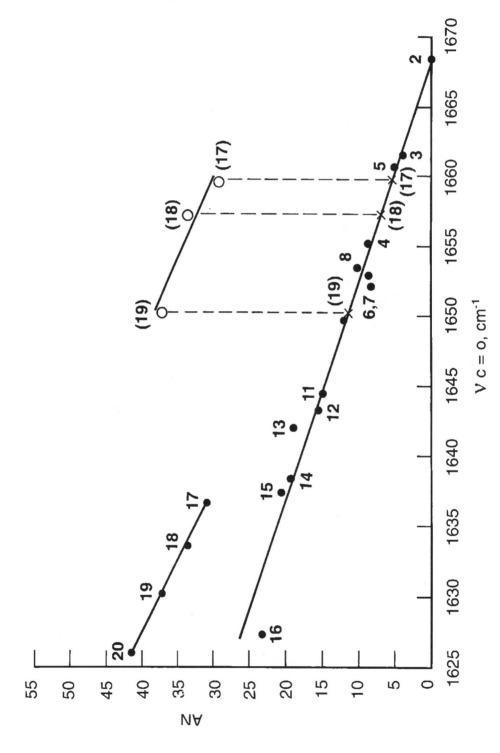

FIGURE 16.7 Plots of νC=O for tetramethylurea in various solvents vs the solvent acceptor number (AN).

TABLE 16.1 IR group frequency correlations for organ carbonates, thiol carbonates, chloroformates, and thiol chloroformates

Compound type	C=O str. cm⁻¹	a.X–C–X str. cm⁻¹	s.X–C–X str. cm⁻¹	C–X str. cm⁻¹	C–Y str. cm⁻¹	a.Y–C–Y str. cm⁻¹	s.Y–C–Y str. cm⁻¹	ABC=O o.p.def. A and B =X or Y cm⁻¹	C=O str. Corrected for Fermi Res. cm⁻¹
(R–O–)2C=O	1739–41	1240–80	~900					785–80	
(R–O–)C=O(–O–C₆H₅)	1754–87	1211–48	~900					~800	
C₆H₅–O–)2C=O	1775–1819	1205–21							
(CH₂–O–)2C=O	1818.3	~1140	~900						1812.5*
(R–S–)2C=O	1640–66					870–80	~572		~1655*
(R–S–)C=O(–S]–C₆H₅)	1649					839	~567		~1664*
(C₆H₅–S–)2C=O	1714–18					827–33	~560		~1690*
(CH₂–S–)2C=O	~1678					880			~1694*
(R–O–)C=O(–S–R)	1702–10			1142–62		~800		~670	
(R–O–)C=O(–S–C₆H₅)	1719–31			1125–451		~800		~670	
(C₆H₅–O–)C=O(–S–R–)	1730–9			1056–1102				650–655	[C=N str. [1649–56]
(–O–CH₂CH₂–S–)C=O	1757			1070 or 1040					
(R–S–)C=O(O–N=C(R')₂	1720–25			1117–19		~800			
(R–O–)C=O(Cl)	1775–80			1139–69				685–87	
(C₆H₅–O–)C=O(Cl)	1784								
(R–S–)C=O(Cl)	1766–72			1113		840–50	579	~560	~1760[CCL₄]* ~1763[vapor]* for CH₃–S–C=O(Cl)
(C₆H₅–S–)C=O(Cl)	1769–75					814–16	595	~560 760–80	N–H str.(not N–H: O=C in dilute soln.)
(R–O–)C=O(NHR')	1732–38	1210–1250							
(R–S–)C=O(NH₂)	1699								a.NH₂ str.,3530 s.NH₂ str.,3414 (not H-bonded in dilute soln.)
(R–S–)C=O(NHR')	1690–95			1170–1230					N–H str.,3431 (in dilute soln.)
(R–S–)C=O(NHC₆H₅)	1652–59			1152–60					N–H str.,3289–3320
(R–S–)C=O(N(R')₂)	1650–60			1220–65					i.p.N–H def.,1517–56
(C₆H₅–S–)C=O(N(R')₂)	1666			1248					
(R–S–)C=O(N(C₆H₅)₂	1670			~1275					
(R)₂NC=O(Cl	1739			1091?		671		660?	C=O str,1763[vapor] for C₂H₅ analog
(C₆H₅)2NC=O(Cl)	1743 1751sh			1138?				661?	

*Corrected for Fermi resonance.

TABLE 16.2   The C=O stretching frequency for ethylene carbonate n1various solvents [1% wt./vol.]

Ethylene carbonate 1 % wt./vol. in different solvents Solvent	cm^{-1}	cm^{-1}	C=O corrected for F.R. cm^{-1}	cm^{-1}	AN
Diethyl ether	1771.2	1818.69	1812.43	1777.46	3.9
Carbon tetrachloride	1771.93	1818.31	1812.45	1777.79	8.6
Carbon disulfide	1769.98	1816.82	1811.73	1775.07	
Toluene	1772.14	1715.24	1810.58	1776.8	
Benzene	1772.67	1714.72	1808.14	1779.25	8.2
Nitrobenzene	1775.06	1807.9	1798.12	1784.84	14.8
Nitromethane	1778.26	1806.91	1799.03	1786.14	
Acetonitrile	1777.9	1807.86	1800.51	1785.25	19.3
Benzonitrile	1775.69	1807.93	1799.46	1784.16	15.5
chloroform	1778.97	1810.27	1801.74	1787.44	23.1
Dimethyl sulfoxide	1774.37	1801.73	1793.54	1782.56	19.3
Methylene chloride	1777.57	1810.16	1803.21	1784.52	20.4
t-Butyl alcohol	1779.23	1806.43	1796	1789.66	[29.3]
Isopropyl alcohol	1778.88	1804.95	1793.53	1790.3	33.5
Ethyl alcohol	1779.14	1804.17	1793.01	1791.3	37.1
Methyl alcohol	1778.77	1804.68	1791.68	1791.77	41.3

TABLE 16.3 Vibrational assignments for methyl and 3-propynyl chloroformate, and FCl(C=O)

Methyl chloroformate cm^{-1}	3-Propynyl chloroformate cm^{-1}	F—C(=O)Cl cm^{-1}	Assignment
1797	1782	1868	C=O str.
1150	1141(conformer 1)		a.C—O—C(=) str.
	1121(conformer 2)		
		1095	C—F str.
954	1014(conformer 1)		s.C—O—C(=) str.
	974(conformer 2)		
822	834(conformer 1)	776	C—Cl str.
	807(conformer 2)		
690	687	667	out-of-plane C=O def.
484	479	501	O—C=O or F—C=O bend
413	362	415	O—C—Cl or F—C—Cl bend
277	255		C—O—R bend
167	~170		C—O—R torsion
3044			a.CH$_3$ str.
3018			a.CH$_3$ str.
2961			s, CH$_3$ str.*[1]
2844			2(s.CH$_3$ bend)*[1]
1453			a.CH$_3$ bend
1434			s.CH$_3$ bend
1202			CH$_3$ rock
	3314		C—H str.
	3012		a.CH$_2$ str.
	2959		s.s.CH$_2$ str.
	2149		CC str.
	1459(conformer 1)		CH$_3$ bend
	1439(conformer 2)		
	1366		CH$_2$ wag
	1366		CH$_2$ twist
	992		CH$_2$ rock
	916(conformer 1_		C—C str.
	937(conformer 2		C—C str.
	687		in-plane C—H bend
	634		out-of-plane C—H bend
	531(conformer 1)		out-of-plane C—C—O bend
	574(conformer 2)		out-of-planeC—C—O bend
	306		out-of-plane C—CC
	222		in-plane C—CC bend

*[1]on Fermi resonance.

TABLE 16.4   IR and Raman data and assignments for acetyl chloride, acetyl-d_3 chloride, and acetyl-d_1 chloride

CH3C(=O)Cl cm⁻¹	CD3C(=O)Cl cm⁻¹	Assignment	Description	CH₂DCC(=O)Cl cm⁻¹ [trans]	Assignment	CH₂DCC(=O)Cl cm⁻¹ [gauche]	Description
3010		v1,v11		2955	v1		s.CH₂ str.
					v1	3020	a.CH₂ str.
	2248	v1		2245	v2		C–D str.
					v2	2955	s.CH₂ str.
					v3	2245	C–D str.
2930		v2		1802	v3		C=O str.
					v4	1802	C=O str.
	2104	v2		1397	v4		CH₂ bend
					v5		CH₂ bend
1807	1812	v3	C=O str.	1254	v5		
					v6	1281	
1421		v4,v12		1011	v6		
					v7	1254	
	1132	v4		908	v7		C–D wag
					v8	1088	
1361		v5		554	v8		C=O bend
	1040	v5		437	v9		C(=O)Cl bend
					v10	852	CD wag+C–C str.
1098		v6		340	v10		CCCl bend
					v11	588	C=O bend
	962	v6		3005	v11		a.CH₂ str.
					v12	487	gamma C=O
953		v7		1281	v12		
					v13	437	C(=O)Cl bend
	818	v7		973	v13		
					v14	340	CCCl bend
594	563	v8		503	v14		gamma C=O
436	438	v9		?	v15	?	
348	317	v10					
3010		v11,v1					
	2257	v11					
1421		v12,v4					
1021		v13					
	887	v13					
514	522	v14					
238	?	v15					

TABLE 16.4a   IR and Raman data and assignments for acetyl chloride and acetyl-d_3 chloride

Mode	$CH_3C(=O)Cl$ Calculated	$CH_3C(=O)Cl$ Observed	Approximate description	Mode	$CD_3C(=O)Cl$ Calculated	$CD_3C(=O)Cl$ Observed	Approximate description
A'	cm^{-1}	cm^{-1}		A'	cm^{-1}	cm^{-1}	
v1	3002	3029	a > CH_3 str.	v1	2232	2280	a.CH_3 str.
v2	2939	2934	s.CH_3 str.	v2	2114	2104	s.CD_3 str.
v3	1780	1822	C=O str.	v3	1777	1820	C=O str.
v4	1441	1432	a.CH_3 bend	v4	1147	1132	s.CD_3 bend+a.CCCl str.
v5	1378	1370	s.CH_3 bend	v5	1031	1040	a.CD_3 bend
v6	1104	1109	CH_3 rock + C−C str.	v6	986	962	a.CCCl +s.CD_3 bend
v7	962	958	C−C str.+CH_3 rock	v7	802	818	CD_3 rock
v8	661	608	C−Cl str.	v8	601	563	CCl str.
v9	469	436	O=C−Cl bend	v9	462	437	O=CCl bend
v10	391	348	CCCl bend	v10	355	317	CCCl bend
A''				A''			
v11	3000	3029	a.CH_3 str.	v11	2227	2280	a.CD_3 str.
v12	1444	1432	a.CH_3	v12	1031	1040	a.CD_3
v13	1040	1029	CH_3 rock	v13	854	877	CD_3 rock
v14	568	514	gamma C=O	v14	516	498	gamma C=O
v15	136	238	CH_3 torsion	v15	100	?	CS_3 torsion

TABLE 16.4b   IR and Raman data and assignments for acetyl-d_1 chloride

$CH_2DC(=O)Cl$ Calculated trans [D-eclipsed] cm^{-1}	$CH_2DC(=O)Cl$ Observed trans [D-eclipsed] cm^{-1}	Approximate description	Mode	$CH_2DC(=O)Cl$ Calculated gauche [H-eclipsed] cm^{-1}	$CH_2DC(=O)Cl$ Observed gauche [H-eclipsed] cm^{-1}	Approximate description
A'			A'			
v1 2962	2972	a.CH_2 str.	v1	3002	3020	a.CH_2 str.
v2 2192	2255	C−D str.	v2	2962	2972	s.CH_2 str.
v3 1778	1820	C=O str.	v3	2189	2255	C−D str.
v4 1418	1408	bend	v4	1780	1820	C=O−H_2 str.
v5 1266	1265	CH_2 wag	v5	1417	1422	CH_2 bend
v6 1033	1020	C−C str.	v6	1300	1290	CH_2 wag
v7 899	909	C−D wag.	v7	1260	1265	CH_2 rock
v8 607	565	C−Cl str.	v8	1085	1088	C−C str.+CH_2 rock
v9 467	437	O=CCl	v9	968	987	C−D bend+C−C str.
v10 381	340	CCCl bend	v10	847	851	C−D bend+C−C str.
A''			A''			
v11 3000	3020	a.CH_2	v11	657	588	C−Cl str.
v12 1306	1290	CH_2 twist	v12	539	489	gamma C=O
v13 974	987	C−D o.p.bend	v13	467	437	O=CCl bend
v14 568	507	gamma C=O	v14	378	340	CCCl bend
v15 120	?	CH_2D torsion	v15	121	?	CH_2D torsion

TABLE 16.4c   IR spectra-structure correlations for carboxylic acid halides in the vapor phase

Carboxylic acid halides	$vC=O$ cm^{-1}	See Text cm^{-1}	$vC-X$ cm^{-1}	gamma C=O cm^{-1}
acetyl fluoride	1869	1032	827	599
acetyl chloride	1818	1108	601	514
acetyl bromide	1818	1095	552	
trichloroacetyl chloride	1814	1023	591	622
hexanoyl chloride	1810	960	577	
octanoyl chloride	1809	965	570	
benzoyl fluoride	1832	1251	1022	
3-fluoro, benzoyl chloride	1790	1255	821	
4-trifluoromethyl, benzoyl chloride	1782.7[*1]	1181	882	
2-fluoro, benzoyl chloride	1794.9[*1]	1198	889	
2-bromo, benzoyl chloride	1801.8[*1]	1200	874	
2,6-dichloro, benzoyl chloride	1802[*1]			

[*1] Corrected for Fermi resonance.

TABLE 16.4d   IR spectra-structure correlations for benzoyl halides in neat phase

Benzoyl halide	perturbed C=O str. cm^{-1}[A]	perturbed overtone cm^{-1}[A]	C=O str. corrected for F.R. cm^{-1}	overtone corrected for F.R. cm^{-1}	complex aryl–C str. cm^{-1}	complex aryl–C–X str. cm^{-1}
fluoride	1812[0.94]	1778[0.75]	1796.9	1793.1	1000	
chloride	1785[0.88]	1735[0.66]	1762.9	1757.1	1205	875
bromide	1775[0.67]	1693[0.10]	1764.4	1703.6	1190	847
3-bromo, bromide	1775[0.84]	1754[0.84]	1764.5	1764.5	1179	875
2-methyl, chloride	1765[0.87]	1710[0.38]	1757.2	1717.8	1203.	870
2-trifluoromethyl, chloride	1803[1.12]	1750[0.38]	1789.6	1763.4	1200	871
2-fluoro, chloride	1788[1.07]	1756[0.94]	1774	1771	1195	884
2-chloro, chloride	1780[0.94	1735[0.45]	1765.4	1749.6	1190	865
2-bromo, chloride	1797[1.13]	1747[0.21]	1789.2	1754.8	1200	864
2-nitro, chloride[*1]	1800[0.70]	1758[0.37]	1785.5	1772.5	1195	870
3-fluoro, chloride	1771[1.03]	1741[1.03]	1756	1756	1245	785
3-chloro, chloride	1765[1.03]	1815[0.15]	1780	1798.8	1200	918
3-bromo, chloride	1750[1.15]	1795[0.59]	1765.3	1779.7	1172	884
3-nitro, chloride[*2]	1752[1.14]	1783[0.74]	1758.3	1776.7	1191	840?
4-methyl, chloride	1775[1.04]	1740[0.85]	1769.3	1745.7	1203	875
4-tert-butyl, chloride	1782[1.13]	1730[0.94]	1758.4	1753.6	1210	874
4-trifluoromethyl, chloride	1775[0.58]	1740[0.55]	1758	1757	1200	879
4-chloro, chloride	1780[1.15]	1732[0.95]	1768	1743.8	1190	860

[*1] $v$asym.NO$_2$, 1530 cm^{-1}; $v$sym.NO$_2$, 1342 cm^{-1}.
[*2] $v$asym.NO$_2$, 1529 cm^{-1}; $v$sym.NO$_2$, 1340 cm^{-1}.

TABLE 16.5   IR data for benzoyl chloride in CS_2 solution in the temperature between 31 and $-70°C$

Benzoyl chloride [CS_2] °C	C=O str. in Fr cm^{-1}	2[C_6H_5-C-Cl str.] in FR cm^{-1}	C_6H_5-C-C] str. cm^{-1}	A[C=O str.] /A[2(C_6H_5-C-Cl str.)]
31	1774.6	1732.5	871.5	4.44
15	1774.5	1732.2	873	4.31
0	1774.5	1732	873	4.17
−10	1774.4	1732	873.5	3.82
−20	1774	1732.1	873.7	3.62
−30	1773.6	1732	873.7	3.53
−40	1773.5	1732	873.9	3.5
−50	1772.4	1732	874.3	3.25
−60	1771.5	1731.7	873.8	3.25
−70	1771.2	1731.3	874.1	3.03
delta C [−101]	delta C=O str. [−3.4]	delta 2[C_6H_5-C-Cl] [−1.2]	delta C_6H_5-C-Cl [−2.6]	delta A[C=O str.] /A[2(C_6H_5-C-Cl)] [−1.41]

TABLE 16.6   Vibrational assignments for S-methyl thiol chloroformate

CH$_3$−S−C(=O)Cl S−C(=O)Cl fundamentals	Infrared 25% solutions in 0.1 KBr cell cm^{-1}	C(=O)Cl$_2$ fundamentals	Infrared cm^{-1}	CH$_3$−S−C(=O)Cl Raman liquid cm^{-1}(R.I.)	Dep. ratio
C=O str.	1767.2(0.862)*	C=O str.	1827	1757(14.9)	0.34
asym.S−C−Cl str.	846.4(0.831)*	asym.Cl−C−Cl str.	847	840(5.2)	0.74
sym.S−C−Cl str.	580.7(0.137)*	sym.Cl−C−Cl str.	567	5881(100.0)	0.09
gamma C=O	571sh	gamma C=O	580		
C=O rock	430.2(0.371)	C=O rock	440	422(43.2)	0.45
S−C−Cl bend	304	Cl−C−Cl bend	285	305(89.8)	0.29
C−S−C fundamentals					
C−S str.	712.8(0.059)			710(50.0)	0.22
C−S−C bend	205			199(5.6)	0.32
C−S−C torsion	126				
CH$_3$−S fundamentals		CH$_3$−S−P−(=O)Cl2 CH$_3$ fundamentals			
asym.CH$_3$ str.	3020.5(0.065_	asym.CH$_3$ str.	3011	3013(2.8)	0.64
sym.CH$_3$ str.	2938.3(0.435)	sym.CH$_3$ str.	2938	2932(36.0)	0.05
asym.CH$_3$ bend	1422.3(0./608)	asym.CH$_3$ bend	1431	1420	
sym.CH$_3$ bend	1316.5(0.605)	sym.CH$_3$ bend	1321	1311(2.0)	0.44
CH$_3$ rock	972.7(0.464)	CH$_3$ rock	972	967(2.5)	0.35
CH$_3$ torsion	?	CH$_3$ torsion	?		

*2% solutions in 0.1 mm KBr cell] [CCl$_4$ solution 3800–1333 cm^{-1}] [CS$_2$ solution 1333–400 cm^{-1}.

TABLE 16.7   IR spectra-structure correlations for carbamic acid: aryl-, alkyl esters

Aryl–NH–C(=O)–O–R	R	C=O str. cm⁻¹	C=O:H–N str. intermolecular*1 cm⁻¹	N–HO=C str. intermolecular*1 cm⁻¹	N–H str. cm⁻¹	2–X:H–N str. intramolecular cm⁻¹	N–H i.p.-bend cm⁻¹	N–H o.p.-bend cm⁻¹	Aryl–N str. cm⁻¹	N–C–O and N–H bend cm⁻¹
Phenyl 2-, 3-, or 4- F, Cl, Br, or C₆H₅ H–CCCH₂	(CH₃)2N(CH₂)₂	1730–1758	3295–3460 1690*2	3409–3461		1504–1546	503–570	1237–1282 624–680*2	1195–1225	1219–1257*2
	H–CC–CH₂	1705–1734				3411–3440				
2–F	H–CC–CH₂					3435				
2–Cl	H–CC–CH₂					3419				
2–Br	H–CC–CH₂					3409				
2–C₆H₅	H–CC–CH₂					3425				
1-Naphthyl	H–CC–CH₂			3325–3340	3441–3461					
2-Naphthyl	H–CC–CH₂			3360	3445					

*1 Disappears upon dilution.

*2 Solid phase.

TABLE 16.8 The carbonyl stretching frequency for 1,1,3,3-tetramethylurea in aprotic solvents [Nyquist data vs Kolling and Wohar data] vs solvent parameters

Solvent	AN[3]	[A] C=O str. Nyquist data cm^{-1}	[B] C=O str. Kolling data cm^{-1}	[C] C=O str. Wohar data cm^{-1}	[A]–[B] cm^{-1}	[A]–[C] cm^{-1}	[B]–[C] cm^{-1}	e[1]	n[2]
Hexane	0	1668.4	1658	1656	10.4	12.4	2	1.88	1.3719
Diethyl ether	3.9	1661.5	1651	1658	10.5	3.5	−7	4.33	1.3494
Methyl tert-butyl ether	[5.0]	1660.7							
Tetrahydrofuran	[8.8]	1655.1	1646		9.1			7.58	1.405
Carbon disulfide	[10.1]	1653,2	1652		1,4			2.64	1.6241
Carbon tetrachloride	8.6	1652.9		1653		−0.1		2,24	1.457
Benzene	8.2	1652.1	1654	1654	−1.9	−1.9	0	2.28	1.5017
Hexamethylphosphoramide	10.6		1643	1649			−6	30	1.457
Acetone	12.5		1646	1648			−2	20.7	1.356
Nitrobenzene	14.8	1644.4	1642.		2.4			34.82	1.55
Benzonitrile	15.5	1643.2		1642			1.2	25.2	1.5257
Acetonitrile	19.3	1638.4	1639	1638	−0.6	0.4	1	46.68	1.4771

[1]e = dielectric constant.
[2]n = refractive index.
[3]AN = solvent acceptor number.

List of Tables

CHAPTER 5

CHAPTER 6

CHAPTER 7

CHAPTER 8

CHAPTER 9

CHAPTER 10

CHAPTER 11

CHAPTER 12

CHAPTER 13

CHAPTER 14

CHAPTER 15

CHAPTER 16

List of Figures

CHAPTER 1

CHAPTER 3

CHAPTER 4

CHAPTER 5

CHAPTER 6

CHAPTER 7

CHAPTER 9

CHAPTER 10

CHAPTER 11

CHAPTER 12

CHAPTER 13

CHAPTER 14

CHAPTER 15

CHAPTER 16

Name index

Subject index

ISBN 0-12-523355-8